Communications
in Computer and Information Science 1314

Editorial Board Members

More information about this series at http://www.springer.com/series/7899

Yongtian Wang · Xueming Li ·
Yuxin Peng (Eds.)

Image and Graphics Technologies and Applications

15th Chinese Conference, IGTA 2020
Beijing, China, September 19, 2020
Revised Selected Papers

 Springer

Editors
Yongtian Wang
Beijing Institute of Technology
Beijing, China

Xueming Li
Beijing University of Posts
and Telecommunications
Beijing, China

Yuxin Peng
Peking University
Beijing, China

ISSN 1865-0929   ISSN 1865-0937 (electronic)
Communications in Computer and Information Science
ISBN 978-981-33-6032-7   ISBN 978-981-33-6033-4 (eBook)
https://doi.org/10.1007/978-981-33-6033-4

This Springer imprint is published by the registered company Springer Nature Singapore Pte Ltd.
The registered company address is: 152 Beach Road, #21-01/04 Gateway East, Singapore 189721, Singapore

Preface

We were honored to organize the 15th Conference on Image and Graphics Technology and Applications (IGTA 2020). The conference was hosted by the Beijing Society of Image and Graphics, China, which was originally planned to be held at the Beijing University of Posts and Telecommunications, China. Due to the influence of COVID-19, the conference was held in virtual format on September 19, 2020. In this special year, we were excited to see the successful continuation of this event.

IGTA is a professional meeting and an important forum for image processing, computer graphics, and related topics, including but not limited to, image analysis and understanding, computer vision and pattern recognition, data mining, virtual reality and augmented reality, and image technology applications.

The IGTA 2020 conference collected 115 technical reports from different countries and regions. Despite the influence of the pandemic, the paper selection remained highly competitive this year. Each manuscript was reviewed by at least two reviewers, and some of them were reviewed by three reviewers. After careful evaluation, 24 manuscripts were selected for oral and poster presentations.

The keynote, oral, and poster presentations of IGTA 2020 reflected the latest progress in the field of image and graphics, and they were collected in this volume of proceedings, which we believe will provide valuable references for scientists and engineers in related fields.

On behalf of the conference general chairs, I would like to express our gratitude to our committee and staff members for all the hard work preparing for this conference under unprecedented difficulties. Thanks go to all authors for their contributions, and to all reviewers who dedicated their precious time. Finally, I would also like to thank our host, professors, and students from Beijing University of Post and Telecommunications, China, who gave great support for the event.

A poem from the Song Dynasty reads: Where the hills and streams end and there seems no road beyond, amidst shading willows and blooming flowers another village appears. We firmly believe that mankind will overcome the pandemic, and we look forward to seeing you face to face at IGTA 2021 next year.

September 2020

Yongtian Wang

Preface

Organization

General Conference Chairs

Yongtian Wang   Beijing Institute of Technology, China
Xueming Li   Beijing University of Post and Telecommunications, China

Executive and Coordination Committee

Huang Qingming   University of Chinese Academy of Sciences, China
Wang Shengjin   Tsinghua University, China
Zhao Yao   Beijing Jiaotong University, China
Chen Chaowu   The First Research Institute of the Ministry of Public Security of P.R.C, China
Liu Chenglin   Institute of Automation, Chinese Academy of Sciences, China
Zhou Mingquan   Beijing Normal University, China
Jiang Zhiguo   Beihang University, China

Program Committee Chairs

Peng Yuxin   Peking University, China
He Ran   Institute of Automation, Chinese Academy of Sciences, China
Ji Xiangyang   Tsinghua University, China
Ma Zhanyu   Beijing University of Post and Telecommunications, China

Organizing Chairs

Liu Yue   Beijing Institute of Technology, China
Hai Huang   Beijing University of Post and Telecommunications, China
Yuan Xiaoru   Peking University, China

Research Committee Chairs

Liang Xiaohui   Beihang University, China
Dong Jing   Institute of Automation, Chinese Academy of Sciences, China
Yang Jian   Beijing Institute of Technology, China

Publicity and Exhibition Committee Chairs

Yang Lei	Communication University of China, China
Zhang Fengjun	Software Institute of the Chinese Academy of Sciences, China
Zhao Song	Shenlan College

Program Committee Members

Eduard Gröller	Vienna University of Technology, Austria
Henry Been-Lirn Duh	La Trobe University, Australia
Takafumi Taketomi	NAIST, Japan
HuangYiping	National Taiwan University, China
Youngho Lee	Mokpo National University, South Korea
Nobuchika Sakata	Osaka University, Japan
Seokhee Jeon	Kyung Hee University, South Korea
Wang Yongtian	Beijing Institute of Technology, China
Shi Zhenwei	Beihang University, China
Xie Fengying	Beihang University, China
Zhang Haopeng	Beihang University, China
Wang Shengjin	Tsinghua University, China
Zhao Yao	Beijing Jiaotong University, China
Huang Qingming	University of Chinese Academy of Sciences, China
Chen Chaowu	The First Research Institute of the Ministry of Public Security of P.R.C, China
Zhou Ming-quan	Beijing Normal University, China
Jiang Zhiguo	Beihang University, China
Yuan Xiaoru	Peking University, China
He Ran	Institute of Automation, Chinese Academy of Sciences, China
Yang Jian	Beijing Institute of Technology, China
Ji Xiangyang	Tsinghua University, China
Liu Yue	Beijing Institute of Technology, China
Ma Huimin	Tsinghua University, China
Zhou Kun	Zhejiang University, China
Tong Xin	Microsoft Research Asia, China
Wang Liang	Institute of Automation, Chinese Academy of Sciences, China
Zhang Yanning	Northwestern Polytechnical University, China
Zhao Huijie	Beijing University of Aeronautics and Astronautics, China
Zhao Danpei	Beijing University of Aeronautics and Astronautics, China
Yang Cheng	Communication University of China, China
Yan Jun	Journal of Image and Graphics, China

Xia Shihong	Institute of Computing Technology, Chinese Academy of Sciences, China
Cao Weiqun	Beijing Forestry University, China
Di Kaichang	Institute of Remote Sensing and Digital Earth, Chinese Academy of Sciences, China
Yin Xucheng	University of Science and Technology Beijing, China
Gan Fuping	Ministry of Land and Resources of the People's Republic of China, China
Lv Xueqiang	Beijing Information Science & Technology University, China
Lin Hua	Tsinghua University, China
Lin Xiaozhu	Beijing Institute of Petrochemical Technology, China
Li Hua	Institute of Computing Technology, Chinese Academy of Sciences, China
Jiang Yan	Beijing Institute of Fashion Technology, China
Dong Jing	Institute of Automation, Chinese Academy of Sciences, China
Sun Yankui	Tsinghua University, China
Zhuo Li	Beijing University of Technology, China
Li Qingyuan	Chinese Academy of Surveying and Mapping, China
Yuan Jiazheng	Beijing Union University, China
Wang Yiding	North China University of Technology, China
Zhang Aiwu	Capital Normal University, China
Cheng Mingzhi	Beijing Institute Of Graphic Communication, China
Wang Yahui	Beijing University of Civil Engineering and Architecture, China
Yao Guoqiang	Beijing Film Academy, China
Ma Siwei	Peking University, China
Liu liang	Beijing University of Posts and Telecommunications, China
Jiang Yan	Beijing Institute of Fashion Technology, China
Liao Bin	North China Electric Power University, China

Contents

Computer Graphics

Virtual Reality and Human-Computer Interaction

Applications of Image and Graphics

**Other Research Works and Surveys Related to the Applications
of Image and Graphics Technology**

Image Processing and Enhancement Techniques

Single Image Super-Resolution Based on Generative Adversarial Networks

Kai Su[1(✉)], Xueming Li[2], and Xuewei Li[3]

[1] School of Digital Media and Design Arts, Beijing University of Posts and Telecommunications, Beijing 100876, China
sukai0805@bupt.edu.cn
[2] Beijing Key Laboratory of Network System and Network Culture, Beijing University of Posts and Telecommunications, Beijing 100876, China
[3] Department of Information and Communication Engineering, Beijing University of Posts and Telecommunications, Beijing 100876, China

Abstract. Traditional super-resolution algorithms are computationally intensive and the quality of generated images is not high. In recent years, due to the superiority of GAN in generating image details, it has begun to attract researchers' attention. This paper proposes a new GAN-based image super-resolution algorithm, which solves the problems of the current GAN-based super-resolution algorithm that the generated image quality is not high and the multi-scale feature information is not processed enough. In the proposed algorithm, a densely connected dilated convolution network module with different dilation rate is added to enhance the network's ability to generate image features at different scale levels; a channel attention mechanism is introduced in the network to adaptively select the generated image features, which improves the quality of the generated image on the network. After conducting experiments on classic test datasets, the proposed algorithm has improved PSNR and SSIM compared to ESRGAN.

Keywords: GAN · Super-resolution · Dilated convolution · Channel attention

1 Introduction

Image resolution here refers to the amount of information stored in the image, generally expressed in pixels per inch of the image, the unit is pixels per inch (PPI), and sometimes people also use the number of pixels in a unit area to measure the resolution, that is, "horizontal dots × vertical dots". The image resolution determines the quality of the image, and the image resolution and the image size together determine the size of the image file. Generally speaking, the higher the resolution of an image, the more information stored in the image, the more comprehensive human perception of the image, so high-resolution images have great significance for human production and life.

However, due to the defects of the hardware equipment and the noise, optics, motion and other factors in the natural scene, the final generated image will lose certain pixel information. In addition, in the process of image transmission, in order to reduce transmission costs or reduce transmission pressure, sometimes the high-resolution

© Springer Nature Singapore Pte Ltd. 2020
Y. Wang et al. (Eds.): IGTA 2020, CCIS 1314, pp. 3–16, 2020.
https://doi.org/10.1007/978-981-33-6033-4_1

image will be processed to reduce the pixel information in the image. Image super-resolution refers to a technique that transforms an image with a small number of pixels per unit area into an image with more pixels per unit area by a certain method, which will increase the amount of information in an image.

2   Related Work

In the field of traditional methods, common super-resolution algorithms consist of interpolation-based methods, degradation model-based methods, and learning-based methods. Traditional methods [1–6] have a large amount of calculation and the image reconstruction effect is not good. With the rise of deep learning, CNN-based method has gradually become the focus of research. From FSRCNN [7], a general form of deep learning-based super-resolution network has been formed, that is, "feature extraction + upsampling" mode. In 2014, Ian proposed GAN [8] in NIPS, which greatly affected the research of machine learning. GAN consists of two parts: generator and discriminator. Its central idea is to use the game of these two parts to train the generator so that it can generate the image we want. After the emergence of GAN, it has exhibited excellent results in many aspects. It can try to avoid people manually designing a priori information of natural images, but by learning to get information hidden in natural images. In addition, it can also cooperate with many excellent CNNs in the past to achieve tasks that were difficult to accomplish before. With further research, the researchers found that GAN's ability to generate image details is far superior to ordinary CNN, so researchers began to apply GAN to image super-resolution problems. Ledig et al. Proposed SRGAN [12], obtained good objective indicators and amazing subjective reconstruction effect. Subsequently, Wang et al. revised the network on the basis of SRGAN and proposed ESRGAN [15], which further improved the network effect and won the first place in the PIRM2018 competition in the super-resolution world.

However, the current GAN-based super-resolution algorithms have some problems. SRGAN simply adds the ResNet [16] structure to the super-resolution network, and does not effectively use the features extracted from the network, which means a lot of effective feature information is wasted. Compared with the simple ResNet structure in SRGAN, the RRDB structure in ESRGAN uses a densely connected structure to connect the convolutions together, so that the feature information between layers is utilized to a certain extent. Although ESRGAN has made further use of features, it still lacks the use of multi-scale feature information.

Based on the above problems, this paper proposes a new GAN-based image super-resolution network. In the proposed network, a densely connected dilated convolution network module with different dilation rate is added to enhance the ability of the network to generate image features at different scale levels. The channel attention mechanism is introduced into the network to adaptively select the features of the generated image, which improves the quality of the generated image of the network.

3 Proposed Method

3.1 Network Structure

The image super-resolution network based on generative adversarial network proposed in this paper is shown in Fig. 1, which includes a generator and a discriminator. During the training process, the low-resolution image I_{low} generates a high-resolution image $I_{restore}$ through the generator G, and then the image $I_{restore}$ generated by the generator and the high-resolution image I_{high} in the training dataset are sent to the discriminator D for processing, which determines the similarity between the restored image and the clear image according to the designed loss function. Discriminator D distinguishes clear high-resolution images and restores the authenticity of high-resolution images as much as possible, and the generator should try to make discriminators unable to distinguish the difference between the two. Both the generator and the discriminator use the Adam optimizer to optimize the parameters. The orange part in the frame diagram is the improved algorithm part proposed in this paper.

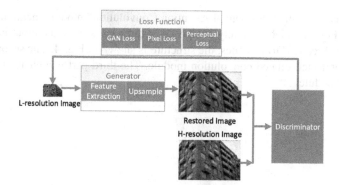

Fig. 1. Network structure.

The image super-resolution network generator based on the adversarial generation network designed in this paper follows the basic structure of excellent super-resolution networks such as SRGAN and ESRGAN, that is, the structure of "residual feature extraction + upsampling". The overall network structure of the generator is shown in the Fig. 2, the initial features are extracted through a 3 × 3 convolution, and then through 15 "residual attention intensive convolution" modules, the selection of the residual module mainly solves the problem that gradient disappears as the network deepens. In the ESRGAN network structure, the number of residual modules is 23. In the network proposed in this paper, due to memory limitations, and in order to keep the parameters of the two networks on the same order of magnitude, the number of residual modules is reduced to 15, On this basis, the test results of the improved network proposed in this paper on each test datasets have still achieved a certain improvement. The part of the network from the beginning to here is the feature extraction module. The output of the feature extraction module and the input before the "residual attention intensive convolution" module are connected by a long skip connection so that the network can learn the residual correction. After this, the network performs upsampling operations through two upsampling modules, which include a nearest neighbor interpolation upsampling function, a 3 × 3 convolution, and a ReLU function, followed by a 3 × 3 convolution and the initial Corresponds to the convolution.

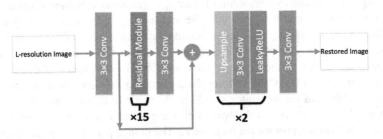

Fig. 2. Generator structure.

The structure of the "Residential Attention Convolution" module mentioned above is shown in Fig. 3, which is mainly composed of three parts. First the input image will pass through three RRDB modules, the structure is shown in Fig. 4. The second part is a densely connected dilated convolution module. The third part is a channel attention mechanism module.

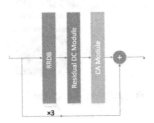

Fig. 3. Residual attention convolution structure.

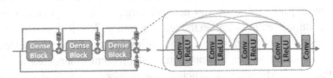

Fig. 4. RRDB structure.

3.2   Densely Connected Dilated Convolution

Dilated convolution, also known as Atrous convolution, was proposed by Yu [17] et al. in 2015 to solve problem of image classification and semantic segmentation tasks. In the process of extracting features, the traditional convolutional neural network will use multiple convolution and pooling methods to increase the receptive field of the network, and then restore the image size through the upsampling operation. Although this method of pooling and de-pooling the feature map multiple times increases the receptive field, it is easy to cause the depth information to contain less and less effective information. Dilated convolution solves this problem well. Dilated convolution introduces the concept of "dilation rate" in the convolution layer, mainly to add holes to the convolution kernel, as shown in Fig. 5.

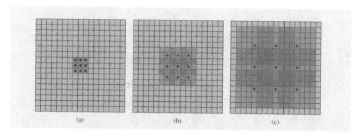

Fig. 5. Dilated convolution

The three graphs above respectively represent the case of dilated convolutions with different dilation rates, where (a) shows the 3 × 3 dilated convolution with an dilation rate of 1, the dilated convolution is actually the original 3 × 3 convolution, (B) indicates a dilated convolution with an dilation rate of 2, the convolution can be regarded as a convolution kernel with a size of 7 × 7, but there are only 9 effective values of the convolution, the amount of convolution in the network is not changed, but the receptive field is expanded to 7 × 7. (C) and (b) are similar, the dilation rate is 4, the receptive field is increased by 15 × 15, and the amount of convolution parameters in the network is not increased.

Because continuous use of the same dilation rate of dilated convolution will bring about a serious grid effect, Wang [18] et al. studied this effect and proposed the HDC (hybrid dilated convolution) dilated convolution design framework. According to the HDC framework, three dilated convolutions with different dilation rate are used to reduce the grid effect in the dilated convolution module. The dilation rate of the three dilated convolutions are 1, 2, 5. The final receiving field of the continuous dilated convolution module designed in this way can fully cover the entire image area without any holes or missing edges. On the other hand, due to the jumping interval of the dilated convolution on the input feature map, the feature maps obtained by the dilated convolution with different dilation rate can be regarded as the feature maps in different scale spaces. The improved super-resolution network proposed in this paper uses the dilated convolution module to realize the extraction of multi-scale features. The dilated convolution with different dilation rate uses the densely connection to connect feature maps in different scale spaces, with this method, features between different layers can be fully reused, and the network parameter amount is not greatly increased.

Due to the densely connected dilated convolution module, the network will generate feature maps of different scales in this part and connect them in the dimension of channel, so that the output of this part of the network contains rich feature map information, so behind this part, the network is connected to a channel attention mechanism module to selectively weight the combination of these feature map information. The specific structure of the dense convolution module is shown in Fig. 6. The convolution size in the figure is 3 × 3 unless otherwise specified.

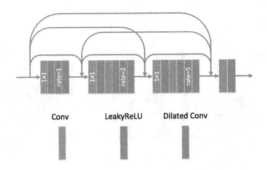

Fig. 6. Densely connected dilated convolution structure.

3.3 Channel Attention Mechanism

The attention mechanism model was proposed by Bahdanau et al. [14]. In 2015. This mechanism simulates the signal processing mechanism of the human brain. When observing an image, humans will quickly scan the entire image and find areas that need focus. In the subsequent observations, humans will devote more attention to this area to obtain effective information, while suppressing useless information. As shown in Fig. 7, this picture visually shows how human attention points are distributed when seeing an image. The red area indicates the more focused image area. The high-frequency information part with details will pay more attention, while distributing less attention to the flat image area.

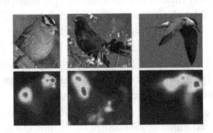

Fig. 7. Attention mechanism. (Color figure online)

The basic principle of the channel attention mechanism is to disperse the network's attention in the input multi-channel feature map, and let the network learn the appropriate attention distribution through training, which is an important channel for the output results. The network will give it high weight to make this part of the feature map work, while the feature maps of other channels will be given low weight. The general channel attention mechanism structure is shown in Fig. 8.

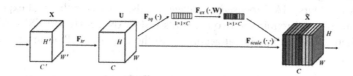

Fig. 8. Channel attention mechanism.

The structure consists of two parts: a feature extraction convolution layer and a channel attention layer. The first several convolution layers of the network are used to convert the input image into a multi-channel feature map, and then pass through the channel attention layer. The channel attention mechanism module first turns each input two-dimensional feature map into a real number through pooling operation. The output of this part is then mapped to the weight values of these channels through a mapping method. The specific operation is to connect two convolutional layers and a Sigmoid function together, and learn this mapping function through network training. Then each channel is multiplied by the corresponding weight to recombine, completing the selective recalibration of all feature channels. It can be seen that in the network without the channel attention mechanism, the network distributes all the feature channel information with the same weight, while in the network with the channel attention mechanism, the feature channels are selectively weighted and combined. This has more practical significance for the network learning. The specific structure of the network channel attention mechanism module in this chapter is shown in Fig. 9.

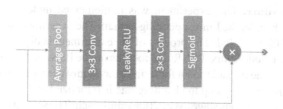

Fig. 9. Channel attention structure.

4   Experiment

4.1   Experiment Preparation

In order to make an equal comparison with the ESRGAN model, the network in this paper uses the same training method as the ESRGAN model, that is, first train a model Ours_PSNR focused on high PSNR, prove the effectiveness of the proposed network structure learning ability. On the basis of this network, join GAN and perceptual loss to train a network Ours_GAN focused on image details, achieving the final experimental effect. In this paper, I retrain the two networks ESRGAN_PSNR and ESRGAN as the reference items for the experiment according to the training method in the ESRGAN paper.

Datasets. Similar to the training method of the ESRGAN model, the experiments in this paper also use the DF2 K datasets to train the PSNR-focused network Ours_PSNR and the DIV2K datasets to train Ours_GAN. DIV2K is a datasets specially prepared for the NTIRE2017 Super Resolution Challenge. There are a total of 1000 images in 2K resolution in the DIV2K datasets, of which 800 are used as the training set, 100 as the verification set, and 100 as the test set. The DF2K datasets is a combination of the DIV2K datasets and the Flick2K datasets, which contains 800 training images of the DIV2K datasets and 2650 resolution images from the Flick website, which is 3450 training images as the training set. There is no test set.

Data Preprocessing. Experimental preprocessing can speed up the training process of deep learning networks and is an indispensable part of deep learning research. The data preprocessing in this chapter includes three steps: image cropping, low-resolution training image generation and data enhancement. The images in the training set are cropped into 480×480 images in 240-pixel steps by random cropping, thereby reducing GPU memory overhead during training. The data set has only high-resolution images but no low-resolution images, so it cannot be trained in pairs, so the next step is to downsample the 480×480 training image generated in the previous step to the size of 120×120 through the Bicubic function in Matlab. To generate the low-resolution image portion of the training set. Then the data set is enhanced by random horizontal flip and $90°$ rotation.

Experiment Procedure. Our experiment was implemented under the PyTorch deep learning framework under the Linux operating system. The experiment was conducted on a GeForce GTX 1080 Ti GPU. The process of the experiment is divided into two steps. First, we generate a model Ours_PSNR that focuses on the PSN. In this step, we only use the generator part of the overall model for training, and only use L_1 pixel loss in terms of loss function. In this training step the learning rate is initialized to 2×10^{-4}, and then is reduced to half of the previous stage every 200k mini-batch. The second step is to train the model Ours_GAN based on image details. In this step, GAN is used for overall training. The model Ours_PSNR obtained in the previous step is used as a pre-training model to accelerate the network convergence speed. The loss function is as follow.

$$L = w_1L_1 + w_2L_2 + w_3L_3 \tag{1}$$

The loss coefficients in each part are: $w_1 = 10^{-2}$, $w_2 = 5 \times 10^{-3}$, $w_3 = 1$, this training step has a total of 400k mini-batch iterations, and the learning rate is initialized to 1×10^{-4}, and reduced to half of the previous stage at the 50k, 100k, 200k and 300k times of mini-batch. The entire training step uses the Adam optimizer for training.

4.2   Experiment Results

This section focuses on comparing the experimental results of the two models in this paper with the existing methods. These existing results include SRCNN [9], VDSR [10], SRGAN, and ESRGAN and other classic networks with milestone significance. The ESRGAN results used in this paper are reproduce the results obtained on the laboratory server. In order to better compare the effects between the models, this chapter also uses Set5, Set14, BSD100, Urban100 to compare several classic datasets, and all are tested when the image zoom factor is 4, the evaluation index is image restoration commonly used peak signal-to-noise ratio PSNR (unit dB) and structural similarity index SSIM. For a fair comparison, all the PSNR and SSIM test results shown are performed on the y channel of the image. The specific experimental results are shown in the following table, where SRResNet and ESRGAN_PSNR in the table are the results of SRGAN and ESRGAN respectively which using only the generator part to train with pixel loss. Bold data indicates the best results in the table.

Table 1. Results compared to other methods.

Method	Data							
	Set5		Set14		BSD100		Urban100	
	PSNR	SSIM	PSNR	SSIM	PSNR	SSIM	PSNR	SSIM
Bicubic	28.42	0.8104	26.00	0.7027	25.96	0.6675	23.14	0.6577
SRCNN [9]	30.48	0.8628	27.49	0.7503	26.90	0.7101	24.52	0.7221
VDSR [10]	31.35	0.8838	28.01	0.7674	27.29	0.7251	25.18	0.7524
DRRN [11]	31.68	0.8888	28.21	0.7721	27.38	0.7284	25.44	0.7638
SRResNet	32.05	0.8910	28.53	0.7804	27.57	0.7354	26.07	0.7839
EDSR [12]	32.46	0.8968	28.80	0.7876	27.71	0.7420	26.64	0.8033
RDN [13]	32.47	0.8990	28.81	0.7871	27.72	0.7419	26.61	0.8028
ESRGAN_PSNR	32.57	0.8996	28.90	0.7888	27.76	0.7427	26.63	0.8042
Ours_PSNR	**32.62**	**0.9000**	**28.91**	**0.7889**	**27.78**	**0.7431**	**26.81**	**0.8073**

As can be seen from Table 1, the image super-resolution algorithm proposed in this chapter has a significant improvement compared to the classic deep learning super-resolution image algorithm SRCNN. On the Set5 datasets, the peak signal-to-noise ratio PSNR is improved by 7.0% compared to SRCNN, and the structural similarity coefficient SSIM is improved by 4.3% compared to SRCNN. Compared with the excellent algorithms SRGAN and ESRGAN proposed in the past two years, the PSNR and SSIM indicators have also been improved, proving the effectiveness of the network structure.

Fig. 10. repetitive high-frequency structures results

It is worth noting that the improved network's indicators on the Urban100 datasets are more obvious. Urban100 datasets consists of images of urban architectural landscapes. The high-frequency repetitive details of the images in the dataset are particularly rich. In the past, super-resolution algorithms often ignored the selective mining of multi-scale feature information, so the repeated high-frequency detail learning was not enough. However, the combination of dilated convolution and channel attention mechanism solved the problem, such as several examples in Fig. 10, where (a) represents a high-resolution image, (b) represents the result of upsampling using bicubic interpolation, and (c) represents the output of the ESRGAN_PSNR model, (d) represents the output of the network trained in this paper based on the L_1 loss function. It can be seen from the comparison that the super-resolution network proposed in this chapter is more accurate in expressing repetitive high-frequency structures.

Fig. 11. GAN-based model results.

Training based on the L_1 loss function enables the network to learn the overall hierarchical structure and pixel distribution of the image. Then, based on this model, the details of the image continue to be learned through the GAN and perceptual loss. The results are as follows. (a) represents the model based on L_1 loss. (b) represents a model based on GAN and perceptual loss, and (c) represents a real high-resolution image. It can be seen from the comparison that the model based on GAN and perceptual loss is closer to the real image than the output result of the model based on the L_1 loss function (Fig. 11).

4.3   Ablation Experiment

This section focuses on the experimental results of the innovative points of the algorithm proposed in this chapter. Since the network in this chapter is improved based on the problems in ESRGAN, one of the best image super-resolution algorithms, it is mainly compared with the experimental results of ESRGAN. The test method in this section is the same as the previous section, and the test is performed on the Set5, Set14, BSD100 and Urban100 datasets.

Figure 12 is a graph of the PSNR calculation on the Set5 datasets of the model at each stage of the training process after the dense dilated convolution module and channel attention mechanism are added to the network. The PSNR value here is calculated in RGB channels, where the blue curve represents the training curve of ESRGAN_PSNR, and the gray curve represents the training curve of the improved network proposed in this chapter. As can be seen from the figure, the network proposed in this chapter is in the process of training Indeed achieved better results.

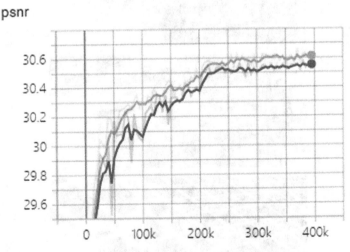

Fig. 12. Dilated convolution + channel attention training diagram. (Color figure online)

Table 2 shows the training results after 100 epoch trainings. The first line shows the output of the ESRGAN_PSNR model. The second line shows the output of the model with the densely connected dilated convolution module added to the ESRGAN_PSNR model. The last line indicates the output result of the model obtained by adding these two methods at the same time. From the table, it can be seen that after the dilated convolution module is added to the network, PSNR (unit dB) and SSIM of the output images have been improved. On this basis, after adding the channel attention mechanism module to the network, the output structure of the network has been further improved, which also proves the effectiveness of the dilated convolution module and channel attention proposed in this paper.

Table 2. Ablation experiment results.

Method	Data							
	Set5		Set14		BSD100		Urban100	
	PSNR	SSIM	PSNR	SSIM	PSNR	SSIM	PSNR	SSIM
ESRGAN_PSNR	32.23	0.8955	28.62	0.7827	27.61	0.7379	26.19	0.7898
Ours_DC	32.28	0.8962	28.65	0.7837	27.64	0.7381	26.23	0.7912
Ours_DC_CA	**32.38**	**0.8965**	**28.74**	**0.7841**	**27.65**	**0.7382**	**26.34**	**0.7932**

5   Conclusion

This paper proposes an image super-resolution algorithm based on GAN. First, this paper points out the problems of recent super-resolution algorithms based on GAN, and then proposes an improved image super-resolution algorithm based on GAN. Finally, this paper introduces the details of the experimental process and analyzes the experimental results to prove the effectiveness of the network.

References

1. Li, Xin, Orchard, Michael T.: New edge-directed interpolation. IEEE Trans. Image Process. **10**(10), 1521–1527 (2001)
2. Giachetti, A., Asuni, N.: Real-time artifact-free image upscaling. IEEE Trans. Image Process. A Publ. IEEE Sig. Process. Soc. **20**(10), 2760–2768 (2011)
3. Carey, W.K., Chuang, D.B., Hemami, S.S.: Regularity – preserving image interpolation. IEEE Trans. Image Process. **8**(9), 1293–1297 (1999)
4. Tsai, R., Huang, T.: Advances in Computer Vision and Image Processing, pp. 317–339. JAI Press Inc., Greenwich (1984)
5. Rhee, S., Kang, M.G.: Discrete cosine transform based regularized high-resolution image reconstruction algorithm. Opt. Eng. **38**(8), 1348–1356 (1999)
6. Nguyen, N., Milanfar, P.: An efficient wavelet-based algorithm for image superresolution. In: Proceedings of the the International Conference on Image Processing, 2000. IEEE (2000)

7. Dong, C., Loy, C.C., Tang, X.: Accelerating the super-resolution convolutional neural network. In: Leibe, B., Matas, J., Sebe, N., Welling, M. (eds.) Computer Vision – ECCV 2016. Lecture Notes in Computer Science, vol. 9906, pp. 391–407. Springer, Cham (2016). https://doi.org/10.1007/978-3-319-46475-6_25

8. Goodfellow, I.J., Pouget-Abadie, J., Mirza, M., et al.: Generative adversarial networks. Adv. Neural. Inf. Process. Syst. **3**, 2672–2680 (2014)

9. Ledig, C., Theis, L., Huszar, F., et al.: Photo-realistic single image super-resolution using a generative adversarial network (2016)

10. Wang, X., Yu, K., et al.: Enhanced super-resolution generative adversarial networks. In: European Conference on Computer Vision Workshop, Munich, Germany. 01 09 2018. https://arxiv.org/pdf/1809.00219.pdf

11. He, K., Zhang, X., Ren, S., et al.: Deep residual learning for image recognition (2015)

12. Yu, F., Koltun, V., Funkhouser, T.: Dilated residual networks. In: 2017 IEEE Conference on Computer Vision and Pattern Recognition (CVPR), Honolulu, HI, pp. 636–644 (2017)

13. Wang, P., Chen, P., Yuan, Y., et al.: Understanding convolution for semantic segmentation. In: 2018 IEEE Winter Conference on Applications of Computer Vision (WACV). IEEE (2018)

14. Bahdanau, D., Cho, K., Bengio, Y.: Neural machine translation by jointly learning to align and translate. Comput. Sci. (2014)

15. Kim, J., Lee, J.K., Lee, K.M.: Accurate image super-resolution using very deep convolutional networks (2015)

16. Tai, Y., Yang, J., Liu, X.: Image super-resolution via deep recursive residual network. In: IEEE Computer Vision and Pattern Recognition. IEEE (2017)

17. Lim, B., Son, S., Kim, H., et al.: Enhanced deep residual networks for single image super-resolution (2017)

18. Zhang, Y., Tian, Y., Kong, Y., et al.: Residual dense network for image super-resolution (2018)

A Striping Removal Method Based on Spectral Correlation in MODIS Data

Zhiming Shang[1(✉)], Wan Jiang[2], Gaojin Wen[1], Can Zhong[1], and Chen Li[1]

[1] Beijing Institute of Space Mechanics and Electricity, Beijing, China
`527064853@qq.com`
[2] National Museum of China, Beijing, China

Abstract. Modis imagery suffers from nonlinear striping in many bands. The nonlinear stripes make great difficulties for conventional linear statistical destriping methods. This paper presents a two-step method for nonlinear stripe removing. Both steps depend on the correlation information between defective band and its reference normal band. First stripes in defective band are divided into different portions with the aid of its high correlation spectrum by GA algorithm automatically and moment matching destriping is performed in each portion independently. Then the residual noise is eliminated by PCA transform and low pass filtering of two high correlation spectrums. Experiment results demonstrate that the proposed algorithm can effectively reduce nonlinear striping in MODIS data.

Keywords: Nonlinear strip · Spectral correlation · GA algorithm · Moment matching

1 Introduction

Modis is a cross-track scanning radiometer with 36 spectral bands and is designed to provide data for studies of the Earth's land, ocean, and atmosphere [1]. Even though MCST MODIS L1B calibration algorithm can effectively handle operational artifacts, stripe noise still remains in many bands images. Without stripe noise correction, this noise will degrade the image quality and introduce a considerable level of noise when the data are processed [2].

There are three types of stripe noises in Modis data: detector-to-detector stripes, mirror side stripes, and noisy stripes. The noise circle of periodic detector-to-detector stripes is also different due to the different number of detectors in different bands. In this letter, we focus on the removal of non-liner detector-to-detector stripes in Modis band 30.

Various methods have been explored to remove striping noise in recent decades. The common methods can be divided into three categories: filtering methods, statistics methods, and optimization-based methods.

Most filtering algorithms construct filters in transformed domain [3], such as fourier transform or wavelet analysis to remove striping noise, while some other algorithms work in spatial domain [4, 5].

© Springer Nature Singapore Pte Ltd. 2020
Y. Wang et al. (Eds.): IGTA 2020, CCIS 1314, pp. 17–28, 2020.
https://doi.org/10.1007/978-981-33-6033-4_2

Histogram matching [6] and moment matching [7] methods are the main representative statistics based methods. Histogram matching attempts to remove the stripe noise by matching the histogram of the contaminated pixels to its reference pixels. While moment matching supposes that the mean and variance of contaminated pixels and its referenced pixels are the same. Since statistics-based methods assume that the contaminated pixels and its reference pixels, which is often spatial proximity normal pixels have similar statistical properties, it can obtain competitive destriping results when the scenes are homogeneous.

The optimization based methods, especially variation methods, introduce regularization methods into the calculation of the image destriping solution model, and obtain the desired image by employing an optimization method. These methods achieve a great success in recent years [8, 9]. Deep learning technologies are also gradually applied to image denoising [10].

Specially multispectral or hyperspectral [11] images can provide redundant data for contaminated bands by its highly correlated normal bands. Based on this fact, a two-step spectral correlation method for striping removal is proposed in this paper.

First stripes in defective band are divided into different portions with the aid of its high correlation spectrum by GA algorithm automatically, and moment matching destriping is performed in each portion independently. Then low pass filtering method based on PCA transform of two high correlation spectrums is proposed to eliminate residual noises in heterogeneous regions. Thus nonlinear and irregular stripes can be fitted in pieces and removed in both heterogeneous and homogeneous regions.

2 The Proposed Destriping Algorithm

As a widely utilized destriping algorithm, moment matching has a better balance of efficacy and efficiency. While it is serious impact by its assumption that contaminated pixels have a linear relationship with normal pixels, which can be described by offset and gain, and this assumption makes it cannot handle situation when nonlinear stripes occur or the land surface is not homogeneous.

Figure 1 shows obviously irregular and nonlinear stripes in Modis band 30 data. This phenomenon generally occurs on images of different gray levels. Dark strip will weaken or disappear in brighter image areas, but will be more pronounced in darker image areas. Bright strip behave just the opposite, as shown in Fig. 1(a), (b). The uniformity of the image also affects the strip. The strips in the homogeneous area are more obvious, while strip in the image area with larger variance is suppressed as shown in Fig. 1(c). Even in a generally homogeneous area, sometimes the nonlinear strips will still remain, as shown in Fig. 1(d), which may be caused by the unstable response of the detector.

Figure 2 shows the comparison of Modis band 30 and 29 images. It can be seen that Modis 30 band suffered serious striping, while band 29 is very clear. The fusion color image is grayish in most areas, except for the striking color stripes caused by band 30 stripes and the gentle color change caused by the spectral difference of band 30 and 29, which proves that these two bands are strong correlated. Therefore, band 29 is selected as the reference band for nonlinear band noise correction.

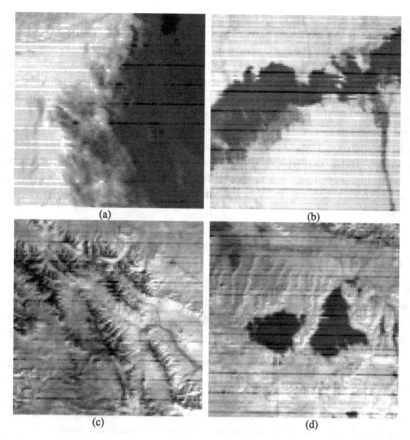

Fig. 1. Nonlinear and irregular stripes in Modis Band 30 image. (a) and (b) shows nonlinear and irregular stripes in different gray levels; (c) shows nonlinear and irregular stripes in heterogeneous image and (d) shows nonlinear in homogeneous image.

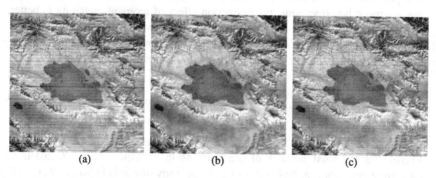

Fig. 2. Comparison of Modis Band 30 and 29 images. (a) is an example image of band 30; (b) is an example image of band 29 and (c) is a fusion image, where the red and blue channels are composed of band 29 image and the green channel is composed of band 30 image. (Color figure online)

To remove nonlinear or irregular stripes of Modis band 30 images, taking Modis band 29 as the reference image, a two-step algorithm based on spectral correlation is proposed in this paper. The basic idea is first to divide a stripe into different portions, and to implement moment matching method in each portion independently. According to the assumption of moment matching, each portion of the stripe line should be linear correlation of its reference portion of normal line. Using this assumption, GA algorithm is employed to find the best cut-off points of the stripe.

After processing, some residual noise still remains, mainly in heterogeneous regions of the image. The noise is not obvious in single band image, and is ignored in many studies. But it is obvious in the fusion color image which is consists of noise band and its high correlation normal band. To eliminate these residual noises, low pass filtering method based on PCA transform of two high correlation spectrums is proposed.

2.1   Auto Piecewise Moment Matching Method Based on Genetic Algorithm(GA)

It can be known from the definition of the correlation coefficient that a linear transformation of a group of numbers will not affect its correlation to other numbers. If the defective stripes of Modis band 30 are only affected by linear noise, its mean and variance change, but its correlation with the corresponding rows of the reference band 29 will not be affected.

Experiment shows that the normal rows of Modis band 30 and it's corresponding rows in band 29 has a very high correlation coefficient, which is about 95%, only affected by different reflectivity of land cover. However, the correlation coefficient between contaminated stripes in Modis band 30 and its corresponding rows in band 29 decreases rapidly to about 70%, due to the presence of non-linear noise.

A piecewise function is proposed to fit this non-linear noise. The stripes are divided into different portions and each portion is assumed to be only affected by linear noise, so the contaminated stripe portions should maintain a high correlation coefficient with its corresponding row portions in its reference band. At the same time, if the stripe portions are more homogeneous, the correction effect of moment matching is better.

So the problem is to find a series of cut-off points which make each portion of defective stripes has a high correlation coefficient with its reference data and make each portion more homogeneous. It can be defined as minimizing function below:

$$J = \frac{std(k)}{\sum_{1}^{N} N(k) \times Corr(k)} \tag{1}$$

Where N is the total number of the portions in a strip, k $(k = 1,2,...n)$ is the sequence number of portion, $N(k)$ stands for the quantity of pixels in portion k, $Corr(k)$ is the correlation coefficient between portion k and its corresponding portion in band 29, and $std(k)$ stands for the standard deviation, which represents the degree of homogeneity of portion k.

Minimizing function J requires searching for the optimal cut-off points in the entire strip. The quantity of cut-off points is large and varies with the noise distribution in different stripes. It is not possible to get an accurate initial value and the unknowns have to be calculated by single function J. Traditional optimization methods such as least squares cannot solve it directly.

Genetic algorithm presents an approach to optimization that is rooted in a model of Darwinian evolution and computational equivalents of selection, crossover and mutation [12] solutions, which has been employed for optimization problems in various domains [13–16]. It takes the advantage of global optimization that operates on groups of multiple individuals simultaneously, and its optimization results are independent of initial conditions, which meet the constraints of solving the objective function J. Therefore, it is employed to deal with the cut-off point position optimizing problems.

The steps involved in the GA method to minimize fitness function J are as follows:

(1) Gene Coding, a vector of cut-off point's positions is set as the gene of GA.

$$X = \{x_1, x_2, x_3 \ldots x_N\} \tag{2}$$

Where N is the quantity of cut-off points, x_1 is forced to be the start position of strip, while x_N is forced to be the end position. As the quantity of cut-off points varies in different stripes, N should be decided automatically during processing with a rather high initial value.

(2) Set initial population as 100.
(3) Sort X to make sure it's element ranking from small to large, and defines the portion P_k as

$$P_k = DLine_{(X_{k-1}, X_k)}(k \in (2, N)) \tag{3}$$

Where $DLine$ is the defective strip of band 30 image, if the length of P_k is less than given threshold, it will be merged with the adjacent portion.

(4) Divide the strip's corresponding normal row in band 29 as Q_k with same cut-off points defined in X.
(5) Minimizing the fitness function J with GA, the correlation coefficient between P_k and Q_k is defined as

$$Corr(k) = \frac{cov_{P_k Q_k}}{s_{P_k} s_{Q_k}} \tag{4}$$

Where $cov_{P_k Q_k}$ represents the covariance between P_k and Q_k, s_{P_k} stands for standard deviation of P_k and s_{Q_k} stands for Q_k.

After X is solved and the final result of cut-off points is calculated, moment matching is implemented in each portion of the strip defined by the cut-off points independently, with its spatial adjacent normal row as the reference. Loop the above steps until the stripes in band 30 are all corrected, and the result image is got.

2.2 Fourier Low-Pass Filtering of Residual Noise of High Correlation Band

After piecewise correction, most of the striping noise is corrected, resulting in extremely high correlation between band 30 and band 29 images, while some residual noise still remains, mainly in heterogeneous images, such as mountains areas.

Assuming that these two bands are completely related, the gray values of the pixels of the two bands will form a straight line in the brightness coordinate system. However, due to the residual noise and the spectral difference, these data will fluctuate around this line slightly.

In the direction of fluctuation, the variance of the data is small while in the direction of the straight line, the image signal will constitute a relatively large variance. This data distribution fits perfectly with PCA's theory, which transforms the image features to a new coordinate system in such a way that the greatest variance of the data comes to lie on the first component, and the second greatest variance on the second component. The residual noise that needs to be further suppressed and spectral differences will be concentrated in the second principal component.

So, PCA transform is done on the fusion image of corrected band 30 and band 29. The first component of the transformed image concentrated the relevant information of these two bands, while the second component contains spectral difference and residual noise.

The problem now is to decompose residual noise from spectral differences. Due to the strong correlation between these two bands, their spectral difference is small and smooth, which is concentrated in the low frequency. On the contrary, the residual noise changes sharply and appears intermittently which is concentrated in the high frequency of second component image. So low-pass filtering of the second component image is done after fourier translation to remove high frequency residual noise.

At last, PCA inverse transform is applied to the filtered image and the final destriping Modis band 30 image is got.

3 Experiment Results

The experiment results are present in this section. Three images of Aqua Modis band 30 which are cropped to 400×400 pixels are employed to evaluate the proposed method. And it is compared with classical moment matching algorithm. Since the stripes are periodic, their positions were determined manually in a single scanning swath of the composite color image, which consist of contaminative image band 30 and its reference image band 29.

Figure 3 consists of not only heterogeneous regions such as mountains, but also homogeneity regions such as lakes. Figure 3(b) shows destriping result of moment matching while Fig. 3(c) shows destriping result of proposed method. For the convenience of visual comparison, detailed regions of lake and mountains are shown in (d)-(f) and (g)-(i). In Fig. 3(e) there are bright strips in the lake after moment matching as a result of overcorrection of original non-linear dark strip while Fig. 3(f) shows a much better visual effects with proposed destriping method.

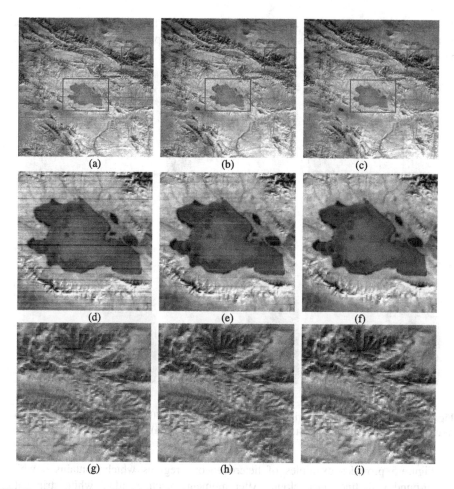

Fig. 3. Destriping results. (a) Original image of Modis band 30 image. (b) Moment matching result. (c) Proposed method result. (d)–(f) detailed regions of (a)–(c). (g–h) detailed regions of (a)–(c), which is composite color image of band 30 and band 29. (Color figure online)

The strip is not obvious in heterogeneous regions of band 30. But in the fusing color image composed of Modis band 30 and band 29, noise is obvious visually, displayed in the form of broken green or purplish red stripes.

After moment matching, there are still intermittent, high-frequency strip residues in heterogeneous mountains regions, as shown in Fig. 3(h). While Fig. 2(i) shows that proposed method successfully suppressed noise in heterogeneous regions. The discrete colored noisy pixels are removed, while the spectral differences between the two bands are retained.

Figure 4 is generally homogeneous, nonlinear strip appears on the dark lake and adjacent land in the picture. The strip is more pronounced in lakes and less pronounced on land. Figure 4(e) shows the result of moment matching in the detailed region of lake and lakeshore. Dark strips in the lake caused by under-correction which are obvious on the pixels left side of the lake, and gradually disappear to the right still exist. Proposed method performs better on this homogeneous region sample in Fig. 4(f), few strip can be noticed visually.

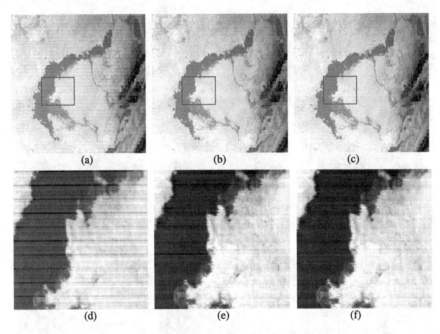

Fig. 4. Destriping results. (a) Original image of Modis band 30 image. (b) Moment matching result. (c) Proposed method result. (d)–(f) detailed regions of (a)–(c).

Figure 5 provides examples of heterogeneous regions which contains a white background with fine black cloud. After moment match, residual white strip with different lengths due to overcorrection appear to the right of the black cloud, as shown in Fig. 5(e). And these strips disappear in Fig. 5(f) which is the result of proposed method.

Figure 6 shows the ensemble-averaged power spectra of the original images and the destriped images of moment matching and proposed algorithm. The horizontal axis represents the normalized frequency, and the vertical axis represents the value of averaged power spectrum [17]. For better visualization, spectral magnitudes above 500 are not plotted. Since the Modis senor has ten detectors, the pulses caused by detector-to-detector stripes are located at the frequencies of 0.1, 0.2, 0.3, 0.4, and 0.5 cycles [2]. In Fig. 6(e) and Fig. 6(h), it is easily recognized that there are still some residues pulses in the result of the moment matching. However these pulses have been strongly removed by proposed method.

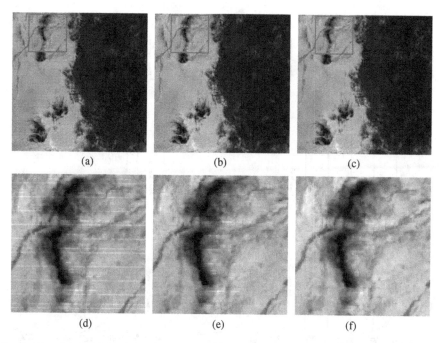

(a)   (b)   (c)

(d)   (e)   (f)

Fig. 5. Destriping results. (a) Original image of Modis band 30 image. (b) Moment matching result. (c) Proposed method result. (d)–(f) detailed regions of (a)–(c), which is composite color image of band 30 and band 29. (Color figure online)

For a quantitative measurement, the quality indices of ICV and NR are employed. ICV [7, 17] is calculated as

$$ICV = \frac{R_a}{R_{sd}} \tag{5}$$

Where the signal component is calculated by averaging the sample pixels as R_a, and the noise component is estimated by calculating the standard deviation of sample pixels as R_{sd}. And ICV index defines the level of stripe noise by calculating the ratio of signal R_a to noise R_{sd}.

According to the definition of ICV, it only works on homogeneous regions. Six 10×10 homogeneous sample pixels from experiment images(Fig. $3 \sim 5$) are selected for ICV evaluation in this letter. The result is shown in Table 1.

Experiment shows proposed method get a higher ICV value than original image and moment matching, especially in sampling areas where is particularly homogeneous, for example in sample 3, ICV values of proposed method is 2.56 times as much as original image and 1.82 times that of moment matching.

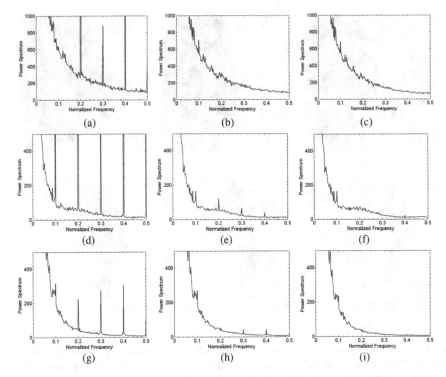

Fig. 6. Power spectra of Modis data. (a)–(c) represents spectral of Fig. 3(a)–(c). (d)–(f) represents spectral of Fig. 4(a)–(c). (g)–(i) represents spectral of Fig. 5(a)–(c)

Table 1. ICV Values of Original and Destriped Modis Data

		Sample 1	Sample 2	Sample 3	Sample 4	Sample 5	Sample 6
Band30	Original	13.73	5.85	46.14	24.87	24.40	40.6767
	Moment	19.86	6.33	64.91	26.49	37.40	86.7686
	Proposed	20.43	6.39	118.43	27.60	46.89	96.7496

NR, the noise reduction ratio, is employed to evaluate the destriped image in frequency domain, it is defined as

$$NR = \frac{N_0}{N_1} \tag{6}$$

where N_0 and N_1 stand for the power of the frequency components produced by the striping noise in the original image and destriped images, respectively [7, 17].

Table 2 shows the NR result of three experiment images. Proposed method also gets a significant increase in NR values than moment matching.

Table 2. NR Values of Original and Destriped Modis Data

		Original	Moment	Proposed
Band30	Figure 3	1	5.04	5.56
	Figure 4	1	15.74	21.90
	Figure 5	1	2.95	4.04

Experiments proof that the proposed method has higher values for both metrics of NR and ICV, and is better than moment matching algorithm in both visual effect and quantitative analysis, which confirms that the proposed method can eliminate striping noise more effectively.

4 Conclusions

The paper presents a two–step method based on spectral correlation for destriping problems. The proposed method is tested on Aqua Modis band 30 images. ICV and NR are employed to evaluate the quality of processed images. Experiments confirmed that proposed algorithm is superior to conventional moment matching in both visual inspection and quantitative evaluation.

References

1. Barnes, W.L., Pagano, T.S., Salomonson, V.V.: Prelaunch characteristics of the Moderate Resolution Imaging Spectroradiometer (MODIS) on EOS-AMl. IEEE Trans. Geosci. Remote Sens. **36**(4), 1088–1100 (1998)
2. Rakwatin, P., Takeuchi, W., Yasuoka, Y.: Stripe noise reduction in MODIS data by combining histogram matching with facet filter. Trans. Geosci. Remote Sens. **45**(6), 1844–1855 (2007)
3. Simpson, J.J., Gobat, J.I., Frouin, R.: Improved destriping of GOES images using finite impulse response filters. Remote Sens. Environ. **52**(1), 15–35 (1995)
4. Cushnie, Janis L., Atkinson, Paula: Effect of spatial filtering on scene noise and boundary detail in thematic mapper imagery. Photogram. Eng. Remote Sens. **51**(9), 1483–1493 (1985)
5. Pan, J.J., Chang, C.I.: Destriping of landsat MSS images by filtering techniques. Photogram. Eng. Remote Sens. **58**, 1417 (1992)
6. Rakwatin, P., Takeuchi, W., Yasuoka, Y.: Restoration of aqua MODIS band 6 using histogram matching and local least squares fitting. IEEE Trans. Geosci. Remote Sens. **47**(2), 613–627 (2009)
7. Shen, H., Jiang, W., Zhang, H., Zhang, L.: A piecewise approach to removing the nonlinear and irregular stripes in MODIS data. Int. J. Remote Sens. **35**(1), 44–53 (2014)
8. Chen, Y., Huang, T.Z., Zhao, X.L., et al.: Stripe noise removal of remote sensing images by total variation regularization and group sparsity constraint. Remote Sens. **9**(6), 559 (2017)
9. Bouali, M., Ladjal, S.: Toward optimal destriping of MODIS data using a unidirectional variational model. IEEE Trans. Geosci. Remote Sens. **49**(8), 2924–2935 (2011)

10. Liu, C., Shang, Z., Qin, A.: A multiscale image denoising algorithm based on dilated residual convolution network. In: Wang, Y., Huang, Q., Peng, Y. (eds.) Image and Graphics Technologies and Applications. IGTA 2019. Communications in Computer and Information Science, vol. 1043, pp. 193–203. Springer, Singapore (2019). https://doi.org/10.1007/978-981-13-9917-6_19
11. Acito, N., Diani, M., Corsini, G.: Subspace-based striping noise reduction in hyperspectral images. IEEE Trans. Geosci. Remote Sens. 49(4), 1325–1342 (2010)
12. Mitchell, M.: An Introduction to Genetic Algorithms, Cambridge, MA (1996)
13. Shang, Z., Wen, G., Zhong, C., Wang, H., Li, C.: UAV Ill-conditioned target localization method based on genetic algorithm. In: 2018 14th International Conference on Natural Computation, Fuzzy Systems and Knowledge Discovery (ICNC-FSKD), Huangshan, China, pp. 530–534 (2018)
14. Bies, Robert R., Muldoon, Matthew F., Pollock, Bruce G., Manuck, Steven., Smith, Gwenn, Sale, Mark E.: A genetic algorithm-based, hybrid machine learning approach to model selection. J. Pharmacokinet Pharmacodyn. 33(2), 196–221 (2006)
15. Wang, J., He, S., Jia, H.: A novel three-dimensional asymmetric reconstruction method of plasma. In: Wang, Y., et al. (eds.) Advances in Image and Graphics Technologies. IGTA 2017. Communications in Computer and Information Sc, vol. 757, pp. 96–103. Springer, Singapore (2018). https://doi.org/10.1007/978-981-10-7389-2_10
16. Qin, W., Fang, Q., Yang, Y.: A facial expression recognition method based on singular value features and improved BP neural network. In: Tan, T., Ruan, Q., Chen, X., Ma, H., Wang, L. (eds.) Advances in Image and Graphics Technologies. IGTA 2013. Communications in Computer and Information Science, vol. 363, pp. 163–172. Springer, Berlin, Heidelberg (2013). https://doi.org/10.1007/978-3-642-37149-3_20
17. Shen, H., Zhang, L.: A MAP-based algorithm for destriping and inpainting of remotely sensed images. IEEE Trans. Geosci. Remote Sens. 47(5), 1492–1502 (2009)

Multi-modal 3-D Medical Image Fusion Based on Tensor Robust Principal Component Analysis

Xiao Sun and Haitao Yin[✉]

College of Automation and College of Artificial Intelligence,
Nanjing University of Posts and Telecommunications, Nanjing 210023, China
{1218053610,haitaoyin}@njupt.edu.cn

Abstract. Multimodal 3-D medical image fusion is important for the clinical diagnosis. The common fusion operation processes each slice of 3-D medical image individually, that may ignore the 3-D information. To extract the 3-D information of multi-modal medical images effectively, we propose a Tensor Robust Principal Component Analysis (TRPCA) based 3-D medical image fusion method. The low rank and sparse components of TRPCA can reveal the 3-D structures of medical image, specially the correlation and differences among the adjacent slices. In addition, we adopt the "3-D weighted local Laplacian energy" rule and the "max-absolute" rule for low-rank components fusion and sparse components fusion, respectively. The developed 3-D weighted local Laplacian energy fusion rule can determine the activity level of 3-D features and preserve 3-D structural information effectively. The experimental results on various medical images, acquired in different modalities, show the performance of proposed method by comparing with several popular methods.

Keywords: Medical image fusion · TRPCA · 3-D weighted local Laplacian energy · Fusion rule

1 Introduction

Medical imaging plays a significant role in the field of clinical medicine, which can show the information of organ, tissue and bone in vision. Due to different medical imaging technology, different devices acquire the tissue and structure information in different modality with respective characteristics, for example the computed tomography (CT) and magnetic resonance imaging (MRI). The medical imaging develops rapidly, but the information provided by a single imaging system is still limited. An economy image processing technique, called as image fusion, can synthesize an image by fusing multi-modal medical images, which can provide more accuracy, robustness and completeness descriptions [1–3].

Medical image fusion is a hot research field in medical image processing. Lots of image fusion algorithms have been proposed. The Multi-Scale Decompositions (MSDs) is commonly used, such as the classical Discrete Wavelet Transform [4], Non-Subsampled Contourlet Transform [5], Shearlet transform [6] and Non-Subsampled Shearlet Transform [7]. The underlying assumption of MSD fusion procedures is that

© Springer Nature Singapore Pte Ltd. 2020
Y. Wang et al. (Eds.): IGTA 2020, CCIS 1314, pp. 29–42, 2020.
https://doi.org/10.1007/978-981-33-6033-4_3

the salient features of multi-modal images can be highlighted in the MSD domain. However, the MSD needs to be carefully designed and its generalizability is still an issue for image fusion. Sparse representation (SR) is also widely applied into image fusion. The main principle of SR is that any input can be formulated as a sparse linear combination of atoms (columns) in a dictionary. Except the primal SR model [8–10], some structured SR based fusion works also be widely studied, such as the group SR [11], joint SR [12] and multiscale SR [13]. The effectiveness of SR based fusion greatly depends on the choice of dictionaries. In addition, the low-rank model [14] with the constraint of rank sparsity, could capture the global image structure, which is also applied into image fusion [15]. However, the existing SR and low rank model based fusion methods aim to fuse the 2-D images, are not effective for 3-D image fusion.

Recalling the 3-D medical image, the 3-D information containing in the adjacent slices is useful for clinical diagnosis. Common individual implementation slice by slice using 2-D fusion methods may result in some artifacts and 3-D information distortion. The 3-D salient information extraction and fusion rule are the core for 3-D image fusion. Based on these considerations, this paper proposes a novel 3-D medical image fusion method using TRPCA. The slices of 3-D medical image are acquired continuously in a specified direction. There is a strong correlation among adjacent slices. TRPCA can capture such 3-D correlations effectively, and the related low rank and sparse parts reflect different 3-D salient features. Furthermore, we design the "3-D weighted local Laplacian energy" rule and "max-absolute" rule for low-rank parts fusion and sparse parts fusion, respectively. The "3-D weighted local Laplacian energy" rule determines the activity level using the 3-D neighborhood information, which can preserve the 3-D spatial structure effectively.

The rest of this paper is organized as follows. In Sect. 2, the related works are briefly introduced. Section 3 gives the proposed procedures. The experimental comparisons are given in Sect. 4. Finally, this paper is concluded in Sect. 5.

2   Related Works

Robust principal component analysis (RPCA) is a popular model in image processing and pattern recognition [16, 17] and achieves lots of advanced results, such as background modeling, shadows removal and image restoration. RPCA aims to decompose the input as a low-rank component and a sparse component. Formally, let $\mathbf{X} \in \mathbb{R}^{n_1 \times n_2}$ be an observed matrix. RPCA formulizes $\mathbf{X}$ as $\mathbf{X} = \mathbf{L} + \mathbf{E}$, and its components can be solved by

$$min_{\mathbf{L},\mathbf{E}} \|\mathbf{L}\|_* + \lambda \|\mathbf{E}\|_1 \quad s.t. \quad \mathbf{X} = \mathbf{L} + \mathbf{E} \tag{1}$$

where $\|\cdot\|_*$ is the nuclear norm, and $\|\cdot\|_1$ denotes the l_1-norm.

Generally, the RPCA model processes the 2-D matrix data. For the medical image, video and hyperspectral images, these images needs to be reshaped as matrix format. To improve the performance of RPCA in the multi-dimensional image decomposition, Lu et al. [18] proposed the tensor version of RPCA, i.e., TRPCA. Similar to the RPCA, TRPCA decomposes the input tensor $\mathcal{X} \in \mathbb{R}^{n_1 \times n_2 \times n_3}$ as a tensor low rank component $\mathcal{L}$ and a tensor sparse component $\mathcal{E}$. The related optimization problem is

$$\min_{\mathcal{L},\mathcal{E}} \|\mathcal{L}\|_* + \lambda \|\mathcal{E}\|_1 \ s.t. \ \mathcal{X} = \mathcal{L} + \mathcal{E} \tag{2}$$

where $\|\mathcal{L}\|_*$ is the tensor nuclear norm. Currently, various definitions of tensor nuclear norm have been proposed, such as in [18, 19]. If $n_3 = 1$, $\mathcal{X}$ reduces to a matrix, then (2) equals to (1). TRPCA can be solved by the Alternating Direction Method of Multipliers (ADMM) algorithm [18]. Algorithm 1 gives the pseudo-codes of ADMM for solving (2).

Algorithm 1: Solve (2) by ADMM

Input: tensor data $\mathcal{X} \in \mathbb{R}^{n_1 \times n_2 \times n_3}$, parameters λ, ρ, μ_0, ε and μ_{max}

Initialize: $\mathcal{L}_0 = 0$, $\mathcal{E}_0 = 0$, $\mathcal{Y}_0 = 0$

While not converged **do**

    Update $\mathcal{L}_{k+1}$ by

$$\mathcal{L}_{k+1} = arg\ min_{\mathcal{L}} \ \|\mathcal{L}\|_* + \frac{\mu}{2} \left\| \mathcal{L} + \mathcal{E}_k - \mathcal{X} + \frac{\mathcal{Y}_k}{\mu_k} \right\|_F^2;$$

    Update $\mathcal{E}_{k+1}$ by

$$\mathcal{E}_{k+1} = arg\ min\ \lambda_{\mathcal{E}} \|\mathcal{E}\|_1 + \frac{\mu}{2} \left\| \mathcal{L}_{k+1} + \mathcal{E} - \mathcal{X} + \frac{\mathcal{Y}_k}{\mu_k} \right\|_F^2;$$

    Update $\mathcal{Y}_{k+1}$ by

    $\mathcal{Y}_{k+1} = \mathcal{Y}_k + \mu_k (\mathcal{L}_{k+1} + \mathcal{E}_{k+1} - \mathcal{X});$

    Update μ_{k+1} by $\mu_{k+1} = min(\rho \mu_k, \mu_{max});$

    Check the convergence conditions

    $\|\mathcal{L}_{k+1} - \mathcal{L}_k\|_\infty \le \varepsilon$, $\|\mathcal{E}_{k+1} - \mathcal{E}_k\|_\infty \le \varepsilon$, $\|\mathcal{L}_{k+1} + \mathcal{E}_{k+1} - \mathcal{X}\|_\infty \le \varepsilon$.

End while

Output: $\mathcal{L}$, $\mathcal{E}$

---

3 Proposed Method

3.1 Procedures of Proposed Method

In this sub-section, the procedures of proposed method will be described. Figure 1 displays the whole framework of proposed method, which contains three major steps: *TRPCA decomposition*, *Decomposed components fusion*, and *Fused image synthesis*. The descriptions of each step are presented as follows.

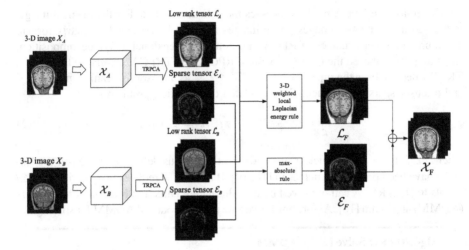

Fig. 1. The framework of proposed method.

1) TRPCA Decomposition

Let $\mathcal{X}_A$ and $\mathcal{X}_B$ be the input multiple 3-D medical images. Through the TRPCA decomposition, we can obtain $\mathcal{X}_A = \mathcal{L}_A + \mathcal{E}_A, \mathcal{X}_B = \mathcal{L}_B + \mathcal{E}_B$. The decomposed components $\mathcal{L}_i, \mathcal{E}_i (i = A, B)$ are calculated by

$$\min_{\mathcal{L}_i, \mathcal{E}_i} \|\mathcal{L}_i\|_* + \lambda \|\mathcal{E}_i\|_1 \; s.t. \; \mathcal{X}_i = \mathcal{L}_i + \mathcal{E}_i, \; i = A, B \tag{3}$$

The above problem (3) is solved by Algorithm 1.

2) Decomposed Components Fusion

The low-rank component of TRPCA reveals the 3-D correlation of adjacent slices, while the sparse component highlights the differences among the adjacent slices. Due to these different characteristics, we employ different rules for different decomposed components. To preserve the 3-D structure information, the 3-D weighted local Laplacian energy rule ($f_{3D-WLLE}$) is designed for combining the tensor low rank components i.e.,

$$\mathcal{L}_F = f_{3D-WLLE}(\mathcal{L}_A, \mathcal{L}_B) \tag{4}$$

The definition of operation $f_{3D-WLLE}(\cdot, \cdot)$ will be given in Sect. 3.2. For the sparse components fusion, the "max-absolute" rule is adopted, i.e.,

$$\mathcal{E}_F(i,j,k) = \begin{cases} \mathcal{E}_A(i,j,k) & if \; |\mathcal{E}_A(i,j,k)| \geq |\mathcal{E}_B(i,j,k)| \\ \mathcal{E}_B(i,j,k) & otherwise \end{cases} \tag{5}$$

3) Fused Image Synthesis

The final fused 3-D image is created as $\mathcal{X}_F = \mathcal{L}_F + \mathcal{E}_F$.

3.2   3-D Weighted Local Laplacian Energy Rule

For fusing the tensor low rank components, we design a specific rule $f_{3D-WLLE}$ which measures the activity level by considering the weighted local Laplacian energy (WLLE) and the weighted sum of 26-neighborhood based modified Laplacian ($\mathcal{W}_{26}$). WLLE is defined as

$$
\begin{aligned}
WLLE(i,j,k) = \sum_{m=-r}^{r} \sum_{n=-r}^{r} \sum_{p=-r}^{r} &\mathcal{W}(m+r+1, n+r+1, p+r+1) \\
&\times \mathbf{L}(i+m, j+n, k+p)^2
\end{aligned}
\tag{6}
$$

where $\mathbf{L}(i,j,k)$ represents the value of $\mathbf{L}$ at location (i,j,k), $WLLE(i,j,k)$ represents the weighted local Laplacian energy of source images at location (i,j,k). $\mathcal{W}$ is a 3-way tensor of size $(2r+1) \times (2r+1) \times (2r+1)$. The entries in $\mathcal{W}$ are set to 2^{3r-d}, where r denotes the radius of tensor $\mathcal{W}$, and d represents the neighborhood distance to the center. Figure 2 is an example of the normalized tensor $\mathcal{W}$ ($r = 1$).

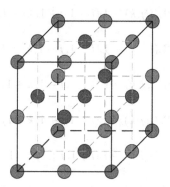

Fig. 2. The normalized tensor $\mathcal{W}$. The red circle is the core of $\mathcal{W}$ with value as 8; the brown circle is the point at distance $d = 1$ from the core, with value as 4; the green circle is the point at distance $d = 2$ from the core, with value as 2; the purple circle is the point at distance $d = 3$ from the core, with value as 1. (Color figure online)

$\mathcal{W}_{26}$ is formulated as

$$\mathcal{W}_{26}(i,j,k) = \sum_{m=-r}^{r} \sum_{n=-r}^{r} \sum_{p=-r}^{r} \mathcal{W}(m+r+1, n+r+1, p+r+1) \\ \times NML_{26}(i+m, j+n, k+p) \tag{7}$$

where $\mathcal{W}$ is with same definition as Eq. (6). NML_{26} in Eq. (7) is defined as

$$
\begin{aligned}
NML_{26}(i,j,k) ={}& |2\mathbf{L}(i,j,k) - \mathbf{L}(i-1,j,k) - \mathbf{L}(i+1,j,k)| \\
&+ |2\mathbf{L}(i,j,k) - \mathbf{L}(i,j-1,k) - \mathbf{L}(i,j+1,k)| \\
&+ |2\mathbf{L}(i,j,k) - \mathbf{L}(i,j,k-1) - \mathbf{L}(i,j,k+1)| \\
&+ \frac{1}{\sqrt{2}} |2\mathbf{L}(i,j,k) - \mathbf{L}(i-1,j,k+1) - \mathbf{L}(i+1,j,k-1)| \\
&+ \frac{1}{\sqrt{2}} |2\mathbf{L}(i,j,k) - \mathbf{L}(i-1,j,k-1) - \mathbf{L}(i+1,j,k+1)| \\
&+ \frac{1}{\sqrt{2}} |2\mathbf{L}(i,j,k) - \mathbf{L}(i,j-1,k+1) - \mathbf{L}(i,j+1,k-1)| \\
&+ \frac{1}{\sqrt{2}} |2\mathbf{L}(i,j,k) - \mathbf{L}(i,j+1,k+1) - \mathbf{L}(i,j-1,k-1)| \\
&+ \frac{1}{\sqrt{2}} |2\mathbf{L}(i,j,k) - \mathbf{L}(i-1,j+1,k) - \mathbf{L}(i+1,j-1,k)| \\
&+ \frac{1}{\sqrt{2}} |2\mathbf{L}(i,j,k) - \mathbf{L}(i-1,j-1,k) - \mathbf{L}(i+1,j+1,k)| \\
&+ \frac{1}{\sqrt{3}} |2\mathbf{L}(i,j,k) - \mathbf{L}(i-1,j-1,k+1) - \mathbf{L}(i+1,j+1,k-1)| \\
&+ \frac{1}{\sqrt{3}} |2\mathbf{L}(i,j,k) - \mathbf{L}(i-1,j+1,k+1) - \mathbf{L}(i+1,j-1,k-1)| \\
&+ \frac{1}{\sqrt{3}} |2\mathbf{L}(i,j,k) - \mathbf{L}(i+1,j+1,k+1) - \mathbf{L}(i-1,j-1,k-1)| \\
&+ \frac{1}{\sqrt{3}} |2\mathbf{L}(i,j,k) - \mathbf{L}(i+1,j-1,k+1) - \mathbf{L}(i-1,j+1,k-1)|
\end{aligned}
\tag{8}
$$

Finally, the fusion rule $f_{3D-WLLE}$ is defined based on $WLLE$ and $\mathcal{W}_{26}$, i.e.,

$$f_{3D-WLLE} : \mathbf{L}_F(i,j,k)$$
$$= \begin{cases} \mathbf{L}_A(i,j,k), & \text{if } WLLE_A^N(i,j,k) + \mathcal{W}_{26_A}^N(i,j,k) \geq WLLE_B^N(i,j,k) + \mathcal{W}_{26_B}^N(i,j,k) \\ \mathbf{L}_B(i,j,k), & \text{otherwise} \end{cases}$$

(9)

where $\mathbf{L}_A$, $\mathbf{L}_B$ and $\mathbf{L}_F$ are low rank components of $\mathbf{X}_A$, $\mathbf{X}_B$ and fused image respectively, $WLLE_A^N = WLLE_A/W\|LLE_A\|_1$, $WLLE_B^N = WLLE_B/\|WLLE_B\|_1$, $\mathcal{W}_{26_A}^N = \mathcal{W}_{26_S}/\|\mathcal{W}_{26_A}\|_1$ and $\mathcal{W}_{26_B}^N = \mathcal{W}_{26_B}/\|\mathcal{W}_{26_B}\|_1$.

4 Experiments

4.1 Experimental Setting

To evaluate the performance of our method, the proposed method is compared with the LRD[1] [20], NSCT-PC[2] [21], MST-LRR[3] [22], RGF-MDFB[4] [23] and 3D-ST [24]. The LRD, NSCT-PC, MST-LRR and RGF-MDFB are implemented on each slice separately. For the LRD, NSCT-PC and RGF-MDFB, the parameters are set as the defaults. For the MST-LRR, the parameters are adjusted for better results. The 3D-ST extracts the 3-D features by the 3D Shearlet transform, and the "average" and "abs-max" rules are used in low-frequencies fusion and high-frequencies fusion, respectively.

We use four quality metrics to objectively assess the performance of tested methods, including the mutual information (Q_{MI}) [25], the gradient-based (Q_G) [25], the image structural similarity (Q_{Yang}) [25] and the weighted sum of edge information based metric (Q_{AB}^F) [26]. The metrics are computed in each slice individually, and then the mean values are reported. For all metrics, higher score demonstrates better fusion performance. All the experiments are programmed by MATLAB 2019a and running on an Intel(R) Core(TM) i7-8750H CPU @ 2.20 GHz Desktop with 8.00 GB RAM.

4.2 Comparison on the Synthetic Data

This subsection studies the performance of tested methods on the synthetic MRI image[5]. The size of selected images is $217 \times 181 \times 181$. Figure 3 shows one slice of sources and fused results. As show in Fig. 3, the contrast of fused image by proposed method is higher than others, and exhibits better visual. Moreover, Table 1 shows that proposed method obtains almost better metrics values than others. It also can demonstrate the superiority of proposed method objectively.

---

[1] The code is available at https://github.com/MDLW.

[2] The code is available at https://github.com/zhiqinzhu123.

[3] The code is available at https://github.com/hli1221/imagefusion_noisy_lrr.

[4] The code is available at https://github.com/Huises/RGF_MDFB.

[5] The data is available at http://www.bic.mni.mcgill.ca/brainweb/

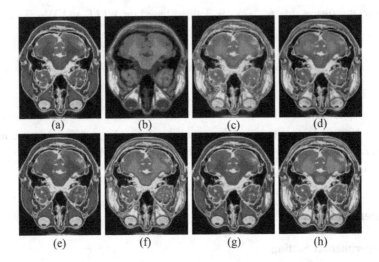

Fig. 3. Visual results of various fusion methods on synthetic MR-T1 and MR-T2 images. (a) MR-T1 image, (b) MR-T2 image, (c) LRD, (d) NSCT-PC, (e) MST-LRR, (f) RGF-MDFB, (g) 3D-ST, (h) Proposed method.

Table 1. Objective metrics results for fusing the synthetic MR-T1 and MR-T2 images

Metrics	LRD	NSCT-PC	MST-LRR	RGF-MDFB	3D-ST	Proposed
Q_{MI}	0.7754	0.6329	1.1253	1.1686	0.9979	**1.2372**
Q_G	0.5881	0.6457	0.7410	0.7389	0.7220	**0.7465**
Q_{Yang}	0.8143	0.7223	0.8988	**0.9000**	0.8723	0.8751
Q_{AB}^F	0.5427	0.6065	0.7129	0.7134	0.6883	**0.7143**

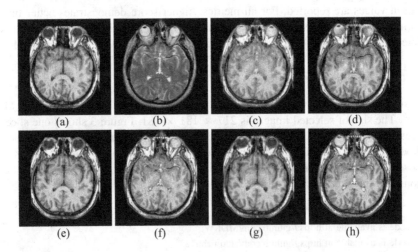

Fig. 4. Visual results of various fusion methods on real MR-T1 and MR-T2 images. (a) MR-T1 image, (b) MR-T2 image, (c) LRD, (d) NSCT-PC, (e) MST-LRR, (f) RGF-MDFB, (g) 3D-ST, (h) Proposed method.

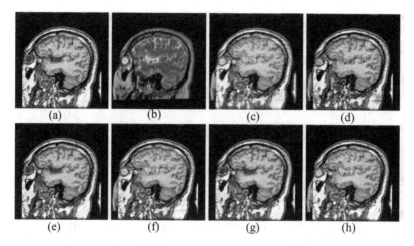

Fig. 5. Visual results of various fusion methods on real MR-T1 and MR-T2 images. (a) MR-T1 image, (b) MR-T2image, (c) LRD, (d) NSCT-PC, (e) MST-LRR, (f) RGF-MDFB, (g) 3D-ST, (h) Proposed method.

4.3   Comparison on the Real Data

In this subsection, six groups of real data[6] acquired by different imaging modalities are investigated to assess the fusion performance in real applications.

1) Fusing MR-T1 and MR-T2 images

Two groups of real MR-T1 and MR-T2 images with size $256 \times 256 \times 81$ are studied. The source image and fused images are shown in Figs. 4 and 5, respectively. As show in Figs. 4 and 5, we can see that the features and detailed information presented in the results of proposed method are much richer than others, especially the enhanced tissues. Overall, the proposed method provides a better visual effect. Table 2 gives the quantitative performance comparisons of results in Figs. 4 and 5. As we can see, the proposed method obtains the best values in most cases in fusing real MR-T1 and MR-T2 images.

2) Fusing MR-T1 and PET Images

In this experiment, two groups of MR-T1 and PET images are tested. Fusing MR-T1 and PET images can provide functional and anatomical information simultaneously. Due to that the PET image is the color image, we use the Y–Cb–Cr transform to change the color space of PET image firstly. Then, the MR-T1 image with the Y-channel of PET image are integrated into a fused component. Finally, the inverse transformation of Y–Cb–Cr is carried out to obtain the fused image.

---

[6] The data is available at http://www.med.harvard.edu/aanlib/home.html.

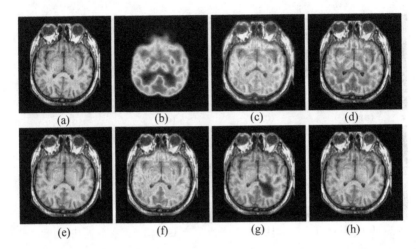

Fig. 6. Visual results of various fusion methods on real MR-T1 and PET images. (a) MR-T1 image, (b) PET image, (c) LRD, (d) NSCT-PC, (e) MST-LRR, (f) RGF-MDFB, (g) 3D-ST, (h) Proposed method.

The tested images are with the size of $256 \times 256 \times 81$. Figures 6 and 7 show two groups of source images and related fused images. By the visual comparison, the proposed method effectively combines the complementary information of sources, and can accurately locate the functional information. Table 3 lists the related indexes values of fused images in Figs. 6 and 7. The proposed method obtains the best scores in four metrics.

Table 2. Objective metrics results for fusing the MR-T1 and MR-T2 images

Images	Metrics	LRD	NSCT-PC	MST-LRR	RGF-MDFB	3D-ST	Proposed
Figure 4	Q_{MI}	0.7534	0.6352	1.0752	0.9575	0.7710	**1.1705**
	Q_G	0.6446	0.6700	0.6796	0.6323	0.5885	**0.7377**
	Q_{Yang}	0.7643	0.6510	0.8474	0.8913	0.5717	**0.9644**
	Q_{AB}^F	0.5988	0.6207	0.6505	0.5538	0.5470	**0.7138**
Figure 5	Q_{MI}	0.6683	0.5877	1.0062	0.9967	0.7843	**1.1347**
	Q_G	0.6059	0.6540	**0.7132**	0.6715	0.6855	0.7005
	Q_{Yang}	0.7339	0.6620	0.7579	0.9265	0.6782	**0.9511**
	Q_{AB}^F	0.5593	0.6036	0.6607	0.5660	0.6450	**0.6735**

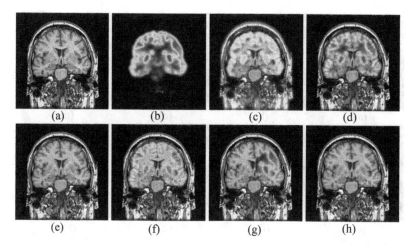

Fig. 7. Visual results of various fusion methods on real MR-T1 and PET images. (a) MR-T1 image, (b) PET image, (c) LRD, (d) NSCT-PC, (e) MST-LRR, (f) RGF-MDFB, (g) 3D-ST, (h) Proposed method.

Table 3. Objective metrics results for fusing the MR-T1 and PET images

Images	Metrics	LRD	NSCT-PC	MST-LRR	RGF-MDFB	3D-ST	Proposed
Figure 6	Q_{MI}	0.5191	0.5278	0.9085	0.9314	0.5708	**1.1803**
	Q_G	0.5474	0.6959	0.7518	0.7255	0.6784	**0.7651**
	Q_{Yang}	0.6204	0.3986	0.8418	0.9292	0.5215	**0.9969**
	Q_{AB}^F	0.5085	0.6587	0.7304	0.6442	0.6449	**0.8481**
Figure 7	Q_{MI}	0.4227	0.5524	0.6396	0.9873	0.5656	**1.1410**
	Q_G	0.5695	0.7588	0.7408	0.8054	0.7612	**0.8155**
	Q_{Yang}	0.6243	0.5300	0.5950	0.9548	0.6118	**0.9556**
	Q_{AB}^F	0.5291	0.7356	0.7346	0.7148	0.7213	**0.7982**

3) Fusing MR-T2 and SPECT Images

The performance on fusing real MR-T2 and SPECT images is studied in this experiment. SPECT image shows the metabolic changes. Fusing MR-T2 and SPECT images can preserve the organ structures, and can locate the color information of SPECT image by tissue structure of MRI image. In addition, the fusion procedures of MR-T2 and SPECT image are implemented as fusing MR-T1 and PET images.

We consider two groups of MR-T2 and SPECT images ($256 \times 256 \times 30$). The visual and quantitative comparisons are presented in Figs. 8, 9 and Table 4, respectively. The visual effects in Figs. 8 and 9, and indexes values in Table 4 demonstrate that the fused images obtained by proposed method are more informative and sharper than other fusion methods.

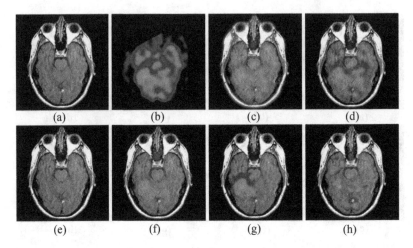

Fig. 8. Visual results of various fusion methods on real MR-T2 and SPECT images. (a) MR-T1 image, (b) SPECT image, (c) LRD, (d) NSCT-PC, (e) MST-LRR, (f) RGF-MDFB, (g) 3D-ST, (h) Proposed method.

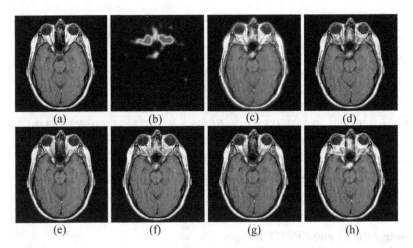

Fig. 9. Visual results of various fusion methods on real MR-T2 and SPECT images. (a) MR-T1 image, (b) SPECT image, (c) LRD, (d) NSCT-PC, (e) MST-LRR, (f) RGF-MDFB, (g) 3D-ST, (h) Proposed method.

Table 4. Objective metrics results for fusing the MR-T2 and SPECT images

Images	Metrics	LRD	NSCT-PC	MST-LRR	RGF-MDFB	3D-ST	Proposed
Figure 8	Q_{MI}	0.6326	0.6801	0.7031	1.0281	0.6408	**1.1447**
	Q_G	0.6916	0.7406	0.7382	0.8290	0.7608	**0.9217**
	Q_{Yang}	0.6562	0.5775	0.4844	**0.9601**	0.5313	0.7748
	Q_{AB}^F	0.6598	0.7181	0.7420	0.7199	0.7193	**0.7843**
Figure 9	Q_{MI}	0.5573	0.6888	0.6996	1.0642	0.6452	**1.2936**
	Q_G	0.6613	0.8117	0.8072	0.8675	0.8350	**0.9630**
	Q_{Yang}	0.6840	0.5358	0.5286	**0.9842**	0.5505	0.7963
	Q_{AB}^F	0.5698	0.8145	0.8165	0.6948	0.7877	**0.9034**

5 Conclusion

This paper proposed a TRPCA based 3-D medical images fusion method. The TRPCA decomposed components are used to capture the 3-D salient features of 3-D medical image Moreover, we adopt "3-D weighted local Laplacian energy" rule and "max-absolute" rule for fusing the low rank components and fusing the sparse components, respectively. By testing on the synthetic data and real data in different imaging modalities, the experimental results demonstrate the superiority of proposed method. In our future work, we will consider the relationship between different imaging methods to further improve proposed method.

Acknowledgements. This research was supported by the National Natural Science Foundation of China under grant No. 61971237 and No. 61501255.

References

1. Chang, L., Feng, X., Zhu, X., et al.: CT and MRI image fusion based on multiscale decomposition method and hybrid approach. IEEE Trans. Image Process. **13**(1), 83–88 (2019)
2. Sauter, A.W., Wehrl, H.F., Kolb, A., et al.: Combined PET/MRI: one-step further in multimodality imaging. Trends Mol. Med. **16**(11), 508–515 (2010)
3. Nastaran, E., et al.: Fusion of quantitative image and genomic biomarkers to improve prognosis assessment of early stage lung cancer patients. IEEE Trans. Biomed. Eng. **63**(5), 1034–1043 (2016)
4. Li, S., Kwok, J.T., Wang, Y.: Using the discrete wavelet frame transform to merge landsat TM and SPOT panchromatic images. Inf. Fusion **3**(1), 17–23 (2002)
5. Zhang, Q., Guo, B.L.: Multifocus image fusion using the non-subsampled contourlet transform. Sig. Process. **89**(7), 1334–1346 (2009)
6. Duan, C., Wang, X.G.: Multi-focus image fusion based on shearlet transform. J. Inf. Comput. Sci. **8**(15), 3713–3720 (2011)
7. Xu, L.P., Gao, G.R., Feng, D.Z.: Multi-focus image fusion based on non-subsampled shearlet transform. IET Image Process. **7**(6), 633–639 (2013)

8. Yin, H.T., Li, S.T., et al.: Simultaneous image fusion and super-resolution using sparse representation. Inf. Fusion **14**(3), 229–240 (2013)

9. Yang, B., Li, S.: Multifocus image fusion and restoration with sparse representation. IEEE Trans. Instrum. Meas. **59**(4), 884–892 (2010)

10. Xuejun, L., Minghui, W.: Research of multi-focus image fusion algorithm based on sparse representation and orthogonal matching pursuit. In: Tan, T., Ruan, Q., Wang, S., Ma, H., Huang, K. (eds.) IGTA 2014. CCIS, vol. 437, pp. 57–66. Springer, Heidelberg (2014). https://doi.org/10.1007/978-3-662-45498-5_7

11. Li, S.T., Yin, H.T., Fang, L.: Group-sparse representation with dictionary learning for medical image denoising and fusion. IEEE Trans. Biomed. Eng. **59**(12), 3450–3459 (2012)

12. Yang, B., Li, S.: Pixel-level image fusion with simultaneous orthogonal matching pursuit. Inf. Fusion **13**(1), 10–19 (2012)

13. Yin, H.T.: Sparse representation with learned multiscale dictionary for image fusion. Neurocomputing **148**, 600–610 (2015)

14. Cheng, Y., Jiang, Z., Shi, J., Zhang, H., Meng, G.: Remote sensing image change detection based on low-rank representation. In: Tan, T., Ruan, Q., Wang, S., Ma, H., Huang, K. (eds.) IGTA 2014. CCIS, vol. 437, pp. 336–344. Springer, Heidelberg (2014). https://doi.org/10.1007/978-3-662-45498-5_37

15. Li, H., Wu, X.J.: Multi-focus image fusion using dictionary learning and low-rank representation. In: Zhao, Y., Kong, X., Taubman, D. (eds.) Image and Graphics. ICIG 2017. Lecture Notes in Computer Science, vol. 10666, pp. 675–686. Springer, Cham (2017). https://doi.org/10.1007/978-3-319-71607-7_59

16. Su, J., Tao, M., Wang, L., et al.: RPCA based time-frequency signal separation algorithm for narrow-band interference suppression. In: IEEE International Geoscience and Remote Sensing Symposium, pp. 3086–3089, IEEE, Fort Worth, TX (2017)

17. Fernando, D.L.T., Michael, J.B.: Robust principal component analysis for computer vision. In: Proceedings 8th IEEE International Conference on Computer Vision, pp. 362–36, Vancouver, BC (2001)

18. Lu, C.Y.: Tensor robust principal component analysis with a new tensor nuclear norm. IEEE Trans. Pattern Anal. Mach. Intell. **42**(4), 925–938 (2020)

19. Zhang, A.R., Xia, D.: Tensor SVD: statistical and computational limit. IEEE Trans. Inf. Theor. **64**(11), 7311–7338 (2018)

20. Li, X., Guo, X., Han, P., et al.: Laplacian re-decomposition for multimodal medical image fusion. IEEE Trans. Instrum. Meas. (2020). https://doi.org/10.1109/TIM.2020.2975405

21. Zhu, Z., Zheng, M., Qi, G., Wang, D., Xiang, Y.: A phase congruency and local Laplacian energy based multi-modality medical image fusion method in NSCT domain. IEEE Access **7**, 20811–20824 (2019)

22. Hui, L., et al. Multi-focus noisy image fusion using low-rank representation. Adv. Comput. Vis. Pattern Recognit. (2018). https://arxiv.org/abs/1804.09325

23. Chen, L.H., Yang, X.M., et al.: An image fusion algorithm of infrared and visible imaging sensors for cyber-physical systems. J. Intell. Fuzzy Syst. **36**(5), 4277–4291 (2019)

24. Xi, X.X., Luo, X.Q., Zhang, Z.C., et al.: Multimodal medical volumetric image fusion based on multi-feature in 3-D shearlet transform. In: 2017 International Smart Cities Conference (ISC2). IEEE (2017)

25. Zheng, L., Erik, B., et al.: Objective assessment of multiresolution image fusion algorithms for context enhancement in night vision: a comparative study. IEEE Trans. Pattern Anal. Mach. Intell. **34**(1), 94–109 (2012)

26. Vladimir, P., Costas, X.: Objective image fusion performance characterization. In: 10th IEEE International Conference on Computer Vision, pp. 1866–1871. IEEE, Beijing (2005)

Accurate Estimation of Motion Blur Kernel Based on Genetic Algorithms

Ziwei Zhou, Pu Huang, Gaojin Wen$^{(\boxtimes)}$, Zhiming Shang, Can Zhong,
Zhaorong Lin, and Jiyou Zhang

Beijing Engineering Research Center of Aerial Intelligent Remote Sensing
Equipment, Beijing Institute of Space Mechanics and Electricity, Beijing
100192, People's Republic of China
76630093@qq.com

Abstract. In this paper, a motion blur kernel estimation method is proposed based on genetic algorithms. Specific individual representation and initialization are developed for optimizing motion blur kernel. Effective fitness function and genetic operators are designed to ensure the convergence of the optimization process. Several traditional methods are compared with the proposed method. Real aerial image data sets are used to test the performance of these methods. It is proved that the proposed method can estimate the blur kernel more effectively and accurately than traditional ones.

Keywords: Motion blur · Remote sensing image · Blur kernel estimation · Genetic algorithms

1 Introduction

Images are an important carrier of people's cognitive world. Aerospace remote sensing cameras provide important image resources for many fields such as national defense, agriculture, commerce and scientific research. In the imaging process of a remote sensing camera, image degradation may occur due to environmental, technical limitations or other effects. Motion blur is a common type of image degradation, which is caused by relative displacement of the camera between the image and observed scene. Image motion deblurring has always been the focus of image restoration research.

Many researchers have conducted research on motion blur simulation [1–9] and achieved many good results. The traditional motion deblurring method [2, 3, 10–16] has changed from the earliest non blind restoration method to blind restoration method in the past 30 years. Recently machine learning methods [17–20] have been developed and show some improvements.

This paper proposes a more-accurate blur kernel estimation method for motion blur images. It can take advantages of many existing blur kernel estimation methods and use adaptive heuristic search method of genetic algorithms to optimize motion blur kernel. It results in solutions with optimal fitness and more accurate motion blur kernels.

The rest of the paper is arranged as follows. Related work is discussed in Sect. 2. The framework of the motion blur kernel estimation method is proposed in Sect. 3. In Sect. 4, experiments and analysis are presented. Finally, a short conclusion is drawn in Sect. 5.

© Springer Nature Singapore Pte Ltd. 2020
Y. Wang et al. (Eds.): IGTA 2020, CCIS 1314, pp. 43–54, 2020.
https://doi.org/10.1007/978-981-33-6033-4_4

2   Related Works

2.1   Traditional Methods

Potmesil M. erected a camera model for simulating motion blur using generalized imaging equations. These equation use optical system transfer function to describe the relationship between an object and its corresponding object point [2]. Boracchi G. derived a new statistical model for the performance of a given de-fuzzification algorithm under arbitrary motion [3]. This model uses random process and Monte Carlo method to represent the recovery performance as the expected recovery error values with some motion randomness descriptors and exposure times. Junseong Park characterized motion blurring by the direction and length of blurring in the cepstrum domain [10]. D. Krishnan introduced an image regularization algorithm [11], which can obtain clear images at a lower cost. Libin Sun proposed a blind blur kernel estimation strategy based on subblock images [12]. D. Krishnan pointed out that all successful MAP and variogram algorithms share a common framework [13], which relied on sparsity enhancement in gradient domains. Weishen Lai proposed an algorithm for estimating the blur kernel from a single image [14]. Jinshan Pan proposed an effective blind image restoration algorithm based on dark channel characteristics [15]. By analyzing the motion blurred light fields in the time domain and Fourier frequency domain, Srinivasan P.S. found the effect of intuitive camera motion on the light field for 3D cameras [16].

2.2   Machine Learning Methods

Seungjun N. proposed a multiscale convolution neural network based on a coarse-to-fine pyramid deblurring method [17], which can restore clear images from blurred images in various end-to-end ways. Orest K. also proposed an end-to-end motion deblurring learning method [18], named DeblurGAN. Tao Xin studied the strategy of gradually restoring clear images at different resolutions from coarse to fine in single image deblurring and proposed a Scale-recurrent network (SRN-DeblurNet) [19]. Leren Han et al. proposed an efficient blind deblurring method based on a data-driven discriminant prior [20].

3   Motion Blur Kernel Estimation Based on Genetic Algorithms

In this section, we present the main framework of our motion blur kernel estimation method based on genetic algorithms, which including individual representation and initialization, fitness function and genetic operators designs.

3.1   Individual Representation

Aiming at the structural characteristics of blur kernels, it is natural to adopt real-number coding method. The two-dimensional blur kernel B(n × n) is adjusted into a one-dimensional vector $G(n^2 \times 1)$ with the sum constraint. The specific formula is as follows:

$$G = (B_{11}, B_{12}, \cdots, B_{1n}, B_{21}, B_{22}, \cdots, B_{2n}, \cdots, B_{nn}) \tag{1}$$

$$\sum_{i=1}^{n}\sum_{j=1}^{n} B_{ij} = 1, \quad 0 \leq B_{ij} \leq 1 \tag{2}$$

Where G represents the encoded individual and B represents 2d matrix of the original blur kernel.

3.2   Initialization of the First Generation Population

Based on available research on image blind restoration, there are a variety of blind restoration algorithms that have a good effect on the estimation of blur kernels. Six different blur kernel estimation methods are selected, respectively: Michaeli's blind deblurring algorithm (BDM) [21], blur kernel estimation from spectral irregularities algorithm (BKESI) [22], kernel original blurred algorithm (KOB) [12], dark channel algorithm (DC) [15], Normalized sparsity blind deblur algorithm (NSBD) [11] and sparse blind deblur algorithm (SBD) [13]. Aim to improve efficiency of the algorithm and speed up convergence, high-quality blur kernels generated by these algorithms are introduced to initialize the first generation population,

In order to facilitate calculation and unified algorithms, the blur kernel is set to 25 × 25, its initialization process is as follows:

(1) Read all blurred images in a folder and the clear images paired with it.
(2) Use the six selected blur kernel estimation methods to calculate blur kernels one by one for the blurred image cyclically.
(3) Use blur kernels to degrade the clear image.
(4) Calculate the peak signal to noise ratio (PSNR) of the degraded image and the original blurred image.
(5) Output the generated blur kernel and compare the PSNR for the 6 methods.
(6) Repeat steps 2 to 5 until all images have been calculated.

In addition, the rest of individuals adopt a random generation method. The initialization value is set to a non-negative real number and normalized to satisfy the special form of the blur kernel. The population size is set to 200.

3.3 Genetic Operators

(1) Crossover Operator

The crossover scattered operator is adopted as crossover operator. The crossover rate is 0.8. The binary vector consists of 1 and 0. If one bit of the vector is 1, then the gene at that position comes from the P1; if it is 0, the gene comes from the P2. The formula is as follows:

$$P_1 = \left(B_{11}^1, B_{12}^1, \cdots, B_{1n}^1, B_{21}^1, B_{22}^1, \cdots, B_{2n}^1, \cdots, B_{nn}^1\right) \tag{3}$$

$$P_2 = \left(B_{11}^2, B_{12}^2, \cdots, B_{1n}^2, B_{21}^2, B_{22}^2, \cdots, B_{2n}^2, \cdots, B_{nn}^2\right) \tag{4}$$

If binary vector is [1, 0, ..., 1, 1, 0, ..., 0, ..., 1], generated child C is:

$$C = \left(B_{11}^1, B_{12}^2, \cdots, B_{1n}^1, B_{21}^1, B_{22}^1, \cdots, B_{2n}^2, \cdots, B_{nn}^1\right) \tag{5}$$

(2) Mutation Operator

Gaussian mutation is used as the mutation operator, which adds a Gaussian distribution random vector Gauss(n2, 1) on the parent generation. Its mean value is 0, the square of variance is the maximum value in B, the mutation rate is set to 0.15. The expression is as follow:

$$C = P + Gauss \tag{6}$$

(3) Normalization Operator

In later generations, due to the uncertainty caused by crossover and mutation, non-negative or sum constraints in genes will not be ensured. So before evaluating the suitability, standardization and non negative process will be carried out:

$$C = \frac{Relu(C)}{\sum_{i=1}^{n^2} Relu(C_i)} \tag{7}$$

Where n is the size of the blur kernel, and the Relu function is as follows:

$$Relu(x) = \begin{cases} x(x > 0) \\ 0(x < 0) \end{cases} \tag{8}$$

3.4 Fitness Function

Multiple blur kernels (shown in Fig. 3) calculated by different algorithms that are similar or slightly different. It is hard to compare these blur kernels directly. Most of the time, the original clear image is not known beforehand. But for a special GOPRO data [17] selected in this article, it happens to have a pair of clear and blurred images of the same scene, so it is feasible to adopt this evaluation method. So, there are two methods to evaluate the blur kernels. The first is to compare the difference between the obtained blurred image and the original blurred image. The second is to compare the difference between the deblurred image and the original clear image. Because the blur and deblur process are not a pair of exactly reciprocal process, we adopt the second method to evaluate the blur kernels. PSNR is the most used indicator to evaluate the difference between two images.

Fitness is a criterion for evaluating individuals in a population for genetic algorithms. The goal of optimization is to obtain a better blur kernel to simulate blur image, so the quality of blur kernel can be evaluated according to the similarity between the degraded image simulated by the blur kernel and the standard degraded image. The fitness function can be set as the PSNR between the degraded image $\hat{g}(x, y)$ simulated by the blur kernel B and the standard degraded image $g(i, j)$. The higher the PSNR, the better the individual B and the better the adaptability.

$$PSNR = 10 log_{10} \left[\frac{255^2 \times M \times N}{\sum_{i=1}^{M} \sum_{j=1}^{N} [g(i,j) - \hat{g}(i,j)]^2} \right] \tag{9}$$

$$\hat{g}(x, y) = f(x, y) \otimes B(x, y) \tag{10}$$

3.5   Parameter Setting

The parameter settings for genetic algorithms optimization are shown in Table 1:

Table 1.   Genetic algorithms parameter table

Population size	Iteration number	Stagnation algebra
200	1000	100
Crossover probability	Mutation probability	Good offspring probability
0.8	0.15	0.05

4   Experiment and Analysis

4.1   Sample Data of Blur Kernel Estimation

In order to better verify the accuracy and make full use of advantages of various traditional blur kernel estimation methods, we use the data set containing clear images and blurred images of the same scene provided by Seungjun N. et al. [17]. This data set uses the GOPRO4 Hero Black camera to shoot 240fps high-speed video, and then synthesize continuous multi-frame images to obtain images with different degrees of blur. The formula to simulate a blurred image is as follows:

$$I = g\left(\frac{1}{T}\int_{t=0}^{T} S(t)dt\right) \simeq g\left(\frac{1}{M}\sum_{i=0}^{M-1} S[i]\right) \tag{11}$$

Where I represents the simulated blurred image, T represents the exposure time, S(t) represents the clear image generated by the camera at time t, M represents the number of clear images sampled, and S[i] represents the i-th image sampled.

Through this method, exposure time of the camera can be improved equivalently, and the process of image motion blur is simulated in principle. There are many images in the GOPRO dataset. Only a few representative clear and corresponding blurred images are selected for display, as shown in the Fig. 2.

As can be seen from Fig. 1, the clear and blurred images of this data set are matched perfectly, and these pictures include various scenes and different degrees of motion blur. It is suitable for the performance comparison experiments of blur kernel estimation algorithm. The selected images are all RGB images with resolution 1280 × 720.

Fig. 1. GOPRO image dataset [17] (Color figure online)

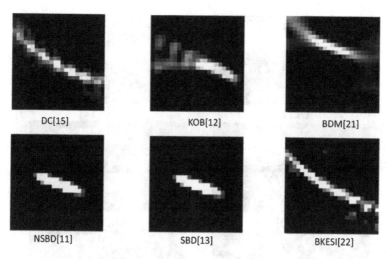

Fig. 2. The estimated blur kernels

4.2   The Rough Estimation Process of Motion Blur Kernel

During the experiment, six algorithms mentioned in Sect. 3.2 are integrated to conduct the rough estimation of the motion blur kernel from the dataset. After this process, each pair of images has six different blur kernels representing its degradation process, which provide an initialization for the accurate estimation of blur kernels. Figure 3 shows an example for blur kernels generated by six motion blur estimation methods.

It can be seen from Fig. 2 that blur kernels generated by these six methods are subjectively close, indicating that these blur kernels have certain correctness. The blur kernels measured by the DC algorithm and BKESI algorithm have a better blur trajectory from a subjective point of view. However, it is not enough to evaluate blur kernels subjectively based on the shape of blur kernel images. It is necessary to use the previously designed evaluation criteria to achieve an objective comparison.

Figure 3 shows images degraded by blur kernels from the original clear image. It is obvious that these images simulated by blur kernels have some certain motion blur effect. The simulated images generated by DC algorithm, BDM algorithm and BKESI algorithm are closer to the original blurred images. The image obtained by KOB appears to be a little excessively blurred. The blur effect of the images generated by NSBD and SBD is much weaker.

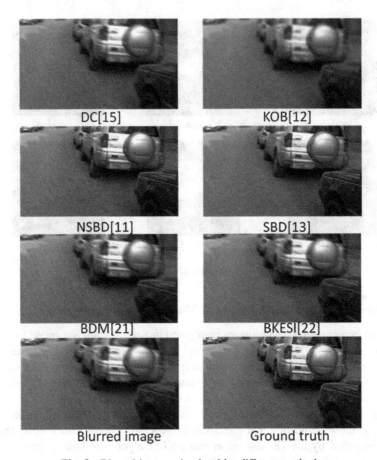

DC[15]   KOB[12]

NSBD[11]   SBD[13]

BDM[21]   BKESI[22]

Blurred image   Ground truth

Fig. 3. Blurred images simulated by different methods

4.3   Blur Kernel Estimation Results

The convergence process of the proposed method is shown in Fig. 4. GAs can significantly improve the estimation accuracy of the blur kernel at the beginning. But with the increase of the number of iterations, the speed of accuracy improvement slows down. After the iteration number reaches 1000, the simulated PSNR value improves to 27.8. Compared with the previous 26.2, there is a significant improvement.

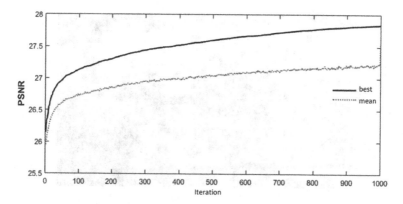

Fig. 4. Convergence process of the proposed method.

Figure 5 is the blur kernel obtained by genetic algorithms. Figure 6 is the corresponding blurred image simulated.

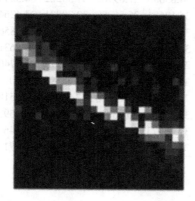

Fig. 5. Blur kernel generated by our method

It can be deduced from Fig. 5 and Fig. 2 that the shape of the blur kernel optimized by our GAs is closer to the DC and BKESI methods. It indicates that the blur kernel obtained by these two algorithm is closer to the optimal solution.

It can be deduced from Fig. 6 and Fig. 3 that the blurred image formed by GAs optimized kernel is closer to the original blur image. A large number of experimental data have been used and the performance of the related methods are compared in Table 2. From an objective perspective, our GAs optimized blur kernels have achieved a greater improvement than the kernels generated by other 6 method mentioned above. Compared to the DC and BKESI methods, PSNR of are improved by 8.63% and 11.13% respectively.

Fig. 6. Blurred image formed by our GAs optimized blur kernel

Table 2. Accuracy comparison of blur kernels

Picture number	BDM	BKESI	KOB	DC	NSBD	SBD	Ours
1	24.71435	34.85688	27.09902	37.30222	24.32481	24.32481	**41.2035**
2	23.42939	35.12684	25.56988	32.29821	24.42644	24.42644	**36.8737**
3	20.69008	34.20246	24.54581	32.54868	25.28563	25.28563	**35.5081**
4	26.29067	35.04271	29.96341	31.43363	26.86314	26.86314	**39.1930**
5	22.03133	23.68857	25.95706	24.53323	21.17407	21.17407	**28.2036**
6	25.32574	32.54563	28.11242	30.66594	24.80795	24.80795	**33.9744**
7	24.98136	29.55461	27.79136	27.06245	22.85184	22.85184	**30.7975**
8	26.63046	38.14687	30.76924	38.38482	29.25976	29.25976	**40.0485**
9	27.03469	37.66373	29.15143	39.82804	29.34631	29.34631	**41.0022**
Avg.	24.56979	33.42537	27.66218	32.67302	25.37111	25.37111	**36.31161**

5 Conclusion

This paper mainly discusses the method of extracting high-quality motion blur kernels of image, and proposes an accurate blur kernel estimation algorithm based on genetic algorithms. Specific individual representation and initialization methods are developed for motion blur kernel. Effective fitness function and genetic operators are designed to ensure the convergence of the optimization process. Several traditional methods are compared with the proposed method. Real aerial image data sets are used to test the performance of these methods. It is proved that the proposed method can estimate the blur kernel more effectively and accurately than traditional ones.

Acknowledgment. The authors gratefully acknowledge the support to this work from all our colleagues in Beijing Engineering Research Center of Aerial Intelligent Remote Sensing Equipment.

References

1. Mei, C., Reid, I.: Modeling and generating complex motion blur for real-time tracking. In: 2008 IEEE Conference on Computer Vision and Pattern Recognition, Anchorage, AK, USA, pp. 1–8. IEEE Press (2008)
2. Potmesil, M., Chakravarty, I.: Modeling motion blur in computer-generated images. J. ACM SIGGRAPH Comput. Graph. **17**, 389–399 (1983)
3. Boracchi, G., Foi, A.: Modeling the performance of image restoration from motion blur. J. IEEE Trans. Image Process. **21**, 3502–3517 (2012)
4. Navarro, F., Francisco, J., Serón, G.D.: Motion blur rendering. J. State Art. **30**, 3–26 (2011)
5. Choi, B.D., Jung, S.W., Ko, S.J.: Motion-blur-free camera system splitting exposure time. J. IEEE Trans. Consum. Electron. **54**, 981–986 (2008)
6. Zhao, H., Shang, H., Jia, G.: Simulation of remote sensing imaging motion blur based on image motion vector field. J. Appl. Remote Sens. **8**, 083539 (2014)
7. Q i P. Correct Generation of Motion Blur Simulation Images. Jinan University (2005)
8. Tiwari, S., Shukla, V.P., Singh, A.K., et al.: Review of motion blur estimation techniques. J. Image Graph. **1**, 176–184 (2013)
9. Kuciš, M., Zemčík, P.: Simulation of camera features. In: The 16th Central European Seminar on Computer, pp. 117–123 (2012)
10. Park, J., Kim, M., Chang, S.K., et al.: Estimation of motion blur parameters using cepstrum analysis. In: 2011 IEEE 15th International Symposium on Consumer Electronics, Singapore, pp. 406–409. IEEE Press (2011)
11. Krishnan, D., Tay, T., Fergus, R.: Blind deconvolution using a normalized sparsity measure (2011)
12. Sun, L., Cho, S., Wang, J., et al.: Edge-based blur kernel estimation using patch priors. In: 2013 IEEE International Conference on Computational Photography, Cambridge, MA, USA, pp. 1–8. IEEE Press (2013)
13. Krishnan, D., Bruna, J., Fergus, R.: Blind deconvolution with non-local sparsity reweighting. In: Computer Science (2013)
14. Lai, W.S., Ding, J., Lin, Y., et al.: Blur kernel estimation using normalized color-line priors. In: 2015 IEEE Conference on Computer Vision and Pattern Recognition, Boston, MA, USA, pp. 64–72. IEEE Computer Society (2015)
15. Pan, J., Sun, D., Pfister, H., et al.: Blind image deblurring using dark channel prior. In: 2016 IEEE Conference on Computer Vision and Pattern Recognition, Las Vegas, NV, USA, pp. 1628–1636. IEEE Press (2016)
16. Srinivasan, P., Ng, R., Ramamoorthi, R.: Light field blind motion deblurring. In: 2017 IEEE Conference on Computer Vision and Pattern Recognition, pp. 2354–2362 (2017)
17. Nah, S., Kim, T.H., Lee, K.M.: Deep multi-scale convolutional neural network for dynamic scene deblurring (2016)
18. Kupyn, O., Budzan, V., Mykhailych, M., et al.: DeblurGAN: blind motion deblurring using conditional adversarial networks (2017)
19. Tao, X., Gao, H., Wang, Y., et al.: Scale-recurrent network for deep image deblurring (2018)
20. Li, L., Pan, J., Lai, W.S., et al.: Learning a discriminative prior for blind image deblurring (2018)

21. Michaeli, T., Irani, M.: Blind deblurring using internal patch recurrence. In: Fleet, D., Pajdla, T., Schiele, B., Tuytelaars, T. (eds.) ECCV 2014. LNCS, vol. 8691, pp. 783–798. Springer, Cham (2014). https://doi.org/10.1007/978-3-319-10578-9_51
22. Goldstein, A., Fattal, R.: Blur-kernel estimation from spectral irregularities. In: Fitzgibbon, A., Lazebnik, S., Perona, P., Sato, Y., Schmid, C. (eds.) ECCV 2012. LNCS, vol. 7576, pp. 622–635. Springer, Heidelberg (2012). https://doi.org/10.1007/978-3-642-33715-4_45

Biometric Identification Techniques

Graph Embedding Discriminant Analysis and Semi-Supervised Extension for Face Recognition

Jiangshan Du[1], Liping Zheng[1], and Jun Shi[2(✉)]

[1] School of Computer Science and Information Engineering,
Hefei University of Technology, Hefei 230009, China
[2] School of Software, Hefei University of Technology, Hefei 230009, China
juns@hfut.edu.cn

Abstract. Considering graph embedding based dimension reduction methods are easily affected by outliers and the fact that there exist insufficient labeled samples in practical applications and thus degrades the discriminant performance, a novel graph embedding discriminant analysis method and its semi-supervised extension for face recognition is proposed in this paper. It uses the intraclass samples and their within-class mean to construct the intrinsic graph for describing the intraclass compactness, which can effectively avoid the influence of outliers to the within-class scatter. Meanwhile, we build the penalty graph through the local information of interclass samples to characterize the interclass separability, which considers the different contributions of interclass samples within the neighborhood to the between-class scatter. On the other hand, we apply the low-rank representation with sparsity constraint for semi-supervised learning, aiming to explore the global low-rank relationships of unlabeled samples. Experimental results on ORL, AR and FERET face datasets demonstrate the effectiveness and robustness of our method.

Keywords: Graph embedding · Low-rank representation · Semi-supervised learning · Dimension reduction · Face recognition

1 Introduction

Over the past decades, dimension reduction has been one research hotspot of face recognition, which aims to discover a low-dimensional representation from original high-dimensional data and thus capture the intrinsic structure of data and finally contribute to classification. Principal Component Analysis (PCA) [1] and Linear Discriminant Analysis (LDA) [2] are the most classic methods. PCA seeks a set of optimal projection directions to maximize the data covariance in an unsupervised way, while LDA uses the label information to construct the within-class scatter and between-class scatter and thus finds the discriminant projection directions. Therefore, LDA is better than PCA in face recognition and other recognition tasks. However, both of them fail to reveal the local structure due to their global linearity.

In order to capture the local information, the manifold learning-based methods have been proposed, which generally assume that the closely located samples are likely to

© Springer Nature Singapore Pte Ltd. 2020
Y. Wang et al. (Eds.): IGTA 2020, CCIS 1314, pp. 57–73, 2020.
https://doi.org/10.1007/978-981-33-6033-4_5

have the similar low-dimensional manifold embeddings. As the most representative manifold learning method, Locality Preserving Projections (LPP) [3] applies k-nearest-neighbor to explore the local structure of each sample. In contrast to LPP, Marginal Fisher Analysis (MFA) [4] uses k_1 nearest intraclass samples to build the intrinsic graph and simultaneously employs k_2 nearest interclass marginal samples to construct the penalty graph. Both of these two graphs can describe the intraclass compactness and the interclass separability respectively. It verifies that the manifold learning-based dimensionality reduction methods can be well explained by graph embedding framework [4]. However, compared with the k_1 nearest intraclass samples and k_2 nearest interclass samples, the distant intraclass and interclass samples are not considered during the process of graph construction. Therefore, it may cause the distant intraclass samples are far from each other and the distant interclass samples are close to each other in the projected feature space, which inevitably goes against classification. To address this problem, Huang et al. proposed the Graph Discriminant Embedding (GDE) [5] to focus on the importance of the distant intraclass and interclass samples in the graph embedding framework. It uses a strictly monotone decreasing function to describe the importance of nearby intraclass and interclass samples and simultaneously applies a strictly monotone increasing function to characterize the importance of distant intraclass and interclass samples. Consequently, the different contributions of nearby and distant samples to the within-class and between-class scatters are fully considered. For the most graph embedding methods, the importance of samples is directly based on the similarity between two samples. Therefore, outliers have the influence to the graph construction and especially affect the intraclass compactness.

On the other hand, considering there exist unlabeled samples in practical application, Cai et al. [6] proposed the Semi-Supervised Discriminant Analysis (SDA) which performs LDA on labeled samples and uses unlabeled samples to build k-nearest-neighbor graph for preserving local neighbor relationships. It effectively applies semi-supervised learning to achieve low dimensional embedding. However, the neighbor parameter k is difficult to define. Inappropriate k even influences the quality of graph construction and thus degrades the semi-supervised discriminant performance. In the past few years, Low Rank Representation (LRR) [7] has been proposed to capture the global structure of samples using low-rank and sparse matrix decomposition. The decomposed low-rank part reflects the intrinsic affinity of samples. In other words, the similar samples will be clustered into the same subspace. However, LRR is essentially unconstrained and is likely to be sensitive to noise for the corrupted data. Therefore, Guo et al. [8] performed the L_1-norm sparsity constraint on low-rank decomposition with the aiming of obtaining the more robust low-rank relationships.

Motivated by the above discussion, we introduce a novel graph embedding discriminant analysis method and its semi-supervised extension for face recognition in this paper. For the labeled samples, we redefine the weights of the intraclass graph through the mean of intraclass samples. In other words, the similarity between two intraclass samples is based on the similarities between them and the mean of intraclass samples, instead of directly using the distance of them. It effectively avoids the influence of outliers to the within-class scatter. Meanwhile we concentrate the different contributions of interclass samples to between-class scatter in view of the idea of GDE. Consequently, we build the improved discriminant criterion to make the intraclass samples as close to each other as possible and interclass samples far from each other. More critically, the constructed graph embedding framework can relieve the influence of outliers to the intraclass compactness. On the other hand, considering there may exist a large number of unlabeled samples in practical application, we apply the low rank representation constrained by L_1-norm sparsity to capture their global low-rank relationships, and then minimizes the error function through preserving the low-rank structure of samples. Finally, we integrate the improved graph embedding discriminant criterion and LRR error function for achieving semi-supervised learning. Experimental results on ORL, AR and FERET face datasets demonstrate the effectiveness and robustness of our method.

The remainder of this paper is organized as follows. Section 2 briefly reviews MFA and GDE. Section 3 introduces the proposed method. Section 4 shows the experimental results and the conclusion is given in Sect. 5.

2   Related Works

Given a set of n training samples $\mathbf{X} = [\mathbf{x}_1, \mathbf{x}_2, \ldots, \mathbf{x}_n]$, containing C classes: $\mathbf{x}_i \in l_c$, $c \in \{1, 2, \ldots, C\}$. By using the projection matrix $\mathbf{A}$, the representation of the data set $\mathbf{X}$ in the low-dimensional space can be represented as $\mathbf{y}_i = \mathbf{A}^T \mathbf{x}_i$. In this section, we will briefly review MFA and GDE.

2.1   Marginal Fisher Analysis (MFA)

As the most representative graph embedding method, MFA [4] builds the intrinsic graph to characterize the compactness of intraclass samples, and constructs the penalty graph to describe the separability of interclass samples. Let $G_{intrinsic} = \{\mathbf{V}, \mathbf{E}, \mathbf{W}^{MFA}\}$ denotes the intrinsic graph, and $G_{penalty} = \{\mathbf{V}, \mathbf{E}, \mathbf{B}^{MFA}\}$ denotes the penalty graph, where $\mathbf{V}$ is the samples, and $\mathbf{E}$ is the set of edges connecting samples. $\mathbf{W}^{MFA}$ and $\mathbf{B}^{MFA}$ are two adjacency matrices identifying the similarity between samples. The weights of these two matrices are defined [4]:

$$\mathbf{W}_{ij}^{MFA} = \begin{cases} 1, & \text{if } \mathbf{x}_i \in N_{k_1}^+(j) \text{ or } \mathbf{x}_j \in N_{k_1}^+(i) \\ 0, & \text{otherwise} \end{cases} \tag{1}$$

$$\mathbf{B}_{ij}^{MFA} = \begin{cases} 1, & \text{if } \mathbf{x}_i \in P_{k_2}(j) \text{ or } \mathbf{x}_j \in P_{k_2}(i) \\ 0, & \text{otherwise} \end{cases} \tag{2}$$

where $N_{k_1}^+(i)$ indicates the k_1 nearest intraclass samples of $\mathbf{x}_i$, and $P_{k_2}(i)$ indicates the k_2 nearest interclass marginal samples of $\mathbf{x}_i$. The intraclass scatter $\mathbf{S}_w^{MFA}$ and interclass scatter $\mathbf{S}_b^{MFA}$ can be formulated [4]:

$$\mathbf{S}_w^{MFA} = \sum\nolimits_{ij} \|\mathbf{y}_i - \mathbf{y}_j\|^2 \mathbf{W}_{ij}^{MFA} = 2tr\left(\mathbf{A}^T \mathbf{X} \mathbf{L}^{MFA} \mathbf{X}^T \mathbf{A}\right) \tag{3}$$

$$S_b^{MFA} = \sum_{ij} ||\mathbf{y}_i - \mathbf{y}_j||^2 \mathbf{B}_{ij}^{MFA} = 2tr\left(\mathbf{A}^T \mathbf{X} \mathbf{L}^{MFA'} \mathbf{X}^T \mathbf{A}\right) \tag{4}$$

where $\mathbf{L}^{MFA} = \mathbf{D}^{MFA} - \mathbf{W}^{MFA}$ and $\mathbf{L}^{MFA'} = \mathbf{D}^{MFA'} - \mathbf{B}^{MFA}$ are the Laplacian matrices; $\mathbf{D}_{ii}^{MFA} = \sum_j \mathbf{W}_{ij}^{MFA}$ and $\mathbf{D}_{ii}^{MFA'} = \sum_j \mathbf{B}_{ij}^{MFA}$ are the diagonal matrices. The discriminant criterion function of MFA can be expressed as [4]:

$$J_{MFA}(\mathbf{A}) = \min_{\mathbf{A}} \frac{S_w^{MFA}}{S_b^{MFA}} = \min_{\mathbf{A}} \frac{tr(\mathbf{A}^T \mathbf{X} \mathbf{L}^{MFA} \mathbf{X}^T \mathbf{A})}{tr(\mathbf{A}^T \mathbf{X} \mathbf{L}^{MFA'} \mathbf{X}^T \mathbf{A})} \tag{5}$$

2.2   Graph Discriminant Embedding (GDE)

The purpose of MFA is to construct two graphs by comprehensively considering the local structure and discriminant structure of samples. However, MFA neglects the importance of the distant samples which may influence the discriminant performance. Therefore, GDE redefines the intrinsic graph $G_{intrinsic}$ and the penalty graph $G_{penalty}$. The weight matrices $\mathbf{W}^{GDE}$ and $\mathbf{B}^{GDE}$ in these two graphs are redefined [5]:

$$\mathbf{W}_{ij}^{GDE} = \begin{cases} e^{-\frac{||\mathbf{x}_i - \mathbf{x}_j||^2}{t}}, & \text{if } \mathbf{x}_j \in N_{k1}^+(i) \text{ or } \mathbf{x}_i \in N_{k1}^+(j), \, l_i = l_j \\ e^{\frac{t}{||\mathbf{x}_i - \mathbf{x}_j||^2}}, & \text{if } \mathbf{x}_j \notin N_{k1}^+(i) \text{ and } \mathbf{x}_i \notin N_{k1}^+(j), \, l_i = l_j \\ 0, & \text{if } l_i \neq l_j \end{cases} \tag{7}$$

$$\mathbf{B}_{ij}^{GDE} = \begin{cases} e^{-\frac{||\mathbf{x}_i - \mathbf{x}_j||^2}{t}}, & \text{if } \mathbf{x}_j \in P_{k2}(i) \text{ or } \mathbf{x}_i \in P_{k2}(j), \, l_i \neq l_j \\ e^{\frac{t}{||\mathbf{x}_i - \mathbf{x}_j||^2}}, & \text{if } \mathbf{x}_j \notin P_{k2}(i) \text{ and } \mathbf{x}_i \notin P_{k2}(j), \, l_i \neq l_j \\ 0, & \text{if } l_i = l_j \end{cases} \tag{8}$$

where the parameter t $(t > 0)$ is used to tune the weight of samples. The intraclass scatter $\mathbf{S}_w^{GDE}$ and the interclass scatter $\mathbf{S}_b^{GDE}$ can be computed [5]:

$$\mathbf{S}_w^{GDE} = \sum_{ij} ||\mathbf{y}_i - \mathbf{y}_j||^2 \mathbf{W}_{ij}^{GDE} = 2tr(\mathbf{A}^T \mathbf{X} \mathbf{L}^{GDE} \mathbf{X}^T \mathbf{A}) \tag{9}$$

$$\mathbf{S}_b^{GDE} = \sum_{ij} ||\mathbf{y}_i - \mathbf{y}_j||^2 \mathbf{B}_{ij}^{GDE} = 2tr\left(\mathbf{A}^T \mathbf{X} \mathbf{L}^{GDE'} \mathbf{X}^T \mathbf{A}\right) \tag{10}$$

where $\mathbf{L}^{GDE} = \mathbf{D}^{GDE} - \mathbf{W}^{GDE}$ and $\mathbf{L}^{GDE'} = \mathbf{D}^{GDE'} - \mathbf{B}^{GDE}$ are the Laplacian matrices; $\mathbf{D}_{ij}^{GDE} = \sum_j \mathbf{W}_{ij}^{GDE}$ and $\mathbf{D}_{ij}^{GDE'} = \sum_j \mathbf{B}_{ij}^{GDE}$ are the diagonal matrices. Therefore, the discriminant criterion function of GDE can be computed [5]:

$$J_{GDE}(\mathbf{A}) = \min_{\mathbf{A}} \frac{S_w^{GDE}}{S_b^{GDE}} = \min_{\mathbf{A}} \frac{tr(\mathbf{A}^T \mathbf{X} \mathbf{L}^{GDE} \mathbf{X}^T \mathbf{A})}{tr(\mathbf{A}^T \mathbf{X} \mathbf{L}^{GDE'} \mathbf{X}^T \mathbf{A})} \tag{11}$$

In GDE method, the similarity defined by monotonic functions can effectively solve the effect of the distant samples on recognition results. However, the similarity of samples is directly defined by the distance of samples, and outliers will have a certain impact on the discriminant performance. Furthermore, GDE computes the similarity for labeled samples without taking unlabeled samples into account.

3   Methodology

3.1   The Proposed Method

As described in Sect. 2.2, GDE does not study the influence of outliers and unlabeled samples on the recognition performance. In this paper, we propose a novel graph embedding discriminant analysis method where the similarity of samples is based on the similarities between them and the mean of intraclass samples. It intends to eliminate the impact of outliers on the within-class scatters and improves the recognition performance.

Let $G_{intrinsic} = \{\mathbf{V}, \mathbf{E}, \mathbf{W}\}$ denotes the redefined intrinsic graph, and $G_{penalty} = \{\mathbf{V}, \mathbf{E}, \mathbf{B}\}$ denotes the penalty graph, and the weight matrices $\mathbf{W}$ and $\mathbf{B}$ are given as follows:

$$
\mathbf{W}_{ij} = \begin{cases} e^{-\frac{||\mathbf{x}_i - \tilde{\mathbf{x}}_i||^2 + ||\mathbf{x}_j - \tilde{\mathbf{x}}_i||^2}{2t}}, & \text{if } l_i = l_j \\ 0, & \text{if } l_i \neq l_j \end{cases} \tag{12}
$$

$$
\mathbf{B}_{ij} = \begin{cases} e^{-\frac{||\mathbf{x}_i - \mathbf{x}_j||^2}{t}}, & \text{if } \mathbf{x}_i \in D_k(\mathbf{x}_j) \text{ or } \mathbf{x}_j \in D_k(\mathbf{x}_i), l_i \neq l_j \\ e^{-\frac{t}{||\mathbf{x}_i - \mathbf{x}_j||^2}}, & \text{if } \mathbf{x}_i \notin D_k(\mathbf{x}_j) \text{ and } \mathbf{x}_j \notin D_k(\mathbf{x}_i), l_i \neq l_j \\ 0, & \text{if } l_i = l_j \end{cases} \tag{13}
$$

where $\tilde{\mathbf{x}}_i$ denotes the mean of within-class samples and $D_k(\mathbf{x}_i)$ denotes the k nearest interclass neighbors of $\mathbf{x}_i$. However, unlike GDE, we redefine the weight matrix $\mathbf{W}$ based on the distance between the sample and the mean of within-class samples instead of directly using the distance of samples. Moreover, we also describe the importance of samples through monotonic functions in view of GDE.

According to the constructed two graphs, the intraclass scatter $\mathbf{S}_w$ and the interclass scatter $\mathbf{S}_b$ can be formulated:

$$
\mathbf{S}_w = \sum_{ij} ||\mathbf{y}_i - \mathbf{y}_j||^2 \mathbf{W}_{ij} = \sum_{ij} ||\mathbf{A}^T\mathbf{x}_i - \mathbf{A}^T\mathbf{x}_j||^2 \mathbf{W}_{ij} = 2tr(\mathbf{A}^T\mathbf{X}\mathbf{L}\mathbf{X}^T\mathbf{A}) \tag{14}
$$

$$
\mathbf{S}_b = \sum_{ij} ||\mathbf{y}_i - \mathbf{y}_j||^2 \mathbf{B}_{ij} = \sum_{ij} ||\mathbf{A}^T\mathbf{x}_i - \mathbf{A}^T\mathbf{x}_j||^2 \mathbf{B}_{ij} = 2tr(\mathbf{A}^T\mathbf{X}\mathbf{L}'\mathbf{X}^T\mathbf{A}) \tag{15}
$$

where $\mathbf{L} = \mathbf{D} - \mathbf{W}$ and $\mathbf{L}' = \mathbf{D}' - \mathbf{B}$ are the Laplacian matrices. $\mathbf{D}_{ij} = \sum_j \mathbf{W}_{ij}$ and $\mathbf{D}'_{ij} = \sum_j \mathbf{B}_{ij}$ are the diagonal matrices. Therefore, the proposed discriminant criterion function is given:

$$J_{Ours}(\mathbf{A}) = \min_{\mathbf{A}} \frac{tr(\mathbf{A}^T\mathbf{X}\mathbf{L}\mathbf{X}^T\mathbf{A})}{tr(\mathbf{A}^T\mathbf{X}\mathbf{L}'\mathbf{X}^T\mathbf{A})} \tag{16}$$

Consequently, our method can avoid the influence of outliers to the within-class scatter, making the intraclass samples close to each other and interclass samples far away from each other in the projected space.

3.2 Semi-Supervised Extension (SSE)

For unlabeled samples, SDA [6] builds the k-nearest neighbor graph on unlabeled samples to preserve local neighbor relationships. However, the neighbor parameter k in SDA is difficult to define. In the past few years, LRR [7] is proposed, and it uses low-rank and sparse matrix decomposition to express the global structure of the samples, so that the low-rank structure can reflect the intrinsic affinity of samples. Nevertheless, LRR is essentially unconstrained, and it may be affected by noise. In this paper, we use the low rank representation with L_1-norm sparsity constraint to capture their low-rank relationships, and minimize the error function through preserving the low-rank structure of samples. For the given training samples $\mathbf{X}$, we refactor the original samples through the data dictionary $\mathbf{D} = [\mathbf{d}_1, \mathbf{d}_2, ..., \mathbf{d}_m] \in R^{d \times m}$. In other words, $\mathbf{X} = \mathbf{D}\mathbf{Z}$, where $\mathbf{Z} = [\mathbf{z}_1, \mathbf{z}_2, ..., \mathbf{z}_n] \in R^{m \times n}$ is the refactor coefficient matrix. Besides, there may exist noise in the distribution of samples, therefore the problem can be modeled as follows [8]:

$$\begin{cases} \min_z \|\mathbf{Z}\|_* + \lambda\|\mathbf{E}\|_{2,1} + \gamma\|\mathbf{Z}\|_1 \\ s.t. \ \mathbf{X} = \mathbf{D}\mathbf{Z} + \mathbf{E} \end{cases} \tag{17}$$

where $\|\mathbf{Z}\|_*$ denotes the kernel norm of $\mathbf{Z}$, that is, the sum of the singular values of the matrix. $\|\mathbf{E}\|_{2,1} = \sum_{j=1}^{n} \sqrt{\sum_{i=1}^{d}(\mathbf{E}_{ij})^2}$ is the $l_{2,1}$-norm of the noise matrix E, which is used to model samples of noise interference. $\|\mathbf{Z}\|_1$ represents the L_1-norm of the matrix, which is used to ensure the sparsity of $\mathbf{Z}$. the parameter λ is used to balance the effects of the noise part; and the parameter γ is to balance the effects of the sparse part on the results. By applying the L_1-norm sparsity constraint and adding the error term to solve noise pollution, we can obtain more robust low-rank relationships.

According to Eq. (16), we can see that $\mathbf{x}_i = \mathbf{X}\mathbf{z}_i + \mathbf{e}_i$, and $\mathbf{e}_i$ is the column vector of the noise matrix $\mathbf{E}$. It is assumed that the $\mathbf{x}_i$ remains the low-rank structure after linearly projection into the low-dimensional feature space, and the influence of noise on classification can be avoided by minimizing the representation error $\mathbf{e}_i$. Therefore, we can minimize the norm of $\mathbf{e}_i$ as follows:

$$\sum_i \|\mathbf{A}^T\mathbf{e}_i\|^2 = \sum_i \|\mathbf{A}^T(\mathbf{x}_i - \mathbf{X}\mathbf{z}_i)\|^2 = tr(\mathbf{A}^T\mathbf{S}_e\mathbf{A}) \tag{18}$$

where $\mathbf{S}_e$ can be computed by:

$$S_e = \sum_i (x_i - Xz_i)(x_i - Xz_i)^T = X(I - Z)(I - Z)^T X^T \tag{19}$$

Finally, we integrate the improved graph embedding discriminant criterion and LRR error function for achieving semi-supervised learning. The discriminant criterion function of our extension method is given as following:

$$J_{Ours-SSE}(A) = \min_A \frac{tr(A^T XLX^T A) + \beta tr(A^T S_e A)}{tr(A^T XL'X^T A)} \tag{20}$$

where the regularization parameter β ($\beta \in (0, 1)$) is used to balance the loss of the discriminant criterion function.

3.3 Algorithm

The dimensionality of face images is often very high, thereby we first use the PCA projection A_{PCA} for dimensionality reduction. For the dataset $X = [X_L, X_U]$ in the PCA subspace, where X_L is the subset of labeled samples and X_U is the unlabeled samples. The proposed algorithm is given as follows:

Step1:	For X_L, construct the two adjacency matrices W and B based on Eqs. (12) and (13), and compute the intraclass scatter S_w and the interclass scatter S_b based on Eqs. (14) and (15).
Step2:	In consideration of unlabeled samples in datasets, compute the reconstruction coefficient matrix Z based on Eq. (17), then compute the matrix S_e based on Eq. (19). Let $a_1, a_2, ..., a_p$ be the eigenvectors associated with the p minimal eigenvalues of the general eigenvalue problem $XL'X^T A = \lambda(XLX^T + \beta S_e)A$, then the optimal projection matrix learned by our extension method can be denoted as $A_{Ours-SSE} = [a_1, a_2, ..., a_p]$, and the final projection is $A = A_{PCA} \cdot A_{Ours-SSE}$.
Step3:	For a test samples x, the representation in the low-dimensional space can be represented as $y = A^T x$, and then we can use a classifier to predict its class.

4 Experiments

In this section, we verify the proposed method with multiple classifiers, i.e. NNC (nearest neighbor classifier), SVM (support vector machine), LRC (linear regression classifier) [9] and LLRC (locality-regularized linear regression classifier) [10], on three public face datasets (ORL, AR and FERET), and particularly evaluates its performance under noise and blur conditions. All the experiments are conducted on a computer with Intel Core i7-7700 3.6 GHz Windows 7, RAM-8 GB, and implemented by MATLAB.

4.1 Experiments on ORL Dataset

The ORL face dataset contains 400 images of 40 people. All images are in grayscale and manually cropped to 92×112 pixels. On ORL face dataset, we select 5 images of each person for training samples and the remaining images for testing. Furthermore, the first r ($r = 2, 3$) images of each person are labeled. The NNC, SVM, LRC and LLRC are picked for classification. Firstly, the dimension of PCA subspace is set to 50. In our method, the value of the interclass neighbor parameter k is 4, and the tuning parameter of adjacency matrices t is set to 2^3. The neighbor parameter k in LPP is 4. For MFA and GDE, the intraclass neighbor parameter k_1 is picked as 1, and the interclass neighbor parameter k_2 is 40.

Table 1 reports the maximal recognition rates of different methods on ORL dataset. It can be seen that the recognition performance of our method is better than other methods in various classifiers, especially GDE. When unlabeled samples are existing in dataset, the L_1-norm sparse constraint is introduced, which makes our extension method is more effective in classifying these samples. Furthermore, the regularization parameter β which is applied for the loss of the discriminant criterion function is critical to the recognition performance, and too large value of β will make the recognition rate degraded. In addition, Fig. 1 shows the recognition rate curves of all methods under different classifiers as the projection axes varies. It can be easily found that the recognition performance of all method outperforms under LLRC, and our extension method achieves the best performance, that further proves the efficiency of our extension method.

Table 1. Maximal recognition rates (%) and corresponding dimensions (shown in the parentheses) of different methods under four classifiers on ORL dataset.

Methods	$r = 2$				$r = 3$			
	NNC	SVM	LRC	LLRC	NNC	SVM	LRC	LLRC
LDA	77.48	77.87	75.12	76.28	80.15	79.56	76.87	78.50
	(25)	(32)	(30)	(28)	(20)	(30)	(35)	(35)
LPP	76.32	77.40	78.54	78.89	80.87	81.32	82.15	82.50
	(50)	(50)	(45)	(45)	(45)	(45)	(40)	(40)
MFA	74.26	77.86	79.20	79.65	81.75	82.55	82.95	83.05
	(25)	(25)	(30)	(30)	(20)	(25)	(25)	(25)
GDE	75.12	78.54	80.25	80.56	82.45	82.87	83.12	84.00
	(30)	(30)	(30)	(35)	(30)	(35)	(30)	(30)
Ours	**76.78**	**79.20**	**80.75**	**81.35**	**82.33**	**83.56**	**83.20**	**84.33**
	(35)	**(30)**	**(35)**	**(35)**	**(35)**	**(30)**	**(35)**	**(35)**
Ours-SSE $\beta = 0.3$	**78.36**	**79.47**	**80.97**	**82.87**	**82.93**	**84.77**	**84.65**	**84.75**
	(35)	**(38)**	**(30)**	**(30)**	**(25)**	**(30)**	**(30)**	**(30)**
Ours-SSE $\beta = 0.4$	**80.15**	**80.96**	**81.37**	**82.48**	**81.74**	**83.87**	**84.48**	**85.46**
	(35)	**(35)**	**(35)**	**(35)**	**(40)**	**(34)**	**(38)**	**(33)**
Ours-SSE $\beta = 0.5$	**80.75**	**81.25**	**82.75**	**82.84**	**82.67**	**83.95**	**84.84**	**85.87**
	(38)	**(34)**	**(35)**	**(35)**	**(38)**	**(37)**	**(37)**	**(35)**
Ours-SSE $\beta = 0.6$	**79.68**	**80.34**	**81.25**	**81.87**	**84.78**	**84.17**	**84.13**	**84.76**
	(35)	**(37)**	**(35)**	**(36)**	**(38)**	**(35)**	**(34)**	**(38)**

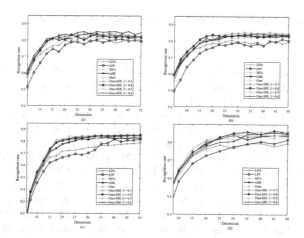

Fig. 1. Recognition rate curves of all methods under different classifiers when the labeled images of each class $r = 3$. (a) NNC; (b) SVM; (c) LRC; (d) LLRC.

4.2 Experiments on AR Dataset

The AR face dataset contains over 3000 face images of 120 people. All images are manually cropped to 50×40 pixels. On AR face dataset, we only use the images without occlusion, and the first 7 images of each person are taken for training and 14–20 images are applied for testing. Furthermore, the first r ($r = 3, 4, 5$) images of each person are labeled. The NNC, SVM, LRC and LLRC are employed for classification. Before the experiments, the dimension of PCA subspace is set to 150. The parameters in the experiments are set as follows: In our method, the value of the interclass neighbor parameter k is 6, and the tuning parameter of adjacency matrices t is chosen as 2^4. The neighbor parameter k in LPP is set to 6. For MFA and GDE, the intraclass neighbor parameter k_1 is picked as 1, and the interclass parameter k_2 is set to 120.

Table 2 reports the maximal recognition rates of different methods on AR dataset. The results indicate that compared with GDE, our method has a small improvement (<1%) under four classifiers, but the recognition performance of our extension method has obvious advantages when unlabeled samples are existing in datasets. Meanwhile, the more labeled samples, the better performance our extension method achieves. In addition, the recognition rate curves of different methods under various classifiers as the projection axes varies are shown in Fig. 2. We can see that LPP performs the worst, but when the regularization parameter β is set to 0.3, the curves in Fig. 2 demonstrate that our extension method performs better than other methods, which proves that the L_1-norm sparse constraint we introduced can eliminate the influence of unlabeled samples on the experimental results.

Table 2. Maximal recognition rates (%) and corresponding dimensions of different methods under four classifiers on AR dataset.

Method	$r = 3$				$r = 4$				$r = 5$			
	NNC	SVM	LRC	LLRC	NNC	SVM	LRC	LLRC	NNC	SVM	LRC	LLRC
LDA	56.18	57.18	56.15	57.33	58.73	59.38	68.56	59.33	61.25	61.55	60.25	61.87
	(125)	(125)	(130)	(110)	(110)	(110)	(110)	(110)	(115)	(115)	(120)	(125)
LPP	49.35	50.12	49.75	51.28	53.65	54.20	55.75	56.15	57.14	58.24	57.35	58.33
	(120)	(120)	(125)	(125)	(115)	(120)	(125)	(120)	(120)	(130)	(125)	(125)
MFA	60.25	61.32	62.55	64.18	63.54	62.87	63.59	65.83	65.55	64.95	64.88	67.75
	(135)	(125)	(130)	(130)	(130)	(125)	(125)	(125)	(140)	(130)	(125)	(125)
GDE	60.88	62.45	63.37	65.28	65.24	63.14	64.87	66.67	67.87	65.25	66.72	68.45
	(130)	(130)	(130)	(130)	(135)	(135)	(130)	(130)	(130)	(130)	(125)	(125)
Ours	**61.75**	**62.45**	**64.78**	**66.54**	**65.76**	**63.84**	**65.87**	**67.75**	**68.74**	**66.48**	**67.50**	**68.87**
	(125)	**(125)**	**(130)**	**(130)**	**(130)**	**(130)**	**(130)**	**(130)**	**(130)**	**(135)**	**(135)**	**(135)**
Ours-SSE $\beta = 0.1$	**63.54**	**65.47**	**66.34**	**67.25**	**66.48**	**66.78**	**67.85**	**69.45**	**68.88**	**67.96**	**68.85**	**69.34**
	(120)	**(130)**	**(125)**	**(130)**	**(125)**	**(130)**	**(135)**	**(130)**	**(125)**	**(135)**	**(125)**	**(125)**
Ours-SSE $\beta = 0.3$	**67.76**	**69.72**	**68.68**	**69.25**	**68.37**	**70.14**	**70.36**	**71.48**	**69.34**	**71.33**	**72.05**	**72.87**
	(110)	**(135)**	**(135)**	**(130)**	**(120)**	**(134)**	**(130)**	**(135)**	**(125)**	**(135)**	**(125)**	**(125)**
Ours-SSE $\beta = 0.5$	**66.87**	**67.26**	**66.28**	**67.39**	**67.46**	**68.24**	**67.52**	**68.25**	**68.90**	**69.26**	**68.98**	**70.34**
	(120)	**(135)**	**(135)**	**(135)**	**(120)**	**(137)**	**(136)**	**(135)**	**(120)**	**(130)**	**(125)**	**(125)**
Ours-SSE $\beta = 0.7$	**65.33**	**66.12**	**65.76**	**66.15**	**66.55**	**67.32**	**67.33**	**68.12**	**67.87**	**68.95**	**68.37**	**68.77**
	(130)	**(130)**	**(130)**	**(135)**	**(125)**	**(125)**	**(125)**	**(130)**	**(130)**	**(130)**	**(125)**	**(125)**

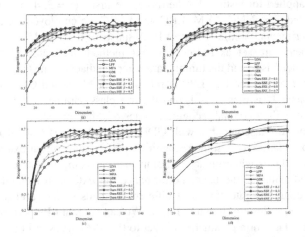

Fig. 2. Recognition rate curves of all methods under different classifiers when the labeled images of each class $r = 5$. (a) NNC; (b) SVM; (c) LRC; (d) LLRC.

4.3   Experiments on FERET Dataset

The FERET face dataset contains over 1000 face images of 200 people. All images are in grayscale and manually cropped to 80 × 80 pixels. On FERET face dataset, the first 4 images of each person are employed for training and the remaining for testing. Similar to the experiment on ORL dataset, the first r ($r = 2, 3$) images are labeled. The NNC, SVM, LRC and LLRC are chosen for classification. First of all, the dimension of PCA subspace is set to 150. The parameters of this experiment are set as follows: In our method, the interclass neighbor parameter k is picked as 3, and the tuning parameter of adjacency matrices t is 2^3. For LPP, the neighbor parameter k is set to 3, and for MFA and GDE, the intraclass neighbor parameter k_1 is set to 1 and the interclass neighbor parameter k_2 is 200.

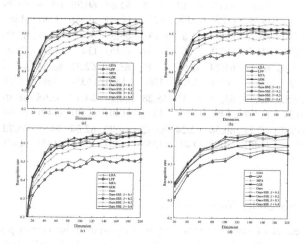

Fig. 3. Recognition rate curves of different methods under different classifiers when the labeled samples of each class $r = 3$. (a) NNC; (b) SVM; (c) LRC; (d) LLRC.

Table 3 reports the maximal recognition rates of different methods on FERET dataset. We can know that the overall recognition rate of all methods in this experiment has decreased, but compared with other methods, our method still has certain advantages. Furthermore, Fig. 3 represents the recognition rate curves of all methods under four classifiers as the projection axes varies. The results show that compared with the other three classifiers, all methods on LLRC will achieve better results, and when the value of β is set to 0.3, in most cases the curve of our extension method reaches the best performance, which also validates the efficiency of our extension method.

Table 3. Maximal recognition rates (%) and corresponding dimensions of different methods under four classifiers on FERET dataset.

Method	r = 2				r = 3			
	NNC	SVM	LRC	LLRC	NNC	SVM	LRC	LLRC
LDA	48.85	47.36	49.32	50.87	55.20	50.36	53.27	55.96
	(120)	(130)	(125)	(125)	(125)	(120)	(125)	(125)
LPP	53.00	49.63	51.76	54.16	56.33	51.48	53.25	56.14
	(125)	(130)	(130)	(120)	(125)	(125)	(120)	(120)
MFA	57.25	54.37	56.84	58.21	59.78	57.87	58.65	60.20
	(130)	(125)	(130)	(120)	(125)	(125)	(120)	(125)
GDE	58.13	57.33	58.67	59.72	59.89	60.32	59.96	60.74
	(125)	(130)	(125)	(125)	(130)	(130)	(125)	(130)
Ours	**58.87**	**58.67**	**59.17**	**60.78**	**61.63**	**60.75**	**63.28**	**63.86**
	(130)	**(125)**	**(130)**	**(125)**	**(130)**	**(125)**	**(130)**	**(125)**
Ours-SSE $\beta = 0.1$	**60.25**	**59.74**	**61.02**	**63.34**	**63.55**	**61.67**	**64.85**	**65.88**
	(135)	**(130)**	**(130)**	**(130)**	**(135)**	**(140)**	**(135)**	**(130)**
Ours-SSE $\beta = 0.2$	**63.11**	**61.24**	**62.95**	**64.72**	**65.67**	**62.57**	**65.28**	**66.25**
	(135)	**(135)**	**(130)**	**(130)**	**(135)**	**(135)**	**(140)**	**(135)**
Ours-SSE $\beta = 0.3$	**63.24**	**65.72**	**64.15**	**65.28**	**66.14**	**64.92**	**65.98**	**66.80**
	(135)	**(140)**	**(140)**	**(135)**	**(140)**	**(140)**	**(140)**	**(140)**
Ours-SSE $\beta = 0.4$	**62.17**	**60.36**	**63.24**	**63.62**	**64.90**	**62.34**	**64.05**	**65.73**
	(140)	**(140)**	**(140)**	**(135)**	**(140)**	**(140)**	**(145)**	**(145)**

4.4   Experiments Under Noise Condition

In view of the effect of noise, in this section, Gaussian noise (mean $\mu = 0$, variance $\sigma = 0.05$), pepper & salt noise (noise density $d = 0.1$) and multiplicative noise (mean $\mu = 0$, variance $\sigma = 0.1$) are applied to face samples on ORL, AR and FERET datasets. Face samples under three different noises on ORL, AR and FERET datasets are given in Fig. 4. The comparative results of the recognition rates of different methods on three datasets are shown in Tables 4, 5, and 6.

(a) Image on ORL   (b) Gaussian   (c) Pepper & salt   (d) Multiplicative   (e) Image on AR   (f) Gaussian   (g) Pepper & salt   (h) Multiplicative   i. Image on FERET   j. Gaussian   k. Pepper & salt   l. Multiplicative

Fig. 4. Face samples under three different noises on ORL, AR and FERET datasets.

It can be seen from Tables 4, 5, and, 6 that under three noise conditions, our method performs better than LDA, LPP, MFA and GDE, due to the improvement in the definition of the similarities of intraclass samples. Especially, since the low rank representation of L_1-norm sparsity constraint is introduced, the recognition rate of our extension method has obvious advantages under three noise conditions, which further proves the anti-noise ability of our extension method. Furthermore, as the number of samples increases and the faces become more complex, our method still has a better and robust performance under three noise conditions.

Table 4. Maximal recognition rates (%) of different methods on ORL dataset under different noise conditions ($r = 3$).

Method	No noise			Gaussian noise			Salt & pepper noise			Multiplicative noise		
	SVM	LRC	LLRC	SVM	LRC	LLRC	SVM	LRC	LLRC	SVM	LRC	LLRC
LDA	79.56 (30)	76.87 (35)	78.50 (35)	77.65 (30)	74.35 (25)	73.28 (27)	74.34 (15)	74.17 (20)	73.56 (35)	76.54 (20)	75.13 (18)	72.84 (26)
LPP	81.32 (45)	82.15 (40)	82.50 (40)	78.56 (35)	76.72 (28)	70.85 (32)	72.25 (30)	73.24 (25)	74.18 (27)	78.32 (25)	72.19 (28)	73.27 (20)
MFA	82.55 (25)	82.95 (25)	83.05 (25)	76.25 (25)	76.56 (30)	74.33 (26)	78.33 (25)	75.83 (28)	76.25 (28)	77.56 (30)	73.44 (30)	72.18 (30)
GDE	82.87 (35)	83.12 (30)	84.00 (30)	75.33 (25)	77.84 (32)	73.28 (25)	77.45 (30)	74.36 (31)	73.57 (32)	76.33 (35)	75.17 (23)	76.34 (25)
Ours	**83.56** (30)	**83.20** (35)	**84.33** (35)	**80.67** (38)	**78.68** (30)	**75.03** (26)	**78.87** (35)	**76.33** (25)	**77.18** (19)	**79.67** (30)	**76.25** (31)	**76.84** (27)
Ours-SSE $\beta = 0.3$	**84.77** **(30)**	**84.65** **(30)**	**84.75** **(30)**	**84.25** **(35)**	**79.34** **(25)**	**76.56** **(30)**	**82.67** **(30)**	**77.54** **(20)**	**78.65** **(30)**	**81.25** **(25)**	**76.87** **(30)**	**77.25** **(35)**

Table 5. Maximal recognition rates (%) of different methods on AR dataset under different noise conditions ($r = 5$).

Method	No noise			Gaussian noise			Salt & pepper noise			Multiplicative noise		
	SVM	LRC	LLRC	SVM	LRC	LLRC	SVM	LRC	LLRC	SVM	LRC	LLRC
LDA	61.55 (115)	60.25 (120)	61.87 (125)	57.50 (120)	54.25 (125)	53.75 (115)	54.72 (125)	53.86 (115)	55.47 (130)	57.36 (120)	57.43 (105)	55.71 (125)
LPP	58.24 (130)	57.35 (125)	58.33 (125)	54.20 (125)	55.18 (100)	53.87 (125)	56.74 (130)	55.74 (125)	55.98 (110)	55.25 (125)	56.33 (105)	54.87 (120)
MFA	64.95 (130)	64.88 (125)	67.75 (125)	59.34 (125)	57.65 (130)	56.73 (125)	58.37 (125)	56.35 (115)	57.76 (130)	57.87 (130)	57.48 (105)	58.67 (120)
GDE	65.25 (130)	66.72 (125)	68.45 (125)	58.78 (125)	56.98 (115)	57.47 (120)	59.25 (130)	56.44 (125)	53.85 (130)	56.74 (135)	56.48 (120)	57.65 (105)
Ours	**66.48** (135)	**67.50** (135)	**68.87** (135)	**60.67** (125)	**58.36** (130)	**58.45** (125)	**61.33** (135)	**57.68** (125)	**56.38** (110)	**58.56** (130)	**58.37** (120)	**58.87** (105)
Ours-SSE $\beta = 0.3$	**71.33** **(135)**	**72.05** **(125)**	**72.87** **(125)**	**61.76** **(135)**	**59.65** **(125)**	**58.76** **(130)**	**63.18** **(130)**	**58.36** **(120)**	**57.75** **(130)**	**59.75** **(125)**	**59.45** **(130)**	**59.79** **(135)**

Table 6. Maximal recognition rates (%) of different methods on FERET dataset under different noise conditions ($r = 3$).

Method	No noise			Gaussian noise			Salt & pepper noise			Multiplicative noise		
	SVM	LRC	LLRC	SVM	LRC	LLRC	SVM	LRC	LLRC	SVM	LRC	LLRC
LDA	50.36	53.27	55.96	54.25	53.75	55.12	48.63	50.13	48.76	50.76	51.25	51.78
	(120)	(125)	(125)	(100)	(125)	(90)	(85)	(95)	(80)	(75)	(115)	(135)
LPP	51.48	53.25	56.14	54.73	54.33	55.36	45.76	48.49	47.36	49.38	52.48	53.76
	(125)	(120)	(120)	(95)	(125)	(130)	(100)	(110)	(120)	(95)	(90)	(120)
MFA	57.87	58.65	60.20	57.65	56.75	54.87	48.56	49.76	50.17	51.45	53.56	52.85
	(125)	(120)	(125)	(90)	(80)	(95)	(95)	(85)	(110)	(80)	(130)	(115)
GDE	60.32	59.96	60.74	58.37	57.67	56.84	46.93	49.35	47.85	52.37	53.64	52.87
	(130)	(125)	(130)	(85)	(120)	(125)	(90)	(120)	(110)	(80)	(120)	(125)
Ours	**60.75**	**63.28**	**63.86**	**59.76**	**58.87**	**57.43**	**49.67**	**51.28**	**50.96**	**54.85**	**56.37**	**54.84**
	(125)	**(130)**	**(125)**	**(110)**	**(115)**	**(110)**	**(80)**	**(125)**	**(100)**	**(75)**	**(120)**	**(100)**
Ours-SSE	**64.92**	**65.98**	**66.80**	**62.43**	**59.17**	**58.25**	**51.27**	**51.45**	**51.73**	**56.84**	**57.74**	**55.28**
$\beta = 0.3$	**(140)**	**(140)**	**(140)**	**(105)**	**(125)**	**(130)**	**(85)**	**(120)**	**(115)**	**(90)**	**(110)**	**(105)**

4.5   Experiments Under Blur Condition

In consideration of the blurred images in dataset, in this section, motion blur (the motion angle is $10°$ and the distance of motion is 5) and defocus blur (the radius is 5) are employed to the face images on ORL, AR and FERET datasets. Figure 5 shows face samples under different blur conditions on ORL, AR and FERET datasets. The comparative results of the recognition rate of all methods on three datasets are shown in Tables 7, 8, and 9.

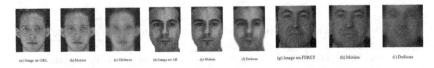

(a) Image on ORL   (b) Motion   (c) Defocus   (d) Image on AR   (e) Motion   (f) Defocus   (g) Image on FERET   (h) Motion   (i) Defocus

Fig. 5. Face samples under two blurs on ORL, AR and FERET datasets.

According to the results in Tables 7, 8, and 9, it can be found that our method has an advantage over other methods in recognition rate under two blur conditions. Moreover, our extension method uses sparse representation to optimize the similarity of intraclass samples, which makes it perform better than GDE. However, as the number of samples increases and the faces become more complex, the recognition performance will be degraded. It can be seen from Table 9 that when the defocused blurred images are existing, the maximal recognition rate of various methods is low, not exceeding 40%, but the performance of our extension method still has a little advantage, which indicates the anti-interference ability of our extension method.

Table 7. Maximal recognition rates (%) of different methods on ORL dataset under different blur conditions.

Method	Original			Defocus blurred			Motion blurred		
	SVM	LRC	LLRC	SVM	LRC	LLRC	SVM	LRC	LLRC
LDA	79.56 (30)	76.87 (35)	78.50 (35)	52.18 (20)	51.37 (25)	53.24 (20)	53.61 (20)	51.48 (25)	53.74 (25)
LPP	81.32 (45)	82.15 (40)	82.50 (40)	50.37 (30)	51.48 (30)	52.65 (25)	43.56 (19)	47.96 (18)	46.47 (25)
MFA	82.55 (25)	82.95 (25)	83.05 (25)	53.65 (35)	52.87 (30)	53.76 (30)	55.78 (20)	54.76 (28)	56.83 (20)
GDE	82.87 (35)	83.12 (30)	84.00 (30)	53.87 (40)	54.73 (35)	53.76 (20)	46.52 (17)	48.75 (20)	50.34 (30)
Ours	**83.56 (30)**	**83.20 (35)**	**84.33 (35)**	**57.34 (35)**	**55.15 (35)**	**54.87 (25)**	**52.96 (25)**	**55.46 (28)**	**57.75 (30)**
Ours-SSE $\beta = 0.3$	**84.77 (30)**	**84.65 (30)**	**84.75 (30)**	**58.36 (25)**	**55.84 (20)**	**55.25 (30)**	**57.87 (22)**	**56.17 (25)**	**58.25 (25)**

Table 8. Maximal recognition rates (%) of different methods on AR dataset under different blur conditions.

Method	Original			Defocus blurred			Motion blurred		
	SVM	LRC	LLRC	SVM	LRC	LLRC	SVM	LRC	LLRC
LDA	61.55 (115)	60.25 (120)	61.87 (125)	49.75 (85)	53.84 (95)	46.72 (100)	50.88 (80)	49.76 (105)	51.74 (110)
LPP	58.24 (130)	57.35 (125)	58.33 (125)	48.26 (100)	50.25 (85)	48.83 (95)	49.20 (90)	48.75 (80)	50.37 (100)
MFA	64.95 (130)	64.88 (125)	67.75 (125)	50.33 (95)	47.26 (90)	51.87 (110)	52.83 (95)	50.78 (85)	51.34 (90)
GDE	65.25 (130)	66.72 (125)	68.45 (125)	49.76 (105)	48.20 (85)	49.25 (100)	51.56 (75)	49.95 (100)	50.83 (95)
Ours	**66.48 (135)**	**67.50 (135)**	**68.87 (135)**	**50.68 (110)**	**51.78 (90)**	**52.50 (105)**	**55.47 (95)**	**52.67 (90)**	**52.86 (90)**
Ours-SSE $\beta = 0.3$	**71.33 (135)**	**72.05 (125)**	**72.87 (125)**	**52.47 (115)**	**52.15 (95)**	**52.73 (105)**	**57.83 (90)**	**53.25 (75)**	**54.28 (105)**

Table 9. Maximal recognition rates (%) of different methods on FERET dataset under different blur conditions.

Method	Original			Defocus blurred			Motion blurred		
	SVM	LRC	LLRC	SVM	LRC	LLRC	SVM	LRC	LLRC
LDA	50.36	53.27	55.96	34.63	33.84	35.25	35.16	34.87	36.73
	(120)	(125)	(125)	(105)	(85)	(90)	(95)	(90)	(85)
LPP	51.48	53.25	56.14	33.75	35.25	35.78	35.78	38.26	36.48
	(125)	(120)	(120)	(110)	(75)	(85)	(105)	(90)	(95)
MFA	57.87	58.65	60.20	35.74	37.26	38.67	36.36	36.18	37.64
	(125)	(120)	(125)	(100)	(76)	(95)	(95)	(90)	(105)
GDE	60.32	59.96	60.74	34.95	38.20	36.47	35.47	36.76	38.78
	(130)	(125)	(130)	(96)	(85)	(105)	(105)	(85)	(100)
Ours	**60.75**	**63.28**	**63.86**	**36.53**	**38.78**	**39.16**	**36.50**	**39.65**	**39.74**
	(125)	**(130)**	**(125)**	**(90)**	**(90)**	**(95)**	**(100)**	**(105)**	**(90)**
Ours-SSE	**64.92**	**65.98**	**66.80**	**36.87**	**39.15**	**39.77**	**36.75**	**40.25**	**41.38**
$\beta = 0.3$	**(140)**	**(140)**	**(140)**	**(110)**	**(95)**	**(100)**	**(95)**	**(110)**	**(105)**

5   Conclusion

In this paper, we propose a novel method to eliminate the influence of outliers in face recognition. It builds the intrinsic graph in the light of the intraclass samples and their with-class mean, and constructs the penalty graph from the perspective of GDE. In addition, a low-rank representation with sparse constraints is introduced to explore the global low rank relationship of unlabeled samples. The experimental results on three datasets under different noise and blur conditions prove the effectiveness and robust of the proposed method.

Acknowledgments. This study was partially supported by the National Natural Science Foundation of China (61906058, 61972128, 61702155), Natural Science Foundation of Anhui Province (1908085MF210, 1808085MF176) and the Fundamental Research Funds for the Central Universities of China (JZ2020YYPY0093).

References

1. Turk, M.A., Pentland, A.P.: Face recognition using eigenfaces. In: Proceedings of the IEEE Computer Society Conference on Computer Vision and Pattern Recognition, pp. 586–591. IEEE Computer Society Press, Los Alamitos (1991)
2. Belhumeur, P.N., Hepanha, J.P., Kriegman, D.J.: Eigenfaces vs. Fisherfaces: recognition using class specific linear projection. IEEE Trans. Pattern Anal. Mach. Intell. **19**(7), 711–720 (1997)
3. He, X.F., Niyogi, P.: Locality preserving projections. In: Proceedings of the Neural Information Processing System Conference, pp. 100–115. MIT Press, Cambridge (2003)

4. Yan, S., Xu, D., Zhang, B., Zhang, H.J., Yang, Q., Lin, S.: Graph embedding and extensions: a general framework for dimensionality reduction. IEEE Trans. Pattern Anal. Mach. Intell. **29**(40), 40–50 (2007)
5. Huang, P., Li, T., Gao, G., Yang, G.: Feature extraction based on graph discriminant embedding and its applications to face recognition. Softw. Comput. **23**(16), 7015–7028 (2018). https://doi.org/10.1007/s00500-018-3340-5
6. Cai, D., He, X., Han, J.: Semi-supervised discriminant analysis. In: 11th IEEE International Conference on Computer Vision, Rio de Janeiro, Brazil, pp. 1–7. IEEE Press (2007)
7. Liu, G., Lin, Z., Yan, S., Sun, J., Yu, Y., Ma, Y.: Robust recovery of subspace structures by low-rank representation. IEEE Trans. Pattern Anal. Mach. Intell. **35**(1), 171–184 (2013)
8. Hai-Hong, G., Jie, M., Jian-Long, L.: Feature extraction based on low rank description. Sci. Technol. Eng. **16**(10), 200–204 (2016)
9. Naseem, I., Togneri, R., Bennamoun, M.: Linear regression for face recognition. IEEE Trans. Pattern Anal. Mach. Intell. **32**(11), 2106–2112 (2010)
10. Brown, D., Li, H.X., Gao, Y.S.: Locality-regularized linear regression for face recognition. In: Proceedings of the 21st International Conference on Pattern Recognition, Tsukuba, pp. 1586–1589 (2012)

Fast and Accurate Face Alignment Algorithm Based on Deep Knowledge Distillation

Biao Qiao, Huabin Wang[(⊠)], Xiaoyu Deng, and Liang Tao

Anhui Provincial Key Laboratory of Multimodal Cognitive Computation,
School of Computer Science and Technology, Anhui University,
Hefei 230031, Anhui, China
wanghuabin@ahu.edu.cn

Abstract. Most existing face key point detection algorithms only consider improving the generalization performance of the model, while ignoring the model efficiency. However, models with good generalization performance usually contain a large number of parameters, which hinders the deployment of face key point detection algorithms on devices with limited resources. In order to overcome this problem, a Fast and Accurate Face Alignment Algorithm is proposed in this paper. First, a teacher network that stacks multiple hourglass modules is designed. Its model structure is robust to meet the positioning accuracy requirements in actual use. Then, a lightweight student network is designed, which has fewer model parameters and faster feature extraction speed, which meets the speed requirements in practical use. Finally, through the knowledge distillation strategy, the knowledge of the teacher network is transferred to the student network, which makes the student network take into account both the model inference speed and the key point positioning accuracy.

Keywords: Deep learning · Knowledge distillation · Face alignment

1 Introduction

Face alignment [1] is a visual task to accurately locate multiple key points on the face. These key points are usually marked around the facial organs, such as the eye sockets, cheeks, upper and lower lips, and so on. Understanding the key points of a person's face is the basis for higher-level applications (such as face recognition [2, 3], sentiment analysis [4, 5], etc.). In recent years, great progress has been made in the research of face alignment. However, it is still challenging to develop a practical face key point detector, as the detection accuracy, processing speed and model size have to be considered at the same time.

In recent years, Convolutional Neural Networks (CNNs) have advanced the progress of human pose estimation and face alignment based on heat map regression.

H. Wang—This work was supported in part by the National Natural Science Foundation of China under Grant 61372137, in part by the Natural Science Foundation of Anhui Province under Grant 1908085MF209, 1708085MF151 and in part by the Natural Science Foundation for the Higher Education Institutions of Anhui Province under Grant KJ2019A0036.

© Springer Nature Singapore Pte Ltd. 2020
Y. Wang et al. (Eds.): IGTA 2020, CCIS 1314, pp. 74–88, 2020.
https://doi.org/10.1007/978-981-33-6033-4_6

Kowalski et al. [6] first introduced the concept of heat map into the face key point detection field, and proposed a deep alignment network (Deep Alignment Network, DAN). DAN is also a multi-stage cascaded convolutional network. Each stage of the network generates a heat map based on the coordinates predicted by the previous stage of the network. Stacked hourglass network (Stacked Hourglass Network, SHN) is a convolutional network consisting of repeated down sampling and up-sampling. Thanks to this symmetrical multi-scale and multi-resolution structure, the network can learn rich semantic information. SHN refreshed the highest level of multiple 2D human pose estimation challenges. [7] first applied SHN to face key point detection. [8] proposed a multi-branch and multi-scale residual block to replace the standard residual block in the hourglass model, studied the parameters and performance of the binary hourglass network. [9] proposed a novel multi-view hourglass model, which attempts to jointly estimate the key points of the front and side faces. Wu et al. [10] designed a boundary-aware algorithm called LAB, which uses eight stacked hourglass networks to predict the boundary heat map of faces and facial features, and then decodes the coordinate information from the boundary heat map. This method can eliminate the ambiguity in the definition of key points, but the large number of parameters makes the calculation resources cost huge. Although these large-scale networks have high accuracy and strong robustness, they all have a common shortcoming, that is, the amount of model parameters is too large, which leads to slow inference speed and cannot be deployed to some edge devices with limited resources.

Recently, the knowledge distillation method proposed by Hinton [11] introduced a novel network training method, which is to train the student network by combining the probability distribution vector output by the teacher network and the manual annotations of the training set. For the input image, the probability distribution vector output by the teacher network has more information than the manual annotation, and the additional auxiliary information can accelerate the training of the student network. The student network jointly learns the probability distribution vector and manual annotation, and removes redundant information while learning the teacher's network knowledge, that is, distilling the teacher's network knowledge to achieve the compression of the model volume. The other is to directly design a lightweight neural network, such as MobileNetV1 [12], which is based on deep separable convolution. Generally speaking, the function of deep separable convolution is to decompose the standard convolution into depth-wise convolution and pointwise convolution. The advantage of this method is that the number of parameters and the amount of calculation can be greatly reduced. MobileNetV2 [13] is an improvement on the basis of V1, which is also a lightweight neural network. V2 proposes an inverted residual structure (Inverted residuals). It first raises the dimension and then reduces the dimension, which enhances the gradient propagation, and significantly reduces the memory occupation during inference. In addition, the RELU6 activation function is used to make the model more robust in the case of low precision. MobileNetV3 [14] adds Squeeze-and-Excitation structure on the basis of V2, and assigns weights on feature maps during training automatically, so as to achieve better results.

In order to solve the problems of the existing human face alignment model, such as large network structure, slow reasoning speed, and high requirements for equipment calculation. This paper proposes a face alignment method based on knowledge distillation. Our model consists of two parts: one part is a stacked hourglass network as the teacher network, which has high positioning accuracy but complicated structure; the other part is a student network with MobileNetV3 as the backbone structure, which has a simplified model and low complexity. Through the deep knowledge distillation method, the output information of the teacher network with high complexity is compressed and extracted through the fully connected layer, then passed to the student network to assist the student network training, and finally a student network model with small scale, high accuracy and fast reasoning speed is obtained.

The main contributions of our method are as follows:

1) A stacked hourglass network is designed as a teacher network. the model structure is robust and can deal with abnormal situations such as partial occlusion, large posture etc.
2) Using deep separable convolution, inverted residuals, SE blocks and other basic modules to design the student network. The model structure is simplified, the reasoning speed is fast, the space occupation is small, and it can be deployed to edge devices.
3) A face alignment method based on deep feature distillation is proposed. The knowledge of the teacher model is distilled through multi-task learning, and the compression of the deep network and feature dimensions is unified to solve the problem that the actual deployment of the deep model requires high computing resources and storage space requirements for edge devices.

2 Related Work

In this section, we first review the traditional face alignment methods, and then briefly introduce the deep learning-based face alignment methods.

In the past few decades, many classical face landmark detection methods have been proposed. The parametric appearance model is represented by active appearance models (AAMs) [15] and constrained local models (CLMs) [16]. This task is accomplished by maximizing the confidence of the local position in the image. Specifically, AAMs and their follow-up studies attempt to jointly model the overall appearance and shape, while CLMs and its variant determine the location of each key point by searching the local model near the key points of the face, and automatically correct the image with shape fitting. In addition, the tree-structured local model (TSPM) [17] uses a deformable local model for simultaneous detection, pose estimation, and landmark positioning. Methods including explicit shape regression (ESR) [18] and supervised descent method (SDM) [19] attempt to solve this problem by regression. These methods usually require an iterative process to find the optimal parameter configuration for a given face, so they are time-consuming and prone to fall into local minima. In addition, due to the limited capacity of the parametric model, these methods have poor robustness to challenges such as occlusion, lighting, and large poses. Besides the model has a large amount of calculation and high complexity.

Before deep learning, cascade regression is a popular method in face alignment, it starts with initializing the face shape, and then refines the face shape in a cascade manner. For each regressor, it learns the mapping function from shape index features to shape increments. The work in [20] proposed the use of random forests to learn local binary features. Due to the sparse binary nature, its speed can reach 3000 frames per second. To reduce the influence of inaccurate shape initializations. In [21] a coarse to fine search method is proposed. It first performs a rough search on the shape pool, and then uses the rough solution to search for the shape more finely. [22] reconstructed the cascade regression scheme into a Cascade Combination Learning (CCL) problem. It divides all training samples into several domains, and each specific domain cascade regressor handles a specific domain. The final shape is composed of multiple predicted shape estimates. Although cascade regression has achieved good performance on public datasets, inaccurate shape initialization, independent regression, and hand-crafted features may still result in sub-optimal results.

The traditional cascade regression method has been replaced by deep learning methods based on CNNs. Sun Yi et al. first introduced CNN into the field of face key point detection [23]. The general idea of this paper is to roughly locate five facial key points, and then optimize the rough location points to get accurate location results. Zhou et al. [24] proposed an improved multi-level deep network to detect key points of faces in a rough to fine way. he innovation lies in: as far as network input is concerned, using CNN to predict the bounding box of the face, instead of using the face image detected by the face detector as the network input. This improvement greatly improves the accuracy of the initial layer location of human face. Zhang et al. [25] combined face key point location and attribute classification modeling (eg. head pose estimation and facial attribute inference) to optimize face key point detection. Zhang et al. [26] used Coarse-to-Fine Auto-encoder Networks (CFAN) to stack multiple auto-encoder networks (SANs) in a cascaded manner. Trigeorgis et al. [27] used a convolutional neural network to extract the features of the average shape of the face, then exploited the recurrent neural network (RNN) to calculate the residual, followed by an adaption of the shape of the previous step using the residual obtained, and looped four times to obtain the final face shape fitting facial key points.

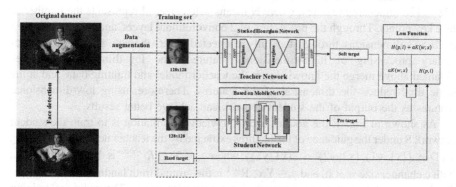

Fig. 1. An overview of deep characteristic distillation proposed. It consists of a teacher network and a student network. The teacher network is a trained network. The student network is a lightweight network to be trained. The output of teacher network is the input of learning network. Through the strategy of knowledge distillation, teachers network supervises students network training.

3 Methodology

3.1 Deep Knowledge Distillation Method

In the knowledge distillation model compression method, a well-trained deep network is regarded as a teacher network, and the output feature map contains rich information. Through a simple fully connection, the neural network can accurately predict the key point information of the input image with a high probability, and the manual annotation only includes the face key points and other attribute information of the input image. Therefore, the information contained in the output characteristics of the teacher network basically includes the information contained in the manual annotations of the input images, and the additional information is kind of auxiliary information, which can help the student network to achieve the best performance in the training process. The student network is a lightweight deep network to be trained, which has fewer parameters than the teacher network, and the forward propagation is faster than the teacher network, but at the same time brings the problem of training difficulties. Due to the existence of auxiliary information, it is easier for the student network to learn from the features output by the teacher network than directly from the original data. Based on this discovery, this paper proposes a deep feature distillation method. Through the design of the loss function, we use the features of the teacher network to guide and supervise the training of the student network, and reduce the dimension of the features of the teacher network. As a result, a more stable and efficient facial key point locator is obtained.

3.2 Model Structure of Deep Knowledge Distillation

The structure of our deep knowledge distillation model is divided into two parts: teacher feature regression and face alignment: 1) Teacher feature regression, the student network directly learns from the output feature maps of the teacher model, thereby transferring the knowledge in the teacher model to the student network, letting the student network acquire the feature representation ability of the teacher network. 2) Face alignment, student network regression from the teacher network to get $N \times M \times 128 \times 128$ dimension features, where N means batch size, M represents the number of face key points. Through the processing of 2 convolutional layers, and then compressed to $M \times 2$ dimensions through the fully connected layer, thereby achieving the secondary processing of high-dimensional features. The low-dimensional features obtained above merge the knowledge in the teacher model and training data, and at the same time reduce the dimension of the features. Therefore, using low-dimensional features as the output of the student network can achieve better results.

As shown in Fig. 1, the goal of deep knowledge distillation is to train the student network S under the guidance of the features extracted by the teacher network T. Training set $D = \{(x1, y1), (x2, y2), \ldots, (x1, y1)\}$, where $x_i \in X \subset R^{c \times w \times h}$ is the face image data with c channel size $w \times h$, and $y \in Y \subset R^{l \times 1}$ is the ground truth landmark of the image. The last layer of student network output a 1×1 dimension vector. The traditional training method optimizes the model parameter w by minimizing the loss function $L_1(w, x, y) = H(p, y)$, where $H(p, y)$ represents the mean square error of p and y.

In this paper, it is proposed that the deep feature distillation method improves the design of the loss function of the student network $S(x,w)$. Based on the original mean square error loss function $H(p,y)$, the new loss function adds the pre-trained teacher network T supervision signal as follows:

$$L_2(w;x) = K(T(x,w'),S(x,w)) = \|T(x,w') - S(x,w)\|_2. \tag{1}$$

Where T is the teacher network, S is the student network, T and S input are x, w is the parameter of the student network, w' is the parameter of the teacher network, $T(x,w')$ is the output of the teacher network, $S(x,w)$ is the output of the student network, The output of the students network and the teacher network have the same dimensions, and the expression of the final loss function is:

$$\begin{aligned} L(w;x,l) &= H(p,y) + aK(w;x) \\ &= H(S(x,w),y) + aK(T(x,w'),S(x,w)). \end{aligned} \tag{2}$$

Where a is the distillation intensity. p is the feature vector of the output of the student network $S(x,w)$. y is the label vector of the input image x. Through the back propagation (BP) algorithm iterative optimization formula (2), the student network S is trained. In order to achieve the purpose of model compression and acceleration, student network S usually selects a lightweight small network, that is, the number of w is far less than the number of w'. At the same time, the teacher network $T(x,w')$ is well trained large network, and its model parameter w' contains a lot of knowledge. Due to the different network structure, the knowledge of teacher network T can not be directly transferred to student network S through parameter learning. It is noted that T is a trained large network, and its output features contain a lot of useful information. Through the fully connected layer, it will be useful. The information is transferred to the student network S, that is, it has more information than the training set D. Let (x, S_T) through formula (1) to supervise the training of student network S, can achieve better results than training directly from training set D. at the same time, let S accept the supervision signal supervision. The supervision signal formula (1) and mean square error loss constitute the loss function formula (2) supervise the training of student network. Finally, the output of student network integrates the information of teacher network and ground truth landmark, The knowledge distillation of teacher network T is realized.

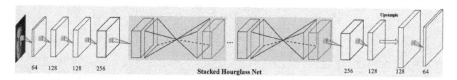

Fig. 2. The network structure of the teacher network. It is mainly formed by stacking four hourglass modules. The model structure is robust and can handle abnormal conditions such as partial occlusion and large poses.

Teacher Network. In order to make the soft target output from the teacher network contain more useful information, we designed a stacked hourglass network as our teacher network. The teacher network architecture is shown in Fig. 2. Input a 128 × 128 face image, through a 7 × 7 convolutional layer with stride 2, padding 3, the original face image is processed into a feature image of 64 × 64, and then three residual blocks [14] are used to increase the dimension of the feature channel, followed by four stacked hourglass networks. The two residual blocks process the feature map to 128 channels, then use the nearest neighbor interpolation up-sampling to increase the spatial resolution to 128 × 128, and finally the residual block and 1 × 1 convolution kernel to generate a heat map. Except for the first 7 × 7 convolutional layer, batch normalization is used before all convolutional layers. ReLU is the activation function. In short, the input of the network is a face image with a size of 128 × 128, and the output are M heat maps and M score maps, where M is the number of key points on the face. Each key point corresponds to a heat map and a score map.

Student Network. In order to reduce the model size of the student network and increase the speed of reasoning, our student network uses the lightweight network MoblienetV3 as the backbone. Compared with MobilenetV2, MobilenetV3 adds SE blocks, Hard-Swish Activation and other modules on its basis, taking into account the speed of model inference while improving network performance. Tables 1 and 2 illustrates the detailed configuration of the student network. From the perspective of the network architecture, the backbone network is simple. Our main purpose is to verify the effectiveness of the deep knowledge distillation method and the loss function of our design. Even a very simple network architecture can achieve the best performance.

Table 1. Structure of student networking

Input	Operate	#exp size	#out	SE	AL	#S
$128^2 \times 3$	Conv2d	–	64	–	Relu	2
$64^2 \times 64$	Conv2d (dw)	–	64	–	Relu	1
$32^2 \times 64$	Bottleneck_1	–	16	True	Relu	2
$16^2 \times 16(x1)$	Bottleneck_2	–	16	True	HSwish	1
$16^2 \times 16(x2)$	Conv2d	–	32	–	Relu	2
$8^2 \times 32(x3)$	Conv2d	–	512	–	Relu	1
(x1, x2, x3)	Full connect	–	136	–	–	–

Table 2. Bottleneck Structure of student networking

	#exp size	#Out	K	#S	SE	AL
Bottleneck_1	128	64	3	2	SE	Relu
	128	64	3	1	True	Relu
	128	64	3	1	True	Relu
	128	64	3	1	True	Relu
	128	64	3	1	False	Relu
Bottleneck_2	128	128	3	2	False	Relu
	128	512	3	1	True	HSwish
	128	512	3	1	True	HSwish
	128	512	3	1	True	HSwish
	128	512	3	1	False	HSwish
	128	512	3	1	True	HSwish
	128	512	3	1	True	HSwish
	16	256	3	1	True	HSwish

Where #exp size represents the expansion channel, #out represents the output channel, K represents the convolution kernel, #S represents the stride, SE represents whether the SE block is used, and AL represents the activation layer.

3.3 Teacher Model and Student Model Analysis

The teacher network T contains 52 convolutional layers in total, and 138×10^6 parameters in total. The student network S uses a lot of deep separable convolutions instead of ordinary convolutions, which greatly reduces the model parameters to 9.5×10^6. That is, the student network S model is compressed by 7.1 times compared to the teacher network T model. Secondly, the teacher network T output feature map is $N \times M \times 128 \times 128$ dimensional, each feature map can represent the position of M feature points and other attribute information, and the student network S output is a $M \times 2$ dimensional feature map, so it can be seen that the output features of teacher network T contains a lot of redundant information.

3.4 Reasoning Complexity

The teacher network T forward propagation needs about 1.53×10^{10} times for feature extraction, while the student network S needs only about 3.68×10^8 times, which increases the speed by 41.6 times and saves 97.34% of the time and computation. Therefore, the time and computational power spent on feature extraction for face alignment are greatly reduced. By using student network S instead of teacher network T, the time needed for feature extraction is effectively reduced.

4 Experimental Results and Analysis

In this experiment, the pre-trained stacked hourglass network model is used as the teacher network, and the proposed deep knowledge distillation method is used to train the student network on the face open data set 300W [15]. The trained student network model is tested and evaluated on the test set, and compared with other state-of-the-art face alignment algorithms in many aspects to verify the effectiveness of this method.

4.1 Experimental Preparation

300-W is the most widely used outdoor data set for 2D face key point detection. In recent years, almost all the work related to 2D face key point detection is bound to do comparative experiments on the 300-W dataset. The 300-W dataset covers facial images of different individuals, rich expressions, strong lighting transformation and various occlusion conditions. The dataset has a total of 3837 images, each containing more than one face, but only one face is labeled for each image. The training set of the 300-W data set contains 3148 face samples, and the test set contains 689 face samples, which are subdivided into a common subset and a challenge subset. In the training process of this paper, a face sample of 68 marked points is used on the 300-W dataset. During dataset pre-processing, we rotate each sample of the training data set from $-30°$ to $30°$ at $5°$ intervals, horizontal inversion, and random occlusion, etc., by introducing a priori knowledge to increase the face image of the data set quantity, thereby improving the generalization performance of the model.

4.2 Evaluation Metrics

We use normalized mean error (NME), area under the curve (AUC) and error rate (ER) to evaluate the error of key points on the face. Specifically, different evaluation metrics are used on different datasets in order to make a fair comparison with existing state-of-the-art methods. For the 300W dataset, we used both the inter-pupillary distance (the distance between the centers of the eyes) and the inter-ocular distance (the distance between the outer corners of the eyes). When the NME of the image is greater than 0.08, it is considered as a detection failure. For the WFLW dataset, we use the inter-eye distance, and when the NME of the image is greater than 0.1, this situation is considered a failure. The NME of the pupil (or inter-ocular) distance is calculated as follows:

$$NME = \frac{1}{M}\sum_{i=1}^{M}\frac{\frac{1}{N}\sum_{j=1}^{N}\left\|p_{i,j}^{pred} - p_{i,j}^{gt}\right\|_2}{\left\|p_{i,l} - p_{i,r}\right\|_2} \tag{3}$$

In the formula (3), M is the number of face pictures in the test set, N is the total number of face key points; $p_{i,l}, p_{i,r}$ is the position of the center of the left eye and the center of the right eye in the $i-th$ face image, $p_{i,j}^{pred}$ is the predicted position of the $j-th$ feature point in the $i-th$ face image, and $p_{i,j}^{gt}$ is the true position of the $j-th$ feature point in the $i-th$ face image. FR is calculated based on the NME value. An image with an NME value greater than ε is considered a failure. The setting of the threshold ε is based on the consensus of most relevant literatures. It is set to 0.08 for the 300-W data set and 0.1 for the WFLW data set. In addition, we use the same threshold setting for AUC.

4.3   Experimental Details

Training Teacher Network. The 128×128 pixel face images and labels in the preprocessed 300W Training set are used as input to the teacher network. Using RMSProp to optimize the network, the initial learning rate is 0.0001, and it drops to 0.00005 after 20 stages. The teacher model uses Pytorch and Nvidia 1080-Ti GPU for training. Save the trained model for the network of trained students.

Training Student Network. The 128×128 pixel face images and labels in the preprocessed 300W training set and the $M \times 128 \times 128$ dimensional features output by the teacher network are used as input to the student network. Among them, face images and tags produce loss $L1$, face images and features produce loss L2, total Loss is the weighted sum of L1 and Loss2. If the distillation intensity $\alpha = 1$ in formula (2) is used during training, then Loss = L1 + L2. Student models are also trained using Pytorch and Nvidia 1080-Ti GPU. In terms of training parameters, the uniform batch size is 128, that is, every 128 data pairs (images, labels, features) form a batch, the initial learning rate is 0.0001, using batch normalization, The Dropout layer accelerates the training of the network and suppresses the occurrence of overfitting. The gradient update uses the Adam optimization algorithm with faster convergence speed.

4.4   Evaluation Results on 300W

The 300W test set consists of a common set and a challenge set. The common test sets are Helen test set and LFPW test set. The challenging dataset is the IBUG dataset. Table 1 shows the results on the 300W dataset. We compare our method with 11 latest face alignment methods, which are CFAN [26], ESR [18], SDM [19], LBF [31], CFSS [21], DAN(6), SHN [7], LAB [10], RAR [30]. Our approach is better than most.

As shown in Table 3, our method is compared with the most advanced method. T-Net is the teacher model, S-Net is the student model, and KD-Net is the model that uses teacher network to train student network through knowledge distillation. Our method achieves 6.56% NME, on the challenge subset. There are various uncontrollable factors in the challenge subset, such as occlusion, large angle pose and extreme expression. It shows that after knowledge distillation, our model can not only compress the model volume, but also keep a low NME. Figure 3 shows some typical output results on a 300W dataset.

Table 3. Landmark detection results on different subsets of the 300-W dataset in terms of the NME averaged over all the test samples.

Method	Common set	Challenging set	Full set
SDM [19]	5.57	15.40	7.52
ESR [18]	5.28	17.00	7.58
CFAN [26]	5.50	16.78	7.69
LBF [31]	4.95	11.98	6.32
CFSS [21]	4.73	9.98	5.76
DAN [6]	4.42	7.57	5.03
RAR [30]	4.12	8.35	4.94
SHN [7]	4.12	7.00	4.68
LAB [10]	3.42	6.98	4.12
T-Net	3.45	6.52	4.08
S-Net	4.73	8.60	5.41
KD-Net (ours)	3.68	**6.56**	**4.10**

Fig. 3. KD-Net's localization of 68 points on several challenging faces in the 300W dataset. We can observe that even under extreme lighting, expression, occlusion and blurry interference, KD-Net can achieve satisfactory results.

4.5   Evaluation Results on WFLW

The facial landmark configuration of this dataset is different from that of 300W dataset. All images in WFLW dataset are manually labeled with 98 points. The dataset contains a subset of various types of challenges, including large poses, lighting, blurring, occlusion, and excessively disturbing backgrounds, and so on. Since WFLW is a newly released dataset, we compare the proposed method with many other methods, including SDM [19], CFSS [21], DVLN [28], LAB [10] and WingLoss [29]. We report NME, failure rate and AUC in six subsets of test set. As shown in Table 4, KD-Net is superior to all other latest methods in terms of NME, failure rate, and AUC.

Table 4. Landmark detection results on different subsets of the WLFW dataset in terms of the NME averaged over all the test samples.

	Method	Full set	Pose	Expression	Illumination	Make up	Occlusion	Blur
Mean error (%)	SDM [19]	10.29	24.10	11.45	9.32	9.38	13.03	11.28
	CFSS [21]	9.07	21.36	10.09	8.30	8.74	11.76	9.96
	DVLN [28]	6.08	11.54	6.27	5.73	5.98	7.33	6.88
	LAB [10]	5.27	10.24	5.51	5.23	5.15	6.79	6.32
	WingLoss [29]	5.51	8.75	5.36	4.93	5.41	6.37	5.81
	T-Net	4.71	8.65	5.24	4.68	4.62	5.80	5.43
	S-Net	6.15	9.79	6.41	5.97	6.45	7.43	6.86
	KD-Net	**4.91**	**8.74**	5.51	**4.69**	**5.28**	**6.16**	**5.73**
Failures rate (%)	SDM [19]	29.40	84.36	33.44	26.22	27.67	41.85	35.32
	CFSS [21]	20.56	66.26	23.25	17.34	21.84	32.88	23.67
	DVLN [28]	10.84	46.93	11.15	7.31	11.65	16.30	13.71
	LAB [10]	7.56	28.83	6.37	6.73	7.77	12.60	6.73
	WingLoss [29]	6.00	22.70	4.78	4.30	7.77	12.60	7.76
	T-Net	5.13	22.48	5.50	3.96	6.90	9.49	6.84
	S-Net	10.75	46.84	11.06	7.22	11.56	16.21	13.60
	KD-Net	**4.29**	**21.84**	7.12	**5.04**	**6.84**	**10.33**	**7.31**
AUC	SDM [19]	0.300	0.022	0.229	0.323	0.312	0.206	0.239
	CFSS [21]	0.365	0.063	0.315	0.385	0.369	0.268	0.303
	DVLN [28]	0.455	0.147	0.388	0.474	0.449	0.379	0.397
	LAB [10]	0.532	0.234	0.495	0.543	0.539	0.449	0.463
	WingLoss [29]	0.550	0.31	0.495	0.540	0.558	0.488	0.419
	T-Net	0.563	0.273	0.526	0.571	0.563	0.478	0.494
	S-Net	0.465	0.157	0.398	0.484	0.459	0.388	0.398
	KD-Net	**0.556**	**0.369**	**0.499**	**0.554**	**0.578**	0.472	**0.474**

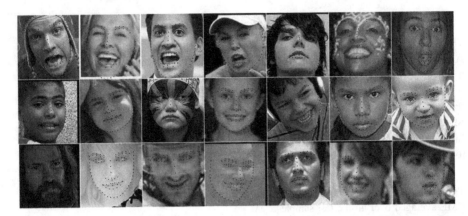

Fig. 4. KD-Net's 98-point localization of several challenging faces in the WFLW dataset. We can observe that even under extreme lighting, expression, occlusion and blurry interference, KD-Net can achieve satisfactory results.

As shown in Table 4, we compare NME, FR and AUC with the existing facial landmark detection algorithm. Among them, T-Net represents the teacher model, S-Net represents the student, and KD-Net is the model that finally uses the teacher network to train the student network through the knowledge distillation method. The proposed method is superior to all other methods. Compared with the teacher network, the student network obtained 7.1 times the parameter reduction, 32 times feature reduction, and 95.1% reasoning complexity reduction. While the model is compressed, the location accuracy of facial landmark on the face is almost the same as the teacher network. The output result of the WFLW data set is shown in Fig. 4.

5   Conclusion

In this paper, a deep knowledge distillation method is proposed, taking the stacked hourglass model as the teacher network and Using the teacher network to guide the training of lightweight student networks, which can ensure that the normalized average error (NME) is 3.68 at the same time realize the compression of teacher model and feature. Experiment results on 300W and COFW dataset show the effectiveness of the proposed method and a face alignment model that takes into account both speed and accuracy is obtained. Since this article aims to speed up the inference speed of the model, the positioning accuracy and the volume of the compressed model, the positioning performance of key points on the partially occluded face image is not ideal. In the next step, we will focus on solving this problem.

References

1. Zhu, X., Lei, Z., Liu, X., et al.: Face alignment across large poses: a 3D solution. In: Proceedings of the IEEE Conference on Computer Vision and Pattern Recognition, pp. 146–155 (2016)
2. Liu, Y., Wei, F., Shao, J., et al.: Exploring disentangled feature representation beyond face identification. In: Proceedings of the IEEE Conference on Computer Vision and Pattern Recognition, pp. 2080–2089 (2018)
3. Sun, Y., Wang, X., Tang, X., et al.: Hybrid deep learning for face verification. In: Proceedings of the IEEE International Conference on Computer Vision, pp. 1489–1496 (2013)
4. Bettadapura, V., et al.: Face expression recognition and analysis: the state of the art. arXiv preprint arXiv:1203.6722 (2012)
5. Guo, Y., Zhao, G., Pietikäinen, M.: Dynamic facial expression recognition with atlas construction and sparse representation. IEEE Trans. Image Process. 25(5), 1977–1992 (2016)
6. Kowalski, M., Naruniec, J., Trzcinski, T., et al.: Deep alignment network: a convolutional neural network for robust face alignment. In: Proceedings of the IEEE Conference on Computer Vision and Pattern Recognition Workshops, pp. 88–97 (2017)
7. Yang, J., Liu, Q., Zhang, K., et al.: Stacked hourglass network for robust facial landmark localisation. In: Proceedings of the IEEE Conference on Computer Vision and Pattern Recognition Workshops, pp. 79–87 (2017)
8. Bulat, A., Tzimiropoulos, G., et al.: How far are we from solving the 2D & 3D face alignment problem? (and a dataset of 230,000 3D facial landmarks). In: Proceedings of the IEEE International Conference on Computer Vision, pp. 1021–1030 (2017)
9. Deng, J., Trigeorgis, G., Zhou, Y., et al.: Joint multi-view face alignment in the wild. IEEE Trans. Image Process. 28(7), 3636–3648 (2019)
10. Wu, W., Qian, C., Yang, S., et al.: Look at boundary: a boundary-aware face alignment algorithm. In: Proceedings of the IEEE Conference on Computer Vision and Pattern Recognition, pp. 2129–2138 (2018)
11. Hinton, G., Vinyals, O., Dean, J., et al.: Distilling the knowledge in a neural network. arXiv preprint arXiv:1503.02531 (2015)
12. Howard, A., Zhu, M., Chen, B., et al.: MobileNets: efficient convolutional neural networks for mobile vision applications. arXiv preprint arXiv:1704.04861 (2017)
13. Sandler, M., Howard, A., Zhu, M., et al.: MobileNetV2: inverted residuals and linear bottlenecks. In: Proceedings of the IEEE Conference on Computer Vision and Pattern Recognition, pp. 4510–4520 (2018)
14. Howard, A., Sandler, M., Chu, G., et al.: Searching for MobileNetV3. In: Proceedings of the IEEE International Conference on Computer Vision, pp. 1314–1324 (2019)
15. Cootes, T., Edwards, G., Taylor, C., et al.: Active appearance models. IEEE Trans. Pattern Anal. Mach. Intell. 23(6), 681–685 (2001)
16. Cristinacce, D., Cootes, T., et al.: Feature detection and tracking with constrained local models. In: BMVC, pp. 929–938 (2006)
17. Zhu, X., Ramanan, D., et al.: Face detection, pose estimation, and landmark localization in the wild. In: 2012 IEEE Conference on Computer Vision and Pattern Recognition, pp. 2879–2886. IEEE (2012)
18. Cao, X., Wei, Y., Wen, F., et al.: Face alignment by explicit shape regression. Int. J. Comput. Vis. 107(2), 177–190 (2014)

19. Xiong, X., De la Torre, F., et al.: Supervised descent method and its applications to face alignment. In: Proceedings of the IEEE Conference on Computer Vision and Pattern Recognition, pp. 532–539 (2013)
20. Ren, S., Cao, X., Wei, Y., et al.: Face alignment at 3000 fps via regressing local binary features. In: Proceedings of the IEEE Conference on Computer Vision and Pattern Recognition, pp. 1685–1692 (2014)
21. Zhu, S., Li, C., Change, L., et al.: Face alignment by coarse-to-fine shape searching. In: Proceedings of the IEEE Conference on Computer Vision and Pattern Recognition, pp. 4998–5006 (2015)
22. Sun, Y., Wang, X., Tang, X., et al.: Deep convolutional network cascade for facial point detection. In: Proceedings of the IEEE Conference on Computer Vision and Pattern Recognition, pp. 3476–3483 (2013)
23. Zhou, E., Fan, H., Cao, Z., et al.: Extensive facial landmark localization with coarse-to-fine convolutional network cascade. In: Proceedings of the IEEE International Conference on Computer Vision Workshops, pp. 386–391 (2013)
24. Zhang, Z., Luo, P., Loy, C.C., Tang, X.: Facial landmark detection by deep multi-task learning. In: Fleet, D., Pajdla, T., Schiele, B., Tuytelaars, T. (eds.) ECCV 2014. LNCS, vol. 8694, pp. 94–108. Springer, Cham (2014). https://doi.org/10.1007/978-3-319-10599-4_7
25. Zhang, J., Shan, S., Kan, M., Chen, X.: Coarse-to-fine auto-encoder networks (CFAN) for real-time face alignment. In: Fleet, D., Pajdla, T., Schiele, B., Tuytelaars, T. (eds.) ECCV 2014. LNCS, vol. 8690, pp. 1–16. Springer, Cham (2014). https://doi.org/10.1007/978-3-319-10605-2_1
26. Trigeorgis, G., Snape, P., Nicolaou, M., et al.: Mnemonic descent method: a recurrent process applied for end-to-end face alignment. In: Proceedings of the IEEE Conference on Computer Vision and Pattern Recognition, pp. 4177–4187 (2016)
27. Wu, W., Yang, S., et al.: Leveraging intra and inter-dataset variations for robust face alignment. In: Proceedings of the IEEE Conference on Computer Vision and Pattern Recognition Workshops, pp. 150–159 (2017)
28. Feng, Z., Kittler, J., Awais, M., et al.: Wing loss for robust facial landmark localisation with convolutional neural networks. In: Proceedings of the IEEE Conference on Computer Vision and Pattern Recognition, pp. 2235–2245 (2018)
29. Xiao, S., Feng, J., Xing, J., Lai, H., Yan, S., Kassim, A.: Robust facial landmark detection via recurrent attentive-refinement networks. In: Leibe, B., Matas, J., Sebe, N., Welling, M. (eds.) ECCV 2016. LNCS, vol. 9905, pp. 57–72. Springer, Cham (2016). https://doi.org/10.1007/978-3-319-46448-0_4
30. Ren, S., Cao, X., Wei, Y., et al.: Face alignment via regressing local binary features. IEEE Trans. Image Process. 25(3), 1233–1245 (2016)

Machine Vision and 3D Reconstruction

3D Human Body Reconstruction from a Single Image

WenTao Xu[1], ChunHong Wu[1(⊠)], Tian Li[1], and Dongmei Fu[1,2]

[1] Key Laboratory of Knowledge Automation for Industrial Processes of Ministry
of Education, School of Automation and Electrical Engineering,
University of Science and Technology Beijing, Beijing 100083, China
cwu@ies.ustb.edu.cn
[2] Shunde Graduate School of University of Science and Technology Beijing,
Fo Shan 528300, China

Abstract. In this paper, a generative adversarial network based on SMPL model is used to reconstruct three-dimensional (3D) human mesh from a single image. The generative adversarial network consists of two parts (generative network and adversarial network). The generative network is made up of a deep residual network and a multilayer perception network. The goal is to generate suitable SMPL model parameters and camera parameters which fit the 2D and 3D position annotation of human skeleton joints. The adversarial network judges whether the shape and pose parameters generated by the generative network match the real human body through a discriminator. The purpose is to guide the generative network output parameters that conform to the laws of human kinematics. A geometric prior of the 3D joint angle limitation of the human body is introduced into the network to further constrain the generated human pose. Experiments verify the effectiveness of the algorithm.

Keywords: SMPL model · 3D reconstruction · Pose and shape estimation · Generative adversarial network

1 Introduction

3D human body reconstruction has a wide range of applications in many fields such as pose estimation, virtual fitting and film production. In the past few decades, researchers explored various approaches like using 3D scanners, depth sensors or multi-camera systems to obtain 3D model of the human body [1–3]. These methods have high accuracy, but with high cost, limited use environment. In recent years, with the requirements for low cost and convenience of human model acquisition, methods that can reconstruct a 3D human body model directly from a single image have appeared. They have the advantages of low cost, simple implementation and have a wider application prospect.

Compared to the traditional method of 3D human body reconstruction, single image lacks sufficient spatial information. Therefore, a parametric human model is required to achieve better reconstruction. The main idea is to fit the parameters of the model from the contour or joint position of the human body. The research starts from the SCAPE [4] model. In the early works, the contours or joint points need to be marked manually as the initial values [5, 6]. With the development of deep neural network, the

© Springer Nature Singapore Pte Ltd. 2020
Y. Wang et al. (Eds.): IGTA 2020, CCIS 1314, pp. 91–100, 2020.
https://doi.org/10.1007/978-981-33-6033-4_7

technology of automatically obtaining human body posture and contour from image is mature. The application of Openpose, Alphapose, Densepose and other methods provides a better foundation for image-based 3D human body reconstruction.

The Skinned Multi-Person Linear (SMPL) [7] model is a parameterized human model proposed by Loper et al. in 2015. Unlike the SCAPE model, SMPL model uses shape and pose parameters to express the 3D surface of human body. It can describe the shape of human muscle stretching and contraction accurately.

In 2016, Bogo et al. [8] proposed the SMPLify method. Based on the automatic detection of 2D joint, the method adds a variety of prior information to fit the SMPL model, and automatically estimates the 3D shape and pose of human body from a single image for the first time. In 2018, Kanazawa et al. [9] proposed an end-to-end network structure based on the SMPL model using the regression method. By introducing the adversarial prior auxiliary training, the human body model parameters were directly regressed from the image features, avoiding the information loss introduced in the multi-step training process. The whole 3D human body can be recovered in real-time even when occlusion or truncation existed. Based on Kanazawa's work, we add 3D joint angle limitations of human body in this paper. With only the 2D and 3D joint label as network supervision, a higher accuracy was achieved.

2 Technical Methods

2.1 Overall Network Structure

Skinned Multi-Person Linear (SMPL) model is a parameterized human model proposed by M. Loper et al. in 2015 [7]. The model includes two basic parameters: shape and pose. Shape parameter ($\beta \in \mathbb{R}^{10}$) contains 10 parameters such as height, weight, body proportions, etc. The pose parameters ($\theta \in \mathbb{R}^{3K}$) are the rotation vectors of the 3D joints (K = 24) which describe how the 3D surface deforms with articulation.

The overall structure of the network we adopted is shown in Fig. 1. The 2D image contains people is put into a **Deep Residual Network (DRN)** to get the image features. The image features are then sent to a **Multi-Layer Perception Network (MLPN)** to generate β, θ parameters of the SMPL model and the camera parameters. After obtaining β, θ parameters, a triangle mesh model (M) of the human body surface can be output through a linear blend skin function. The 3D joint coordinates (X) can be obtained by linear regression from the human mesh model. The 2D joint coordinates (x) on the image can be further calculated by camera projection. The DRN and the MLP network together forms the generative network, as shown in the Fig. 1. The aim of this generative network is to generate appropriate SMPL model parameters ($\hat{\beta}$, $\hat{\theta}$) and camera parameters ($\hat{s}$, $\hat{t}$) so that the 3D and 2D human body joints calculated on the basis of these parameters are in line with the corresponding annotation.

Since indirect supervision was used in our method, there will inevitably some ambiguities existed in the training process, so that an adversarial network is introduced. To guide the training of the generative network, a discriminator is used in judging whether its output conforms to the laws of human body kinematics. The geometric prior of 3D joint angle limitation of the human body is further introduced in the network to constrain the generated human pose.

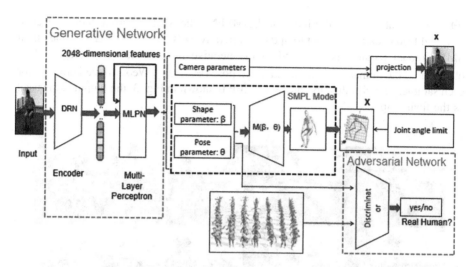

Fig. 1. Block diagram of the overall network structure

2.2 Generative Network

Generative network consists of Deep Residual Network (DRN) and Multi-Layer Perception Network (MLPN) as shown in Fig. 2. The DRN uses ResNet-50 [10], which is usually trained in the ImageNet classification task to encode the input image. In this paper we directly remove the last layer of the pre-trained ResNet model to extract image features Φ from the input image. The MLPN consists of three fully connected layers, as shown in the right part of Fig. 2. It is responsible for generating SMPL model parameters (shape parameter β, pose parameter θ) and camera parameters (translation matrix $t \in \mathbb{R}^2$ and scale $s \in \mathbb{R}$). The output is an 85-dimensional vector $\Theta = \{\theta, \beta, t, s\}$, $\Theta \in \mathbb{R}^{85}$. The pose parameters in the SMPL model need to meet the orthogonal conditions. Since it is difficult to regress the accurate pose parameter directly, an **Iterative Error Feedback (IEF)** method is introduced in the MLPN's training [11]. The output from the DRN (image feature Φ) and the SMPL parameter

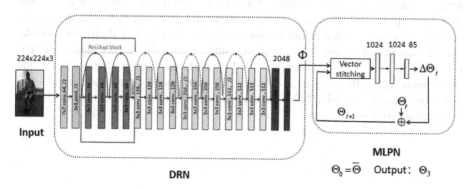

Fig. 2. Generative network structure

(Θ_t) are concatenated as the input of the MLPN. The MLPN output parameter error $\Delta\Theta_t$ and feed back to the input to update Θ_t iteratively ($\Theta_{t+1} = \Theta_t + \Delta\Theta_t$). The initial SMPL model parameters Θ_0 used in the iterative process are the average values obtained by the human body in the standard pose ($\Theta_0 = \overline{\Theta}$). We set three iterations for each sample. The visualization process of IEF is shown in Fig. 3, the output Θ_3 is used as the final output.

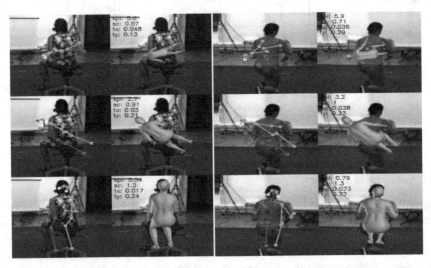

Fig. 3. Visualization of the IEF process. The left and right side shows the output skeleton and reconstructed 3D mesh for two different objects. From top to bottom are the outputs of the three iterations. The output is getting closer to the human body in the image as the iterative process.

2.3   Adversarial Network

The adversarial network uses discriminator as shown in Fig. 4 to guide the training. During the training process, we put the parameters of the real human body and the model parameters that generated from the generative network into the discriminator together. The discriminator identifies β and θ parameters separately. For β, two full connected layers with 10 and 5 neurons output 1D authentication data. For θ, the rotation vector of each joint is transformed into a rotation matrix by Rodrigues formula. The rotation matrix of the 23 joints (root joint is removed) are first input to a convolutional layer of a 32 consecutive 1 x 1 convolution kernels, the output of the convolution layer is then sent to 23 discriminators corresponding to 23 joints and a two fully connected layers with 1024 neurons respectively. The final output is the discrimination result of all joints.

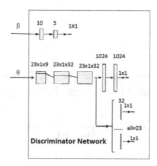

Fig. 4. Adversarial network structure

2.4   3D Joint Angle Limitation

Due to the information loss when projecting from 3D space to 2D images, some wrong 3D postures will be generated in the training process. As shown in the Fig. 5a, the first and second columns shows the normal and abnormal 3D pose of the human body respectively although they have the same 2D projection as shown in Fig. 5b. In response to this case, we introduce joint constraint loss [12] to further constrain the impossible 3D pose.

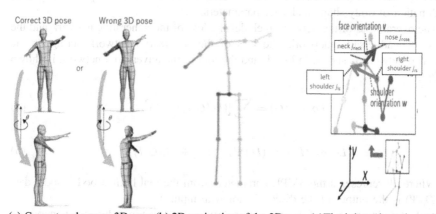

(a) Correct and wrong 3D pose (b) 2D projection of the 3D pose (c)The joint orientation priori

Fig. 5. Joint orientation constraint

Figure 5c gives a schematic diagram of joint orientation constraints, where j_{nose}, j_{neck}, j_{ls}, j_{rs} denote the 3D coordinates of human nose, neck, left shoulder and right shoulder joints, respectively. Define the vector pointing from the neck to the nose as the human body orientation vector, upward is the positive direction. Define the vector pointing from the right shoulder to the left shoulder as the shoulder orientation vector, from right to left is the positive direction, as shown in Fig. 5c.

$$v = [v_x, v_y, v_z] = j_{nose} - j_{neck} \tag{1}$$

$$w = [w_x, w_y, w_z] = j_{ls} - j_{rs} \tag{2}$$

According to the prior knowledge that the head should face forward, the angle between the vectors v and w should satisfy the following equation:

$$\sin\alpha = \frac{v_z w_x - v_x w_z}{\|w\|\|v\|} \geq 0 \tag{3}$$

2.5   Loss of the Network Model

To train the generative adversarial network,we use the least square Gans [13] for its stability. Use G to represent the generative network and D for the adversarial network. $G(I)$ is the output of the generated network for input image I. The result of the i-th adversarial discriminator is $D_i(G(I))$. The output value of the discriminator is between [0, 1], which gives a probability of whether the identified parameter is true or not. During the training process, the adversarial network except the discriminator gives out a value close to 0 to discriminating the generated SMPL parameters. While the generative network hopes that the discriminator gives out a value close to 1 for the generated SMPL parameters. The overall network strikes a balance between the adversarial network (D) and the generative network (G). The aim is to generate a better combination of body shape and pose parameters.

The overall loss of the network includes the loss of the generative network and the loss of the adversarial network. the loss of the adversarial network acting on the generated network is shown in Eq. 4, and the loss of the adversarial network is shown in Eq. 5.

$$Loss_{adv}(G) = \sum_i [(D_i(G(I)) - 1)^2] \tag{4}$$

$$Loss_{adv}(D_i) = [(D_i(\Psi) - 1)^2] + [D_i(G(I))^2] \tag{5}$$

Where, Ψ represents the SMPL parameters from the real human body model data set, $D_i(\Psi)$ is the output of the discriminator with input Ψ.

In the training process of the generation network, we also introduce the supervision of the loss of the 3D and the 2D joint, and the limitation of the joint angle. The loss of the entire generation network is shown in Eq. 6:

$$Loss(G) = \lambda_1 Loss_{joints} + \lambda_2 Loss_{reproj} + \lambda_3 Loss_{angle} + Loss_{adv}(G) \tag{6}$$

Here, $Loss_{joints}$, $Loss_{reproj}$, $Loss_{angle}$ represent **loss of the 3D joint, loss of the 2D joint** and **loss of the joint orientation constraint** respectively. The calculation of these losses will be described later. λ_1, λ_2, λ_3 control the relative importance of each part.

Loss of the 3D Joint

The labeled 3D human joint points are used to guide the output of the generated network, 3D joint loss can be calculated as Eq. 7:

$$Loss_{joints} = \sum_{i=0}^{K} \left\| X_i - \hat{X}_i \right\|_2^2 \tag{7}$$

X_i and $\hat{X}_i$ are the annotated and the predicted coordinates of the i-th 3D joint respectively, K is the number of 3D joints.

Loss of the 2D Joint

The projection from 3D joints to 2D image plane can be calculated as:

$$\hat{x}_i = s \prod \hat{X}_i + t \tag{8}$$

$\prod$ represents the orthographic projection. s, t are camera parameters, $\hat{x}_i$ is the predicted 2D coordinates of point i. The 2D joint loss can be calculated as

$$Loss_{reproj} = \sum_{i=0}^{K} \left\| v_i(x_i - \hat{x}_i) \right\| \tag{9}$$

x_i and $\hat{x}_i$ are the annotated and the predicted coordinates of the i-th 2D joint respectively, $v_i \in \{0, 1\}^K$ represents the visibility of the 2D joint. $v_i = 1$ means visible, 0 means invisible.

Loss of the 3D Joint Angle Constraint

Based on the prior knowledge that the right shoulder should appear to the right side of the human head, as shown in Fig. 5c, the loss function of the 3D joint angle constraint is represented as:

$$Loss_{angle} = \max(0, -\sin\alpha) = \max\left(0, \frac{v_x w_z - v_z w_x}{\|w\| \|v\|}\right) \tag{10}$$

3 Experimental Results

Five datasets were used in our experiment: Human3.6 M [14], MPI_INF_3DHP [15], UP_3D [16], 3dpw [17], SURREAL [18]. Among them, Human3.6 M and MPI_INF_3DHP are used to 2D and 3D joints supervised, which are used for geneative network training. UP_3D, 3dpw and SURREAL are the SMPL model datasets of real human body, which are used for adversarial training. So divided the training datasets into two parts. We build our network on the Tensorflow deep learning framework, use the Adam optimizer to optimize the generative network and the adversarial network. We trained 80 epoch datasets for 5 days on a Intel Core i7-8700K CPU, single GeForce GTX 1080Ti GPU, 12-GB RAM device.

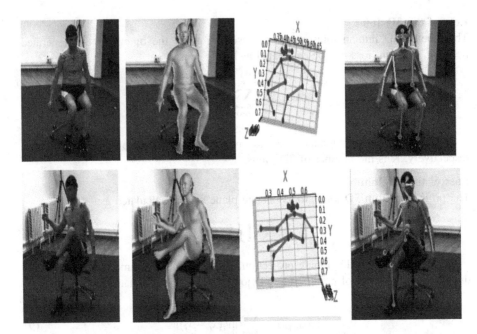

Fig. 6. Human body 3D reconstruction results

Figure 6 shows the reconstruction results from our network. The sample is from Human3.6M datasets. The first column in the figure are the input images. The second column shows the output 3D human model projection onto the image plane, and the third column shows the 3D human skeleton regressed from the SMPL model. The last column shows the 2D skeleton projected from 3D human skeleton.

As far as we known, there is no datasets suitable for evaluating the accuracy of 3D human reconstruction from images, so in this paper only the position error of the reconstructed 3D joint is evaluated. Since Human3.6M has a unified protocols for evaluation human pose, Human3.6M is used as the evaluation dataset. Table 1 gives the comparison of ours and HMRs [9] on dataset Human3.6M using protocols 1 [19] and 2 [8] under the same condition. The average joint position error (MPJPE) we obtained is less than that of HMR under the same conditions,which indicates the effectiveness of adding joint angle limitation in ours work. Besides, our method is

Table 1. Comparison with HMR method

Method	Protocol 1 MPJPE	Protocol 2 MPJPE	Reconstruction time(s)
HMR [9]	141.27 mm	131.03 mm	0.140 s
Ours	**127.67 mm**	**129.26 mm**	**0.126 s**

slightly faster than the HMR in terms of time performance. Table 2 gives the comparison of the reconstruction error with other 3D joint acquisition methods on Human3.6 M. It should be mentioned that our method not only can obtain 3D joints but also can reconstruct 3D human body model in real time.

Table 2. Comparison with other similar works

Method	Reconst. error (mm)
Akhter & Black [20]	181.1 mm
Ramakrishna et al. [21]	157.3 mm
Zhou et al. [22]	106.7 mm
Lassner et al. (direct prediction) [16]	93.9 mm
Ours	**88.6 mm**

4 Conclusion

In this paper, a method of direct reconstructing 3D human body from a single RGB image is achieved. Our work has the following characteristics:

1) A network based on the SMPL model is adopted with reference to HMR [9]. In order to train the whole network in the correct direction, an adversarial prior is introduced by judging whether the generated 3D human body conforms to the laws of human kinematics. Our method can directly return 3D human body mesh model from a 2D image, avoiding information loss in the multi-step training process.
2) We use the 2D and 3D position annotations of the joints that are common in the existing datasets as the network supervision and has a wider applicability.
3) Compared with the reference HMR [9], we add 3D joint angle constraints in the training process, which improves the reliability and accuracy of the reconstructed human body.

Our next goal is to introduce human models such as SMPL-H, SMPL-X, SMPLify-X [23] under the existing network framework. These models add more hand and face joints, which can make the reconstruted human body more realistic.

Funding.. Scientific and Technological Innovation Foundation of Shunde Graduate School, USTB (No. BK19CE017)

References

1. Dou, M., et al.: Fusion4D: real-time performance capture of challenging scenes. ACM Trans. Graph. **35** (2016). https://doi.org/10.1145/2-897824.2925969
2. Yu, L.: Automatic three-dimensional reconstruction of human body based on synchronous camera array (2016)
3. Yu, T., et al.: DoubleFusion: real-time capture of human performances with inner body shapes from a single depth sensor. In: IEEE CVPR 2018 (oral) (2018)

4. Anguelov, D., Srinivasan, P., Koller, D., Thrun, S., Rodgers, J., Davis, J.: SCAPE: shape completion and animation of people. ACM Trans. Graph. (TOG) **24**(3), 408–416 (2005)
5. Guan, P., Weiss, A., Balan, A.O., Black, M.J.: Estimating human shape and pose from a single image. In: CVPR (2009)
6. Sigal, L., Balan, A., Black, M.J.: Combined discriminative and generative articulated pose and non-rigid shape estimation. In: NIPS (2008)
7. Loper, M., Mahmood, N., Romero, J., Pons-Moll, G., Black, M.J.: SMPL: a skinned multi-person linear model. ACM Trans. Graph. **34**, 248:1–248:16 (2015)
8. Bogo, F., Kanazawa, A., Lassner, C., Gehler, P., Romero, J., Black, Michael J.: Keep it SMPL: automatic estimation of 3D human pose and shape from a single image. In: Leibe, B., Matas, J., Sebe, N., Welling, M. (eds.) ECCV 2016. LNCS, vol. 9909, pp. 561–578. Springer, Cham (2016). https://doi.org/10.1007/978-3-319-46454-1_34
9. Kanazawa, A., Black, M.J., Jacobs, D.W., Malik, J.: End-to-end recovery of human shape and pose. In: CVPR (2018)
10. He, K., Zhang, X., Ren, S., Sun, J.: Identity mappings in deep residual networks. In: Leibe, B., Matas, J., Sebe, N., Welling, M. (eds.) ECCV 2016. LNCS, vol. 9908, pp. 630–645. Springer, Cham (2016). https://doi.org/10.1007/978-3-319-46493-0_38
11. Carreira, J., Agrawal, P., Fragkiadaki, K., Malik, J.: Human pose estimation with iterative error feedback. In: IEEE Conference on Computer Vision and Pattern Recognition, CVPR (2016)
12. Kudo, Y., Ogaki, K., Matsui, Y., Odagiri, Y.: Unsupervised Adversarial Learning of 3D Human Pose from 2D Joint Locations (2018)
13. Mao, X., Li, Q., Xie, H., Lau, R.Y.K., Wang, Z., Smolley, S.P.: Least squares generative adversarial networks (2016)
14. Ionescu, C., Papava, D., Olaru, V., Sminchisescu, C.: Human3.6 M: large scale datasets and predictive methods for 3D human sensing in natural environments. PAMI **36**(7), 1325–1339 (2014)
15. Tung, H.-Y.F., Harley, A.W., Seto, W., Fragkiadaki, K.: Adversarial inverse graphics networks: Learning 2D-to-3D lifting and image to image translation from unpaired supervision. In: IEEE International Conference on Computer Vision, ICCV (2017)
16. Lassner, C., Romero, J., Kiefel, M., Bogo, F., Black, M.J., Gehler, P.V.: Unite the people: closing the loop between 3D and 2D human representations. In: CVPR (2017)
17. von Marcard, T., Henschel, R., Black, Michael J., Rosenhahn, B., Pons-Moll, G.: Recovering accurate 3D human pose in the wild using IMUs and a moving camera. In: Ferrari, V., Hebert, M., Sminchisescu, C., Weiss, Y. (eds.) ECCV 2018. LNCS, vol. 11214, pp. 614–631. Springer, Cham (2018). https://doi.org/10.1007/978-3-030-01249-6_37
18. Varol, G., et al.: Learning from synthetic humans. In: Proceedings of the IEEE Conference on Computer Vision and Pattern Recognition, pp. 109–117 (2017)
19. Rogez, G., Weinzaepfel, P., Schmid, C.: LCR-Net: localization-classification-regression for human pose. In: IEEE Conference on Computer Vision and Pattern Recognition, CVPR, July 2017 (2017)
20. Akhter, I., Black, M.J.: Pose-conditioned joint angle limits for 3D human pose reconstruction. In: CVPR (2015)
21. Ramakrishna, V., Kanade, T., Sheikh, Y.: Reconstructing 3D human pose from 2D image landmarks. In: Fitzgibbon, A., Lazebnik, S., Perona, P., Sato, Y., Schmid, C. (eds.) ECCV 2012. LNCS, vol. 7575, pp. 573–586. Springer, Heidelberg (2012). https://doi.org/10.1007/978-3-642-33765-9_41
22. Zhou, X., Zhu, M., Leonardos, S., Daniilidis, K.: Sparse representation for 3D shape estimation: a convex relaxation approach. PAMI **39**, 1648–1661 (2016)
23. Pavlakos, G., Choutas, V., Ghorbani, N.: Expressive body capture: 3D hands, face, and body from a single image (2019)

Image/Video Big Data Analysis
and Understanding

Abnormal Crowd Behavior Detection Based on Movement Trajectory

Fujian Feng[1,2(✉)], Han Huang[1], Haofeng Zhu[1], Yihui Liang[3],
and Lin Wang[2]

[1] College of Software Engineering, South China University of Technology,
Guangzhou 510006, China
[2] College of Data Science and Information Engineering,
Guizhou Minzu University, Guiyang 550025, China
fujian_feng@gzmu.edu.cn
[3] Zhongshan Institute, University of Electronic Science and Technology
of China, Zhongshan 528400, China

Abstract. Abnormal crowd behavior detection on surveillance videos plays an important role in social security and is an important research in computer vision. This work detects crowd abnormal behaviors according to motion trajectory. Specifically, the presented algorithm discovers abnormal direction and abnormal crossing of pedestrians, and uses a novel backtracking method to accurately mark the movement trajectory of abnormal behaviors in the crowd. Experiments were performed on the MOT16 challenge dataset to verify the performance of the proposed method. The experiment shows that the proposed method can effectively detect the abnormal direction, abnormal crossing and movement trajectory of pedestrian with the abnormal behavior.

Keywords: Movement trajectory · Abnormal crowd behavior · Multi-target tracking · Abnormal trajectory · Abnormal direction

1 Introduction

Video surveillance has been widely used in public places, such as labor reform workshops, station entrances and exits, key security control areas, detecting abnormal crowd behavior is an import task, because some hidden safety hazards may cause by the behavior (such as changing the direction of movement in violation of rules and violations of boundaries [1]), which need to be found and processed. In recent years, with the improvement of urban monitoring systems, the government is eager to detect the possible abnormal behavior in real-time from surveillance video through computers in order to reduce the cost of manual processing. Thus, the study of crowd abnormal behavior based on video sequences has become an important computer vision task [2]. Most existing video surveillance systems only detect or track pedestrians in the scene, and rarely further detect and analyze the behavior of pedestrians. However, the purpose of monitoring is to detect and warn pedestrians of abnormal behavior or events in real-time. Therefore, it is of great significance to perform abnormal crowd behavior detection in surveillance video systems.

© Springer Nature Singapore Pte Ltd. 2020
Y. Wang et al. (Eds.): IGTA 2020, CCIS 1314, pp. 103–113, 2020.
https://doi.org/10.1007/978-981-33-6033-4_8

In the last decade, abnormal behavior analysis has made great progress in the field of human behavior. For example, Lahiri et al. [3] calculated the magnitude and direction of the motion using optical flow estimation, then compared with a fixed threshold to determine whether that particular frame is abnormal. It can only detect then crowd abnormal behavior. Direkoglu et al. [4] calculated the angle difference computed between optical flow vectors, and used a one-class SVM to learn normal crowd behaviors. Then sample deviated significantly from the normal behavior was detected as abnormal crowd behaviors. Yuan et al. [5] designs an abnormality detector based on a statistical hypothesis test. Besides, abnormal events are identified as ones containing abnormal event patterns while possessing high abnormality detector scores. Samsudin et al. [6] used a self-adaptive social force model (SFM) as a classifier to classify the features extracted by optical flow method, which can describe crowd behaviors based on the interaction forces between individuals. Gupta et al. [7] used deep convolutional neural network (CrowdVAS-Net) to extract motion and appearance feature representations from video frames. Then a random forest classifier was performed to classify frames into normal or abnormal categories. Most of the existing researches modeled normal events as probabilistic models and classified events with lower probabilistic scores as abnormal events. There are few studies on the special modeling of crowd abnormal directions and abnormal cross-boundary behaviors to determine crowd abnormal behavior.

Movement trajectory is the core difficulty of crowd abnormal behavior detection, it is also the key to the determination of abnormal direction. Many efforts have been made in this field. Liu et al. [8] modeled real-world crowd movement recognition problem as a multilabel classification problem, which took crowd trajectories as features, and used a number of well-designed agent-based motion models (AMMs) to describe the crowd movement. Xu et al. [9] used iterative k-means clustering to find out possible trajectories in the scene and thereby abnormal trajectories which those that do not fit well in any trajectories were identified detected as abnormal trajectory. Wang et al. [10] extracts pedestrian features based on multi-feature fusion and match the same pedestrian between adjacent frames to track the pedestrian. Liu et al. [11] used local outlier factor algorithm to judge whether the sample points of the trajectories are abnormal, it overcome the problem whose trajectory length is not uniform. Although many researches have presented in the analysis of crowd abnormal behavior, most of them don't use it as a key for the analysis of crowd abnormal behavior.

In this paper, we introduce a crowd abnormal behavior detection method based on the motion trajectory. We focus on the abnormal behavior that the pedestrian cross a gate in an abnormal direction. This method only uses the local part of the gate of the monitoring images while multiple object tracking. We use Yolo v3 [12] as detector and Deep-sort [13] as the multiple object tracker to locate pedestrians, determine the pedestrian direction by calculating the motion vector between two or more frames, to determine the abnormal pedestrian direction. We use the location of pedestrians between frames to determine whether the pedestrian is located on both sides of the gate in two frames. If true, he or she will be considered as crossing the gate. Backtracking is used to find the trajectory of the pedestrian which is considered both abnormal direction and crossing. The pedestrian trajectory is traced back by backtracking video using the SiamRPN [14] single object tracker, which improves the accuracy of trajectory detection. The running time of pedestrian abnormal behavior detection is reduced, because only part of the monitoring screen near the entrance is used to detect. Finally, we conducted the experiments on MOT16 dataset [15] by customizing the position of the gate. Experiments verify the effectiveness of the proposed method.

The rest of the paper is organized as follows. Section 2 introduces the algorithms of the multiple object tracking based on detection. Section 3 proposes crowd abnormal behavior detection method, including the abnormal direction detection, abnormal crossing detection and trajectory traceback. Section 4 details our experiments design and results analysis. Section 5 concludes this paper.

2   Multiple Object Tracking Based on Detection

In this section, we will introduce the related work of multiple object tracking (MOT). MOT aims at locating multiple objects of interest and giving the same identities for the same object while the tracking [16]. We will focus on multiple object tracking based on detection and worked on single camera scenario. Figure 1 shows the flow of multiple object tracking based on detection. It needs a detector and a tracker. The detector gets a frame from video stream and detect the object and then both frame the detection result will be send to the tracker to get the multiple object trajectory. The details of the detector and tracker we use will introduce separately.

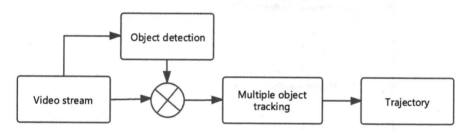

Fig. 1. Flow chart of multiple object tracking based on detection

Yolo v3 is adopted as the detector. Giving an input image, it can output the locations of interest, which is the pedestrian in our method. It works using a trained convolution neural network (CNN).

Deep-sort is used as the tracker. The locations of the pedestrians is given by the detector as input. It uses a standard Kalman filter [17] with constant velocity motion and linear observation model to calculate the possible location of every pedestrian. Then pre-trained CNN is applied to compute bounding box appearance descriptors to find the most possible location of the pedestrian. After that, it can associate the position of pedestrians in the previous and subsequent frames to achieve the purpose of tracking.

As the basis of abnormal behavior detection methods. Multiple object tracking gives the location and motion trajectory of every pedestrian, which will be used to do abnormal detection and analysis.

3   Abnormal Behaviour Detection of the Crowd

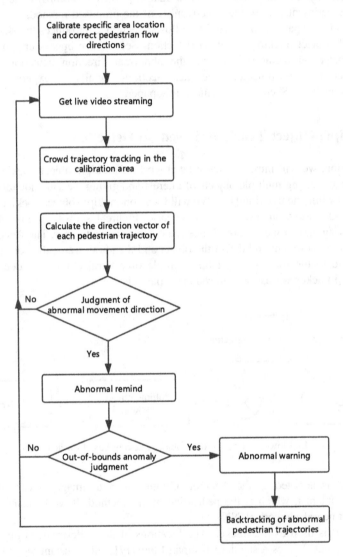

Fig. 2. System design of crowd abnormal detection

In this section, we will introduce abnormal crowd behaviour detection algorithm. Figure 2 shows the flow of the algorithm. First input the location of the gate, calibrate the specific area location to do multiple object detection and correct pedestrian flow direction to initial the algorithm. In our method, the specific area location is a rectangle with the gate as the center. Its length is the same as the gate's length, and the width is W_g pixels. Second, it does get video streaming frame by frame and do multiple objects tracking to get the location of pedestrians. Third, it calculates the motion vector determine if there has any pedestrian walking by abnormal direction. If not, detect the next frame, else give warning. Then the algorithm detects from which has been detected as moving though abnormal direction to find whether these is someone crossing the gate, and give abnormal crossing alarm. At the end, backtracking the pedestrians which was found abnormal crossing.

3.1   Abnormal Direction

If someone wants to cross the gate abnormally, he or she must firstly have an abnormal direction. In this section, we will introduce the direction of movement. Firstly, we should consider the representation of pedestrian. In order to simplify the model, the position of the pedestrian is represented as the coordinates of the center point of the pedestrian. So, the location P of pedestrian in a 2D video is a two-dimensional coordinate point $(P^t_{i,x}, P^t_{i,y})$, where $P^t_{i,x}$ is the x-axis coordinate and $P^t_{i,y}$ is the y-axis coordinate.

In direction calculation method, the influence of jitter caused by the detector should be taken into consider, which will cause the location of pedestrian jitters between frames. Since the Deep-sort tracker mentioned in Sect. 2 has used Kalman filter to fight against jitter, we can simply calculate the motion vector by subtract the position of this frame from the position of the previous N_d frames:

$$D^t_i = P^t_i - P^{t-N_d}_i \tag{1}$$

where P represents the location of pedestrian, t represents the t-th frame and i represents the i-th pedestrian, and N_d is constant throughout the detecting process which must be set in a reasonable interval. A small N_d can't do well in fight against the jitter since the time between two location P^t_i and $P^{t-N_d}_i$ is small, so the change of location dose not cause by movement but jitter. A big N_d may hide the direction change of pedestrian.

Then we can compare the direction D^t_i with normal direction D_N:

$$f = \begin{cases} 1, & D^t_i \cdot D_N \geq 0 \\ 0, & D^t_i \cdot D_N < 0 \end{cases} \tag{2}$$

where 1 represents normal detection and 0 represents abnormal detection. If the angle between two vectors is less than 180°, the point multiplication is positive, otherwise it is negative. So, we can use the sign of the point multiplication between D^t_i and normal direction D_N to judge normal or abnormal direction.

3.2   Abnormal Crossing

After found the pedestrians with abnormal direction, we should judge whether they have crossed abnormally.

Abnormal crossing is someone who crossing the gate with an abnormal direction. In Sect. 3.1 we have found pedestrian that with abnormal direction, then we can judge whether they have crossing the gate. We have known the position of the gate, represents as two points $A(x_1, y_1)$ and $B(x_2, y_2)$ $(x_1 < x_2)$, and we can get the straight line equation l_g of the gate:

$$Ax + By + C = 0, \quad x_1 < x < x_2 \tag{3}$$

where $A = y_2 - y_1$, $B = x_1 - x_2$, $C = y_1(x_2 - x_1) - x_1(y_2 - y_1)$.

Then we can judge whether i-th pedestrian has crossed the gate between t-th frame and $(t + 1)$-th frame from location $P_i^t \left(P_{i,x}^t, P_{i,y}^t \right)$ and $P_i^{t+1} \left(P_{i,x}^{t+1}, P_{i,y}^{t+1} \right)$:

$$f' = \begin{cases} 1, & \left(AP_{i,x}^t + BP_{i,y}^t + C \right) \left(AP_{i,x}^{t+1} + BP_{i,y}^{t+1} + C \right) \geq 0 \\ 0, & \left(AP_{i,x}^t + \dot{B}P_{i,y}^t + C \right) \left(AP_{i,x}^{t+1} + BP_{i,y}^{t+1} + C \right) < 0 \end{cases} \tag{4}$$

where 1 represents normal detection and 0 represents abnormal detection. If two points P_i^t and P_i^{t+1} is on the same side of the l_g, the signs after bringing the point into the formula will be same, and the product is greater than 0. Otherwise the product is less than 0. As a result, the pedestrian crossing abnormally was discovered.

3.3   Trajectory Traceback

The trajectory of pedestrian who crossing abnormally is important in security. So we should find out the trajectory from the whole monitoring area. We use single object tracker and backtracking video stream to record the trajectory. Next, we will introduce a trajectory backtracking algorithm by details.

Algorithm 1 outlines our trajectory backtracking algorithm. As input we provide the bounding box BB_i^t of the i-th pedestrian that has been judged as abnormal crossing at time t and saved video stream $VS = \{v_t, v_{t-1}, \ldots, v_{t-N_v}\}$, where v_t is a video frame at time t and N_v is max video frame to saved. In lines 1 and 2 we initialize the single object tracker and trajectory set. In line 3 to 8, we update single object tracker with video frame v_{t-i}, and get the bounding box BB_i^{t-i}, and calculate the center of BB_i^{t-i} and update the trajectory set T, which we return after completion in line 10. In lines 7 and 8, if the single object tracker cannot find the object, then break.

Algorithm 1 Trajectory Backtracking

Input: pedestrian   *i-th*   bounding   box BB_i^t ,   saved   video
stream$VS = \{v_t, v_{(t-1)}, \dots, v_{t-N_v}\}$

1: Initialize the single object tracker with BB_i^t and v_t

2: Initialize set of trajectories $TT \leftarrow \emptyset$

3: **for** $i \in \{1, \dots, Nv\}$ **do**

4:     $BB_i^{t-i} \leftarrow$ single_object_tracker_update (v_{t-i})

5:     $C_i^t \leftarrow$ center (BB_i^t)

6:     $T \leftarrow T \cup \{C_i^t\}$

7:     **if** single object tracker misses

8:         **break**

9:     **end if**

10: **end for**

11: **return** T

---

4 Experiments and Analysis Results

In this section, we assess the performance of our abnormal crowd behavior method on the MOT16 dataset.

4.1 Experimental Setup

Experiment assesses the performance of our abnormal crowd behavior detection method on the MOT16 challenge dataset [15]. Two video MOT16-03 and MOT16-04 is chosen in the experiment, in which have many pedestrians walk on different directions. This makes is suitable for our experiments to detect direction and crossing. Then the gate, the normal direction (D_N) are defined for every scene respectively. The design of three scenes is given followed:

- Scene A: dataset: MOT16-04, gate location: A (250, 230), B (250, 520), D_N: (1, 0)
- Scene B: dataset: MOT16-04, gate location: A (100, 260), B (750, 260), D_N: (0, 1)
- Scene C: dataset: MOT16-03, gate location: A (100, 260), B (750, 260), D_N: (0, 1).

Also, constant W_g introduced in Sect. 3 was set to 100, constant N_d introduced in Sect. 3.1 was set to 20, and constant N_v introduced in Sect. 3.3 was set to 2000.

There are two groups of experiments were conducted.

(1) The first experiment analyzed of the ability of discovering the pedestrians who have crossing abnormally.

(2) The second experiment analyzed the time saving of our method to use local region for multiple object tracking, which is introduced in Sect. 3.

We deployed the algorithms under comparison on the same computational environment. The experimental computer was equipped with one Intel(R) Xeon(R) CPU E5-2680 2.40 GHz, 16 GB memory, one GTX 2080 GPU and Ubuntu 16.04 OS.

4.2   Experimental Analysis

Two metrics were used in the experiments:

- Accuracy: the right pedestrian number that our method found divide all pedestrians number with manual annotation.
- Frame per second (FPS): the number of frames that the tracker processes in one second, where pedestrian number only represents the pedestrians that cross abnormally. FPS can be used to measure the speed of the tracker. A higher FPS means that the tracker takes less time to process a frame.

Table 1. Abnormal crossing result on three scenes

Scene	Detect number	Annotation number	Accuracy
A	1	1	100%
B	4	4	100%
C	17	18	94.44%
Sum	22	23	95.65%

The accuracy of our method is shown in Table 1. Detect number is total number of pedestrians crossing abnormally that our method detected, Annotation number is the total number of pedestrians crossing abnormally with manual annotation. As shown in Table 1, the accuracy of our method is high. The pedestrian does not have been detected is because the tracker does not track correctly while he crossing the gate.

Table 2. Multiple objects tracking FPS comparison on three scenes

Scene	FPS on local area tracking	FPS on whole monitoring area tracking
A	12.54	6.11
B	10.31	6.11
C	7.34	5.96
Average	10.06	6.06

The FPS is shown in Table 2. FPS on local area tracking means that the input frame of the tracker is cropped and only use the local area of the gate, FPS on whole monitoring area tracking means that the input frame of the tracker is the whole monitoring area. It can be seen that our method increases the FPS of the multiple object tracker when tracking on the local area, which means that our method can make the tracker take less time to process a frame. Also, the size of the local area influences the FPS. If the local area is a small part of the whole monitoring area, this method can significantly increase the FPS.

Scene	Abnormal direction detection	Abnormal crossing detection	Trajectory backtracking

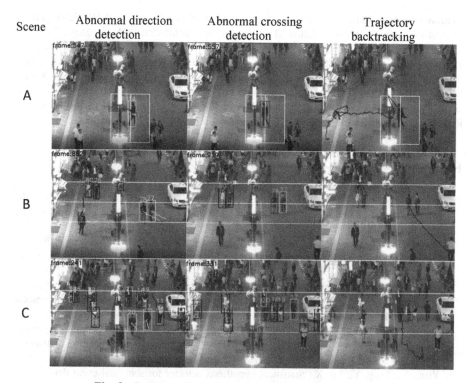

Fig. 3. Detect result on three scenes. (Color figure online)

Figure 3 shows the detect result on scene A, B and C separately. The pedestrian with normal direction was marked with a blue rectangle, with abnormal direction was yellow, and with abnormal crossing was red. The dark green trajectory is the backtracking trajectory.

In conclusion, the experiments verify the proposed method can accurately and efficiently detect the abnormal direction, abnormal crossing of pedestrian with the abnormal behavior. After finding abnormal behavior, the method can backtracking the trajectory in the whole monitoring area.

5 Conclusion

In this study, we proposed an abnormal crowd behavior detection method based on movement trajectory in local region. Due to shrink the track region, we decrease the time cost. However, the trajectory is not completed, so we use backtracking to get back the whole trajectory of pedestrian with abnormal behavior. Experiments verify the accuracy and efficiency of the method. Yet, the method performs poorly in the scene with high-density pedestrian. In future work, we will focus on the study on abnormal behavior of high-density pedestrian.

Acknowledgment. Science Technique Department of Guizhou Province (QKHJC [2019]1164), Funded Research Project of Guizhou Minzu University (GZMU[2019]QN03), Major Research Project of Innovation Group of Guizhou Province (QJHKY[2018]018), Zhongshan Science and Technology Research Project of Social welfare (2019B2010), National Natural Science Foundation of China (61876207), Guangdong Natural Science Funds for Distinguished Young Scholar (2014A030306050), International Cooperation Project of Guangzhou (201807010047), Guangzhou science and technology project (201802010007) and Guangdong Province Key Area Research and Development Program (2018B010109003).

References

1. Rohit, K., Mistree, K., Lavji, J.: A review on abnormal crowd behavior detection. In: 2017 International Conference on Innovations in Information, Embedded and Communication Systems (ICIIECS), pp. 1–3. IEEE (2017)
2. Hong, B., Yufang, S., Bo, X.: Video based abnormal behavior detection. In: Proceedings of the 2011 International Conference on Innovative Computing and Cloud Computing, pp. 32–35 (2011)
3. Lahiri, S., Jyoti, N., Pyati, S., Dewan, J.: Abnormal crowd behavior detection using image processing. In: 2018 Fourth International Conference on Computing Communication Control and Automation (ICCUBEA), pp. 1–5. IEEE (2018)
4. Direkoglu, C., Sah, M., O'Connor, N.E.: Abnormal crowd behavior detection using novel optical flow-based features. In: 2017 14th IEEE International Conference on Advanced Video and Signal Based Surveillance (AVSS), pp. 1–6. IEEE (2017)
5. Yuan, Y., Feng, Y., Lu, X.: Statistical hypothesis detector for abnormal event detection in crowded scenes. IEEE Trans. Cybern. **47**(11), 3597–3608 (2016)
6. Samsudin, W.N.A.W., Ghazali, K.H.: Crowd behavior monitoring using self-adaptive social force model. Mekatronika **1**(1), 64–72 (2019)
7. Gupta, T., Nunavath, V., Roy, S.: CrowdVAS-Net: a deep-CNN based framework to detect abnormal crowd-motion behavior in videos for predicting crowd disaster. In: 2019 IEEE International Conference on Systems, Man and Cybernetics (SMC), pp. 2877–2882. IEEE (2019)
8. Liu, W., Lau, R.W., Wang, X., Manocha, D.: Exemplar-AMMs: recognizing crowd movements from pedestrian trajectories. IEEE Trans. Multimed. **18**(12), 2398–2406 (2016)
9. Xu, S., Ho, E.S., Aslam, N., Shum, H.P.: Unsupervised abnormal behaviour detection with overhead crowd video. In: 2017 11th International Conference on Software, Knowledge, Information Management and Applications (SKIMA), pp. 1–6. IEEE (2017)
10. Wang, X., Song, H., Cui, H.: Pedestrian abnormal event detection based on multi-feature fusion in traffic video. Optik **154**, 22–32 (2018)
11. Liu, W., Xing, W., Qi, J.: Crowd behaviors analysis and abnormal detection in structured scene. In: DMSVLSS, pp. 50–56 (2017)
12. Redmon, J., Farhadi, A.: YOLOv3: an incremental improvement. arXiv preprint arXiv:1804.02767 (2018)
13. Wojke, N., Bewley, A., Paulus, D.: Simple online and realtime tracking with a deep association metric. In: 2017 IEEE international conference on image processing (ICIP), pp. 3645–3649. IEEE (2017)
14. Li, B., Yan, J., Wu, W., Zhu, Z., Hu, X.: High performance visual tracking with Siamese region proposal network. In: Proceedings of the IEEE Conference on Computer Vision and Pattern Recognition, pp. 8971–8980. IEEE (2018)

15. Milan, A., Leal-Taixé, L., Reid, I., Roth, S., Schindler, K.: MOT16: a benchmark for multi-object tracking. arXiv preprint arXiv:1603.00831 (2016)
16. Fan, L., et al.: A survey on multiple object tracking algorithm. In: 2016 IEEE International Conference on Information and Automation (ICIA), pp. 1855–1862. IEEE (2016)
17. Kalman, R.E.: A new approach to linear filtering and prediction problems (1960)

Full Convolutional Color Constancy
with Attention

Tian Yuan[✉] and Xueming Li[✉]

Beijing University of Posts and Telecommunications, Beijing, China
white-sky@qq.com, lixm@bupt.edu.cn

Abstract. Traditional color constancy algorithms solved the problem by modeling the statistical laws of objects and illumination colors. We [1] designed a CNN-based color constancy algorithm using FCNs with an adding pooling, which explained that the semantic information of each local region is different, whose contributions to global illumination color are different. The adding pooling can automatically merge all local patches into a global layer regarding their different weights of contributions, only the features in the last convolutional layer are adaptively weighted. Attention can adaptively learn different weights for all features, which can tell the network where to pay attention and improves the performance of features. This paper improves the color constancy network's ability by using attention modules without significantly increasing the size of model and the running time of network. On standard benchmarks, our network outperforms the previous state-of-the-art algorithms.

Keywords: Color constancy · Attention · FCN

1 Introduction

Color constancy is an important image processing algorithm, which can remove the influence of the illumination on the color rendering of pictures and only retain the inherent color of the objects. The images that processed by color constancy algorithms can be applied better to other computer vision algorithms such as semantic segmentation, image classification, and object detection. Methods for color constancy fall mainly into three categories: static methods, gamut-based methods and learning-based methods. In our previous work, we have proposed a full convolutional neural network architecture, it can take the entire image as input, so that the network can learn from different local image patches at the same time, which can avoid distortion and improve prediction accuracy, then fuse all local information automatically using a new designed pooling layer named adding pooling, which can automatically merge all local patches in the last convolutional layer into a global layer regarding their different weights of contribution to the global illumination color, but only the feature in the last convolutional layer are adaptively weighted.

Attention can be used to solve the problem above. The importance of attention has been widely studied [2–7]. Attention can not only tell the network where to pay attention, but also improves the performance of features. This paper improved the color constancy FCN's accuracy by using attention in computer vision without significantly

© Springer Nature Singapore Pte Ltd. 2020
Y. Wang et al. (Eds.): IGTA 2020, CCIS 1314, pp. 114–126, 2020.
https://doi.org/10.1007/978-981-33-6033-4_9

increasing the size of model and the running time of network. We designed a channel attention module and a spatial attention module, then, designed two methods that mix channel attention and spatial attention by using series mode.

2 Related Work

Attention originates from the study of human vision. Due to the bottleneck of information processing, human will selectively pay attention to a part of information while ignoring other information, this mechanism is usually called attention. In order to make judicious use of the limited visual information processing resources, people need to focus on specific parts of the visual area. For example, people usually pay attention to and deal with some important words while reading. In summary, there are two main aspects of attention: selecting key parts of input that need to be paid attention, allocating limited information processing resources to the key parts.

Attention is a methodology without strict mathematical definition. Traditional local image feature extraction, saliency detection and sliding window methods can be regarded as attention. In neural network, attention is usually implemented by an additional network which is connected to the original one, the model is end-to-end, means the attention module can be trained synchronously with the original network. Attention can choose some parts of input rigidly or assign different weights to different parts of input.

Attention can be divided into four categories: soft attention based on item, hard attention based on item, soft attention based on location and hard attention based on location. The input forms of item-based attention and location-based attention are different. The input of item-based attention is a sequence containing explicit items, or a sequence containing explicit items which requires additional preprocessing steps to be generated. The input of location-based attention is a feature map, and all targets can be specified by their location. Soft attention's result is the probability weights of all candidate vectors, which is differentiable for the input, the whole network can be optimized using gradient method. If just some candidate vectors are selected from the candidates discretely, it is called hard attention, the whole system is not differentiable for the input, and is generally optimized by reinforcement learning. Table 1 shows these four forms of attention.

Table 1. Four forms of attention.

		Item-based	Location-based
Hard attention	Input	Item sequence	Complete feature map
	Operation	Select some items discretely	Select some sub-area discretely
	Optimization	Reinforcement learning	Reinforcement learning
Soft attention	Input	Item sequence	Complete feature map
	Operation	Weighted sum of all terms	Transform
	Optimization	Gradient descent	Gradient descent

This paper mainly focuses on soft attention. Attention has been successfully applied to machine translation. The goal of machine translation is to translate the source language into the target language, each word in target language is not equivalent to every word in source language, but has strong correlation with some words, and weak correlation or no correlation with other words. The function of attention is to allocate the attention of target words to different words in source language according to different weights. Self-attention is used in Transformer [8], which has greatly improve the accuracy and parallelism of the translation model. Unlike traditional attention, the query and key of self-attention belong to same domain, they calculate the attention allocation between different positions in the same statement (image), then, extract the features of this statement (image).

Soft attention in computer vision mainly includes self-attention and scale attention. Non-local [9] is the first algorithm which uses self-attention in computer vision tasks to capture long-range dependencies between different locations in images or videos. Self-attention has achieved great success in visual tasks. It has solved CNN's local receptive field problem, so that each position can obtain the global receptive field, and the characteristics of each position are obtained by the weighted sum of all positions. In computer vision, there are a lot of pixels, using all locations to calculate the attention of each location will lead to huge computation and display memory costs. GC Net [10] is an improvement and a simplified version of Non-local, the authors find that the content of the attention map calculated by different query of Non-local is very similar, so the attention module only needs to learn a global attention map, which can greatly reduce the computational overhead. Self-attention simply treats the image as a sequence, neglecting the relative position relationship between different pixels, so the obtained attention map loses the structural information of images, an improvement direction for self-attention is to add relative position information or absolute position information coding [11, 12]. Scale attention is based on the response of each position in the feature map. In classification task, the greater the response of a location in the feature map, the greater its impact on classification results, and the greater the importance of this location, according to the intensity of response, different positions of the feature map can be enhanced or suppressed, so that attention can be allocated in spatial or channel dimensions. SE Net [13] has a far-reaching influence in the field of computational visual attention, which is essentially channel attention.; SK Net [14] considers the multi-scale receptive field on the basis of SE Net, so that the network can adaptively adjust the size of the receptive domain according to the multi-scale of input information; GE Net [15] extends the channel attention of SE Net to spatial dimension, and uses the second-order pooling function to gather information; CBAM [16] designed a channel attention module and a spatial attention module, combines them in a serial way. Residual Attention Network [17] uses bottom-up top-down to get spatial attention; SGE Net [18] is a lightweight spatial attention module, and its parameters are almost negligible. Scale attention is based on every position in the feature map, it enhances the significant areas and suppresses the non-significant areas, which is more similar to the characteristics of human visual system than self-attention.

Attention is generally divided into three modules: information gather module, information excites module and feature fusion. The framework is shown in formula (1):

$$z_i = F\left(x_i, \delta\left(\sum_{j=1}^{N_p} \alpha_i x_j\right)\right) \tag{1}$$

$\sum_{j=1}^{N_p} \alpha_i x_j$ represents information gather module, this module gathers the features of all positions together by the weighted average determined by the value α_i, the global average pooling layer in SE Net, the global average pooling layer and global maximum pooling layer in CBAM, are all information aggregation modules; $\delta(\cdot)$ represents information excite module, it can process features by transformation to capture the correlation of features in channel dimension or spatial dimension, which is generally composed of convolutional layers, fully connected layers, activation functions and normalization layer, the MLP in SE Net belongs to information excite module, a bottleneck is generally designed in this module to reduce parameters; $F(\cdot, \cdot)$ represents fusion function, which is used to fuse the global context features into each position of the original feature map.

Channel attention and spatial attention proposed in this paper are designed according to this framework.

3   Attention Structure Design

3.1   Channel Attention Mechanism

CNN's feature map can learn the features of input images, the feature maps of the same layer learn different features. It is generally believed that when the feature map is resized to the size of original image, the network has learned the locations where the feature values are relatively large. For example, in face recognition, for a CNN feature map, the first channel may has learned the features of eyes, the second channel may has learned the features of nose and mouth, the third channel may has learned the features of hair, the fourth channel may has learned the features of background area, etc. According to prior knowledge, when human doing face recognition tasks, eyes are most effective to distinguish different characters, and eyebrows, noses and mouths are also conducive to distinguish, but their effects are slightly weaker than eyes, cheek areas are weaker than facial features, hair has a greater effect on sex recognition, but its effect is weaker than face areas, and the features of the background is noise. Different channels of the same layer can learn different features and have different effects on network learning. However, the next layer does not treat the different channels differently, it connects them equally, which will cause the lack of effective information. Channel attention can adaptively learn different weights for different channels, so that the network can adaptively find more useful channels and gives them higher weights, then gives lower weights to noisy channels.

Suppose that the dimension of a CNN feature map is C × H × W, while H × W represents the spatial size, H represents the height and W represents the width, and C represents channel number of this feature map. Channel attention is to learn a C × 1 1 attention vector for original feature map, the essence of this attention vector is the weighted value of each channel, it can be broadcast multiplied or broadcast added to the original feature map so that each channel will have distinct weights. Figure 1 shows the channel attention module proposed in this paper, in which the lower part is the information gather module and the upper part is the information excite module.

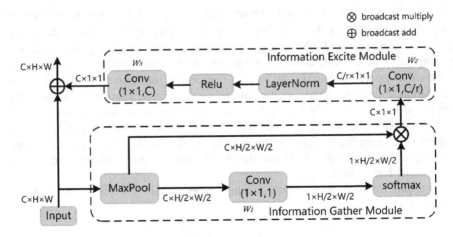

Fig. 1. Channel attention mechanism.

This channel attention module conforms to the attention framework proposed above:

(1) Information gather module: using 1×1 convolution W_1 and softmax function to get feature transformation results, then, global context features are obtained by global attention pooling via matrix multiplication, the maximum pooling is used to reduce the feature resolution and parameter amount;

(2) Information excite module: using 1×1 convolution W_2 and W_3 to transform the features, layer norm is added between the two convolutional layers to make the network converge easily.

(3) Feature fusion: integrating global context feature into original feature map by broadcast addition.

The computational procedure of this channel attention is shown in formula (2):

$$z_i = x_i + W_3 Relu\left(LN\left(W_2 \sum_{j=1}^{N_{\tilde{p}}} \frac{e^{W_1 \tilde{x}_j}}{\sum_{m=1}^{N_{\tilde{p}}} e^{W_1 \tilde{x}_m}} \tilde{x}_j \right) \right) \qquad (2)$$

x_i represents each feature point in original feature map, $\tilde{x}_j$ represents each feature point after maximum pooling, $N\tilde{p}$ represents the number of feature points after maximum pooling, here we use 2×2 maximum pooling, which will make the number of points become 1/4 of the original one. $\alpha_i = \sum_{j=1}^{N_{\tilde{p}}} \frac{e^{W_1 \tilde{x}_j}}{\sum_{m=1}^{N_{\tilde{p}}} e^{W_1 \tilde{x}_m}}$ means the weights of global attention pooling, $\delta(\cdot) = W_3 Relu(LN(W_2(\cdot)))$ represents information excite module with bottleneck design. In this channel attention module, global attention pooling realizes information gather; information excite module with bottleneck design studies the correlation of channels, here, r = 16; Broadcast addition realizes feature fusion.

3.2   Spatial Attention Mechanism

In computer vision, spatial attention is more intuitive than channel attention. Spatial attention obtains the weights of each spatial pixel of original feature map directly in spatial dimension, that is, to find an attention map with the same spatial size as the original feature map, this attention map can be regarded as the weighted values of original feature map in each position. Spatial attention learns a $1 \times H \times W$ attention map for original feature map, the essence of this attention map is the weighted value of each spatial position of the original feature map, it can be broadcast multiplied or broadcast added to the original feature map, which will make each spatial position in original feature map has different weight. Figure 2 shows the spatial attention proposed in this paper:

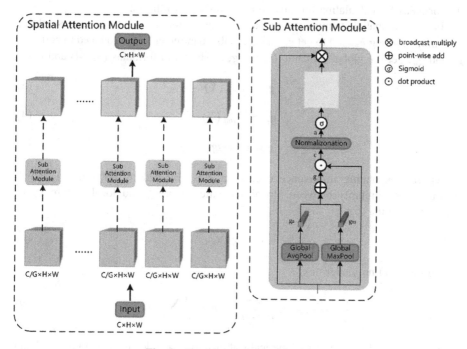

Fig. 2. Spatial attention mechanism.

This spatial attention module conforms to the attention framework proposed above, the original feature map is divided into G groups in channel dimension, and the number of channels in each group is C/G. each group performs the following operations:

(1) Information gather module: the global attention weights are obtained by the weighted sum g of the global average pooling result g_a and the global maximum pooling result g_m;

(2) Information excite module: we generate the corresponding importance coefficient for each feature, which is obtained by dot product that measures the similarity between the global semantic feature g and local feature x_i to get c, then, normalize c over spatial dimension to get $\hat{c}$, then, introduce a pair of parameters γ, β for each coefficient $\hat{c}_i$, which scale and shift the normalized value, after this we will get a, sigmoid function is used to obtain the global context feature $\sigma(a)$, namely spatial attention;

(3) Feature fusion: integrating feature attention into original feature map by broadcast multiplication.

The purpose of grouping channels is to prevent weak generalization due to too many channels, it will be more accurate if each group learns a sub attention module, here, G = 4. For each group of features, suppose $X = \{x_{1...m}\}$, $x_i \in R^{\frac{C}{G}}$, m = H × W, the formulas for calculating this spatial attention are as follows:

For X, our spatial attention use global average pooling $F_{gap}(\cdot)$ to get g_a, use global maximum pooling $F_{gmp}(\cdot)$ to get g_m, the learnable parameter α is introduced to combine g_a and g_m, get the global semantic feature g, as shown in formulas (3), (4) and (5):

$$g_a = F_{gap}(X) \tag{3}$$

$$g_m = F_{gmp}(X) \tag{4}$$

$$g = g_a + \alpha g_m \tag{5}$$

The corresponding importance coefficient c is obtained by dot product, which measures the similarity between global semantic feature g and local feature x_i, as shown in formula (6):

$$c_i = g \cdot x_i \tag{6}$$

Formula (7) normalizes c_i:

$$\hat{c}_i = \frac{c_i - \mu_c}{\sigma_c + \epsilon}, \mu_c = \frac{1}{m}\sum_j^m c_j, \sigma_c^2 = \frac{1}{m}\sum_j^m (c_j - \mu_c)^2 \tag{7}$$

ϵ is a minimal constant for numerical stability. A pair of learnable parameters γ and β are introduced for $\hat{c}_i$, where γ is used for scaling, β is used for shifting, then get a_i:

$$a_i = \gamma \hat{c}_i + \beta \tag{8}$$

The sigmoid function $\sigma(\cdot)$ is used for a_i in spatial dimension, and it is fused to the original feature by broadcast multiplication:

$$\hat{x}_i = x_i \cdot \sigma(a_i) \tag{9}$$

The parameters of this spatial attention module are only α, β and γ, which are very lightweight.

3.3   Mixed Attention Mechanism

The channel attention and spatial attention proposed above can be directly added to CNN networks, and these two attention mechanisms can also be added at the same time, that is, single attention module can be replaced by mixed attention module. There are two ways to mix attention modules: in serial and in parallel. CBAM has proved that series mechanism is more effective than parallel mechanism, because when using parallel method, channel attention and spatial attention have no interaction at all, they calculate attention vector of original feature map respectively as two independent branches, then these two attention vectors are both fused to original feature map. However, when using the series method, the feature map goes through one kind of attention first, then another, the feature map processed by first attention is more effective than the one without attention, and channel attention and spatial attention are coupled in series. This paper, we use series method to mix channel attention and spatial attention. When mixing, two kinds of mixing methods are used: channel-spatial attention module and spatial-channel attention module, as shown in Fig. 3.

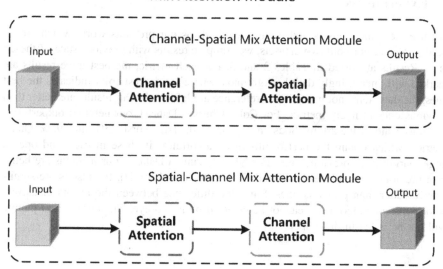

Fig. 3.   Mix attention mechanism.

3.4   Network Structure

Attention is an independent module, which can be added to any part of CNN and learn together with original network. In our previous work [1], we have proposed an FCN structure using VGG Net or SqueezeNet [19] to solve color constancy problem, here we add attention to our proposed SqueezeNet-based FCN to improve the accuracy. SqueezeNet is a lightweight CNN structure, who's base module is fire module, in our previous work, we use SqueezeNet to reduce the number of parameters. In this paper, we replace the ReLU functions in SqueezeNet with Swish functions and add attention module after each fire module to improve the accuracy of this FCN structure, part of this FCN structure is shown in Fig. 4, the attention module can be channel attention, spatial attention or mix attention.

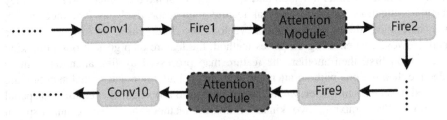

Fig. 4. Part of this FCN structure.

4   Experiments

We use the same datasets and training methods as our previous work, which are no need to repeat. For both two datasets, we compare results with previous state-of-the-art ones. Results are listed in Tables 2 and 3. For each metric, the best three results are marked with increasingly dark backgrounds, where the darkest color indicates the best ones, the data items not given in the reference are left blank. Our results are better than previous ones in many metrics. For Color Checker Dataset, our network outperforms previous state-of-the-art methods in Mean, Med, Tri, Worst 25% and 95% Quant metrics, which obtain the best results of all algorithms in these metrics, and open a large margin from previous ones. For NUS 8-Camera Dataset, our methods are better than previous ones in Mean, Tri and G.M metrics, and for Tri, the results are significantly better than previous ones. Since the difference between the results of current color constancy algorithms cannot be detected by naked eyes, the comparison picture will not be listed in this paper.

Table 2. Results on Color Checker Dataset.

Method	Mean	Med	Tri.	Best 25%	Worst 25%	95% Quant
White-Patch [20]	7.55	5.68	6.35	1.45	16.12	-
Edge-based Gamut [21]	6.52	5.04	5.43	1.90	13.53	-
Gray-World [22]	6.36	6.28	6.28	2.33	10.58	11.30
1st-order Gray-Edge [23]	5.33	4.52	4.73	1.86	10.03	11.00
2nd-order Gray-Edge [23]	5.13	4.44	4.62	2.11	9.26	-
Shades-of-Gray [24]	4.93	4.01	4.23	1.14	10.2	11.90
Bayesian [25]	4.82	3.46	3.88	1.26	10.49	-
General Gray-World [26]	4.66	3.48	3.81	1.00	10.09	-
Intersection-based Gamut [21]	4.20	2.39	2.93	0.51	10.70	-
Pixel-based Gamut [21]	4.20	2.33	2.91	0.50	10.72	14.10
Natural Image Statistics [27]	4.19	3.13	3.45	1.00	9.22	11.70
Bright Pixels [28]	3.98	2.61	-	-	-	-
Spatio-spectral (GenPrior) [29]	3.59	2.96	3.10	0.95	7.61	-
Cheng et al. 2014 [35]	3.52	2.14	2.47	0.50	8.74	-
Corrected-Moment (19 Color) [30]	3.50	2.60	-	-	-	8.60
Exemplar-based [31]	3.10	2.30	-	-	-	-
Corrected-Moment (19Color)* [30]	2.96	2.15	2.37	0.64	6.69	8.23
Corrected-Moment (19 Edge) [30]	2.80	2.00	-	-	-	6.90
Corrected-Moment (19 Edge)* [30]	3.12	2.38	2.59	0.90	6.46	7.80
Regression Tree [32]	2.42	1.65	1.75	0.38	5.87	-
CNN [33]	2.36	1.98	-	-	-	-
CCC (dist+ext) [34]	1.95	1.22	1.38	0.35	4.76	5.85
DS-Net (HypNet+SelNet) [36]	1.90	1.12	1.33	0.31	4.84	5.99
AlexNet-FC4 [37]	1.77	1.11	1.29	0.34	4.29	5.44
SqueezeNet-FC4 [37]	1.65	1.18	1.27	0.38	3.78	4.73
AddingPooling-VGG [1]	1.62	1.14	1.25	0.38	3.70	4.87
AddingPooling-SqueezeNet [1]	1.58	1.16	1.23	0.34	3.60	4.42
Channel Attention	1.54	1.08	1.18	0.36	3.59	4.64
Spatial Attention	1.56	1.09	1.20	0.34	3.60	4.59
Channel-Spatial Attention	1.56	1.10	1.18	0.33	3.64	4.62
Spatial-Channel Attention	1.57	1.10	1.22	0.34	3.58	4.40

Table 3. Results on NUS 8-Camera Dataset.

Method	Mean	Med.	Tri.	Best 25%	Worst 25%	G.M
White-Patch [20]	10.62	10.58	10.49	1.86	19.45	8.43
Edge-based Gamut [21]	8.43	7.05	7.37	2.41	16.08	7.01
Pixel-based Gamut [21]	7.70	6.71	6.90	2.51	14.05	6.60
Intersection-based Gamut [21]	7.20	5.96	6.28	2.20	13.61	6.05
Gray-World [22]	4.14	3.20	3.39	0.90	9.00	3.25
Bayesian [25]	3.67	2.73	2.91	0.82	8.21	2.88
Natural Image Statistics [27]	3.71	2.60	2.84	0.79	8.47	2.83
Shades-of-Gray [24]	3.40	2.57	2.73	0.77	7.41	2.67
Spatio-spectral (ML) [29]	3.11	2.49	2.60	0.82	6.59	2.55
General Gray-World [26]	3.21	2.38	2.53	0.71	7.10	2.49
2nd-order Gray-Edge [23]	3.20	2.26	2.44	0.75	7.27	2.49
Bright Pixels [28]	3.17	2.41	2.55	0.69	7.02	2.48
1st-order Gray-Edge [23]	3.20	2.22	2.43	0.72	7.36	2.46
Spatio-spectral (GenPrior) [29]	2.96	2.33	2.47	0.80	6.18	2.43
Corrected-Moment* (19 Edge) [30]	3.03	2.11	2.25	0.68	7.08	2.34
Corrected-Moment* (19 Color) [30]	3.05	1.90	2.13	0.65	7.41	2.26
Cheng et al. 2014 [35]	2.92	2.04	2.24	0.62	6.61	2.23
CCC (dist+ext) [34]	2.38	1.48	1.69	0.45	5.85	1.74
Regression Tree [32]	2.36	1.59	1.74	0.49	5.54	1.78
DS-Net (HypNet+SelNet) [36]	2.24	1.46	1.68	0.48	6.08	1.74
AlexNet-FC4 [37]	2.12	1.53	1.67	0.48	4.78	1.66
SqueezeNet-FC4 [37]	2.23	1.57	1.72	0.47	5.15	1.71
Adding Pooling-VGG [1]	2.20	1.71	1.81	0.59	4.67	2.20
Adding Pooling-SqueezeNet [1]	2.10	1.56	1.68	0.52	4.56	1.68
Channel Attention	2.13	1.54	1.68	0.53	4.71	1.69
Spatial Attention	2.07	1.56	1.48	0.48	4.57	1.64
Channel-Spatial Attention	2.10	1.53	1.67	0.51	4.63	1.66
Spatial-Channel Attention	2.10	1.52	1.66	0.51	4.62	1.65

5   Conclusion

In this paper, we designed a channel attention module and a spatial attention module respectively, which can improve the performance of color constancy network without increasing the size of model and the running time of network. Then we designed two series methods that mixed the channel attention and spatial attention, which can further improve the performance of color constancy network. Our work can make images better used in computer vision algorithms.

References

1. Yuan, T., Li, X.: Full convolutional color constancy with adding pooling. In: 2019 IEEE 11th International Conference on Communication Software and Networks (ICCSN), pp. 666–671. IEEE (2019)
2. Mnih, V., Heess, N., Graves, A., et al.: Recurrent models of visual attention. In: Advances in Neural Information Processing Systems (2014)
3. Ba, J., Mnih, V., Kavukcuoglu, K.: Multiple object recognition with visual attention. arXiv preprint arXiv:1412.7755 (2014)
4. Bahdanau, D., Cho, K., Bengio, Y.: Neural machine translation by jointly learning to align and translate. arXiv preprint arXiv:1409.0473 (2014)
5. Xu, K., Ba, J., Kiros, R., et al.: Show, attend and tell: neural image caption generation with visual attention. In: Computer Science, pp. 2048–2057 (2015)
6. Gregor, K., Danihelka, I., Graves, A., et al.: DRAW: a recurrent neural network for image generation. In: Computer Science, pp. 1462–1471 (2015)
7. Jaderberg, M., Simonyan, K., Zisserman, A.: Spatial transformer networks. In: Advances in Neural Information Processing Systems, pp. 2017–2025 (2015)
8. Vaswani, A., Shazeer, N., Parmar, N., et al.: Attention is all you need (2017)
9. Wang, X., Girshick, R., Gupta, A., et al.: Non-local neural networks (2017)
10. Cao, Y., Xu, J., Lin, S., et al.: GCNet: non-local networks meet squeeze-excitation networks and beyond. arXiv preprint arXiv:1904.11492 (2019)
11. Bello, I., Zoph, B., Vaswani, A., et al.: Attention augmented convolutional networks. arXiv preprint arXiv:1904.09925 (2019)
12. Parmar, N., Ramachandran, P., Vaswani, A., et al.: Stand-alone self-attention in vision models (2019)
13. Jie, H., Li, S., Albanie, S., et al.: Squeeze-and-Excitation Networks[J]. IEEE Trans. Pattern Anal. Mach. Intell. **PP**(99), 1 (2017)
14. Li, X., Wang, W., Hu, X., et al.: Selective kernel networks (2019)
15. Hu, J., Shen, L., Albanie, S., et al.: Gather-excite: exploiting feature context in convolutional neural networks (2018)
16. Woo, S., Park, J., Lee, J.-Y., Kweon, I.S.: CBAM: convolutional block attention module. In: Ferrari, V., Hebert, M., Sminchisescu, C., Weiss, Y. (eds.) ECCV 2018. LNCS, vol. 11211, pp. 3–19. Springer, Cham (2018). https://doi.org/10.1007/978-3-030-01234-2_1
17. Wang, F., Jiang, M., Qian, C., et al.: Residual attention network for image classification. In: Proceedings of the IEEE Conference on Computer Vision and Pattern Recognition, pp. 3156–3164 (2017)
18. Li, X., Hu, X., Yang, J.: Spatial group-wise enhance: improving semantic feature learning in convolutional networks (2019)
19. Iandola, F.N., Han, S., Moskewicz, M.W., et al.: SqueezeNet: AlexNet-level accuracy with 50x fewer parameters and <0.5 MB model size. arXiv preprint arXiv:1602.07360 (2016)
20. Brainard, D.H.: Analysis of the retinex theory of color vision. J. Opt. Soc. Am. A **3**(10), 1651–1661 (1986)
21. Barnard, K.: Improvements to Gamut mapping colour constancy algorithms. In: Vernon, D. (ed.) ECCV 2000. LNCS, vol. 1842, pp. 390–403. Springer, Heidelberg (2000). https://doi.org/10.1007/3-540-45054-8_26
22. Buchsbaum, G.: A spatial processor model for object colour perception. J. Franklin Inst. **310**(1), 1–26 (1980)
23. van de Weijer, J., Gevers, T., Gijsenij, A.: Edge-based color constancy. IEEE Trans. Image Process. **16**(9), 2207–2214 (2007)

24. Finlayson, G.D., Trezzi, E.: Shades of gray and colour constancy. In: Color and Imaging Conference, vol. 2004, no. 1, pp. 37–41. Society for Imaging Science and Technology (2004)
25. Gehler, P.V., Rother, C., Blake, A., et al.: Bayesian color constancy revisited. In: 2008 IEEE Computer Society Conference on Computer Vision and Pattern Recognition (CVPR 2008), Anchorage, Alaska, USA, 24–26 June 2008. IEEE (2008)
26. Barnard, K., Martin, L., Coath, A., et al.: A comparison of computational color constancy algorithms-Part II: experiments with image data. IEEE Trans. Image Process. 11(9), 985–996 (2002)
27. Gijsenij, A., Gevers, T.: Color constancy using natural image statistics and scene semantics. IEEE Trans. Pattern Anal. Mach. Intell. 33(4), 687–698 (2011)
28. Joze, H.R.V., Drew, M.S., Finlayson, G.D., et al.: The role of bright pixels in illumination estimation. In: Color and Imaging Conference, vol. 2012, no. 1, pp. 41–46. Society for Imaging Science and Technology (2012)
29. Chakrabarti, A., Hirakawa, K., Zickler, T.: Color constancy with spatio-spectral statistics. IEEE Trans. Pattern Anal. Mach. Intell. 34(8), 1509–1519 (2012)
30. Finlayson, G.D.: Corrected-moment illuminant estimation. In: Proceedings of the IEEE International Conference on Computer Vision, pp. 1904–1911 (2013)
31. Joze, H.R.V., Drew, M.S.: Exemplar-based color constancy and multiple illumination. IEEE Trans. Pattern Anal. Mach. Intell. 36(5), 860–873 (2014)
32. Cheng, D., Price, B., Cohen, S., et al.: Effective learning-based illuminant estimation using simple features. In: Proceedings of the IEEE Conference on Computer Vision and Pattern Recognition, pp. 1000–1008 (2015)
33. Bianco, S., Cusano, C., Schettini, R.: Single and multiple illuminant estimation using convolutional neural networks. IEEE Trans. Image Process. 26(9), 4347–4362 (2017)
34. Barron, J.T.: Convolutional color constancy. In: Proceedings of the IEEE International Conference on Computer Vision, pp. 379–387 (2015)
35. Cheng, D., Prasad, D.K., Brown, M.S.: Illuminant estimation for color constancy: why spatial-domain methods work and the role of the color distribution. JOSA A 31(5), 1049–1058 (2014)
36. Shi, W., Loy, C.C., Tang, X.: Deep specialized network for illuminant estimation. In: Leibe, B., Matas, J., Sebe, N., Welling, M. (eds.) ECCV 2016. LNCS, vol. 9908, pp. 371–387. Springer, Cham (2016). https://doi.org/10.1007/978-3-319-46493-0_23. IEEE International Workshop on Information Forensics and Security, pp. 234–239. IEEE (2012)
37. Hu, Y., Wang, B., Lin, S.: FC4: fully convolutional color constancy with confidence-weighted pooling. In: Proceedings of the IEEE Conference on Computer Vision and Pattern Recognition, pp. 4085–4094 (2017)

Simplifying Sketches with Conditional GAN

Jia Lu, Xueming Li$^{(\boxtimes)}$, and Xianlin Zhang$^{(\boxtimes)}$

School of Digital Media and Design Arts, Beijing University of Posts
and Telecommunications, Beijing, China
{lvjia,lixm,zxlin}@bupt.edu.cn

Abstract. In the process of creating paintings by artists, the first and foremost step is to outline the overall structures and lines of the object on paper, and then convert it into simplified sketches by scanning or redrawing. Traditional CNN-based sketch simplification models often ignore the overall semantic structure of the sketch, and have limited simplification effect on sketch with shadows and a large number of messy lines. In this paper, we present a novel sketch simplification model based on conditional GAN which learned to extract the semantic features of important lines in the original sketch and generate clear simplified sketches. We achieved this by introducing self-attention mechanism and perceptual loss. The results of multiple groups of user researches show that the simplified sketch generated by our model is superior to state-of-the art models in all aspects such as image aesthetics, line simplicity and content consistency.

Keywords: Sketch simplification · Conditional GAN · Self-attention mechanism · Perceptual loss

1 Introduction

Sketch is a communication tool commonly used by human, which is natural, concise and abstract. The drawing of sketches is usually the first step when artists conceive their works. Since the final artwork needs to be continuously processed and perfected on the basis of sketches, the importance of sketches for artistic creation is self-evident. As creative inspiration is usually fleeting, artists need to quickly record their creative concepts and ideas on paper, and determine the overall structure of the work, rather than spending time on processing sketch details. As a result, the lines of hand-drawn sketches are usually messy and rough. Given the original rough sketch, artists need to carry out refined secondary processing to make it a clear and simplified line drawing, including deleting and modifying the rough lines in the sketch. Predictably, this process is cumbersome and time-consuming.

In order to facilitate the creation of artists, researchers have proposed many approaches for automatically simplifying sketches. A common approach is to gradually modify the sketch lines during the drawing process: such as the marker-based re-parameterization method [1], cubic bézier curve fitting [2], and stepwise merge algorithm based on proximity and topology [3]. However, most of these step-by-step drawing approaches rely on vectorized stroke orders of sketches, and are difficult to apply to sketches already drawn on paper.

© Springer Nature Singapore Pte Ltd. 2020
Y. Wang et al. (Eds.): IGTA 2020, CCIS 1314, pp. 127–139, 2020.
https://doi.org/10.1007/978-981-33-6033-4_10

Recently, a sketch simplification model based on convolutional neural networks [4] has been proposed. It takes scanned hand-drawn pencil sketches as input and obtains simplified sketches using mean-square error (MSE) loss. This model requires a large amount of supervised data for training, however, dataset containing pairs of rough sketches and their corresponding simplified sketch is relatively scarce. To solve this problem, Simo-Serra et al. [5] proposed another sketch simplification model combining supervised learning and unsupervised learning. This model uses unpaired sketches as enhanced data to improve the learning ability and robustness of the original model. Compared with the original model, the sketch simplification performance of this model has been significantly improved. However, it still relies heavily on high-quality supervised data. Moreover, although the quality of simplified sketches is improved, the convergence rate of the network is greatly reduced.

In this paper, we proposed a novel sketch simplification model based on conditional GAN which learned to extract the semantic features of important lines in the original sketch and generate clear simplified sketches. The sketch simplification network using U-Net structure effectively combines the multi-layer features of the sketch, and imposes precise constraints on the global structure of the sketch by introducing the self-attention mechanism. The loss function, consisting of the adversarial loss, perceptual loss and content loss, ensures that the simplified sketch is clean and consistent with the real sketch on the human perception level. Moreover, we build a brand new sketch simplification dataset consisting of supervised sketch data for model training. The results of multiple groups of user researches show that the simplified sketch generated by our model is superior to state-of-the art models in all aspects such as image aesthetics, line simplicity and content consistency.

2 Related Work

2.1 Sketch Simplification

For a long time, teaching computer to automatic simplify rough sketches has been a hot topic since it reduces tedious manual work for most artists. Traditional sketch simplification approaches are usually based on tracking [6], refinement [7] and fitting regression [8] of lines. However, these approaches can only handle sketches in binary format. In order to deal with sketches drawn by artists on paper, many works have been proposed, such as a system [9] that converts pixel sketches into vector sketches recognizable by CAD system. This system uses the pattern analysis technology based on Gabor filter to simplify sketches. Another sketch simplification approach [10] is based on gradient field and emphasis filtering to simplify lines of sketches. However, most of these approaches are difficult to deal with the complex gradient information generated by sketch strokes, thus cannot accurately separate the most important structural framework from the large number of irrelevant details in the sketch.

Recently, a sketch simplification model [4] based on deep full convolution network is proposed. Different from traditional sketch simplification approaches, this model takes the scanned hand-drawn sketch as input, and output the simplified sketch through a series of convolution operators trained by MSE loss. However, this model requires a large amount of supervised data for training. More importantly, since training sketches are different from the true rough sketches, models would generate simplified sketches with shadows and blurs. Researchers proposed another framework [5] that uses unpaired sketches as enhanced data to improve the performance in real usage cases based on this work. However, this approach suffers from the convergence rate, and keeps a lot of unnecessary lines in the output sketch.

2.2 Conditional GAN

Generative Adversarial Network (GAN) [11] proposed by Goodfellow et al. is one of the most popular generative models in the field of machine learning. It is a deep generative model using unsupervised data for training. In the GAN approach, the generator and the discriminator are trained jointly using game theory. The goal of the generator is to generate as many samples as possible that correspond to the real data distribution, while the goal of discriminator is to distinguish the real sample from the generated sample as far as possible. The generator is able to generate realistic images from random inputs after the joint training [12]. However, GAN cannot control the mode of data being generated, especially when the number of pixels in the training image is large.

Mirza and Osindero proposed Conditional GAN(CGAN) [13], which is GAN with additional constraints. CGAN introduces conditional variables in both generator and discriminator to guide the process of data generation. These variables can be based on a variety of information, such as category labels [14], partial data for image recovering [15], and data from different modality. CGAN is an improvement of turning unsupervised GAN into supervised model, which has proved to be effective and widely used in various image generation work. Therefore, in this paper, we proposed a sketch simplification model based on CGAN to improve the performance of simple GAN for sketch generation.

3 Architecture

3.1 Overview

The overall network architecture of our sketch simplification model is shown in Fig. 1, which is mainly composed of two parts. The first part is a CGAN network consisted of the simplification network S and the discriminator network D, and the second part is a pre-trained VGG19 network. CGAN converges after sufficient iterative training using paired rough sketches and clean sketches as supervised training data. The simplification network S is responsible for simplifying rough sketches into clean sketches, while the discriminator D is responsible for determining whether the paired rough sketches and clean sketches are derived from the ground truth or the output of the simplification network. During the training process, both of them are constantly optimizing based on the adversarial loss L_{ADV}. The pre-trained VGG19 network extracts the high-dimensional features of the simplified sketch output by the simplification network and the ground truth separately, and obtains the perceptual loss L_{VGG} by calculating the difference between them. Besides, we take the mean square error L_{MSE} between the simplified sketch output by our model and the ground truth into consideration to ensure the pixel similarity of them. The details of the loss function will be introduced in the following sections of this chapter.

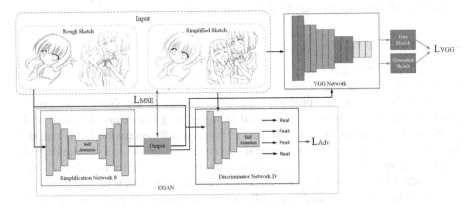

Fig. 1. Schematic diagram of our sketch simplification model.

3.2   CGAN

As shown in Fig. 2, the CGAN network in our model is consisted of the simplification network S and the discriminator network D. Both of them can observe the rough sketches which are to be simplified. Given the rough sketch x and random noise z, the goal of the simplification network S is to learn the mapping {x, z} → y, where y denotes the simplified sketch. The details of the CGAN network will be discussed as follows.

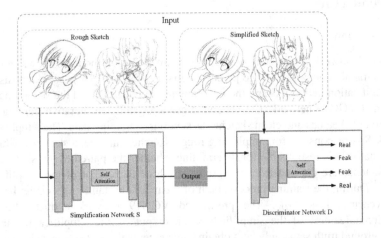

Fig. 2. Architecture of CGAN in our model.

Simplification Network Based on U-Net. The simplification network in our model is a fully convolutional neural network based on U-Net structure [16]. As shown in Table 1, the simplification network is mainly consisted of downsampling convolutional layers and upsampling convolutional layers. We use downsampling convolutional layers with stride of 2 instead of the max pooling operation to implement downsampling of feature maps. Similarly, we also use the upsampling convolutional layers with stride of 1/2 to decode the feature map. Besides, we use a large number of flat convolutional layers with stride of 1 to deepen our network, so as to extract important line features from the sketch. If the convolutional layer with the same size of the output feature map is regarded as the same level, our sketch simplification network contains 4 levels. In order to make full use of the low-dimensional detail features and high-dimensional semantic features of the input sketch, we add skip connections between downsampling convolutional layers and upsampling convolutional layers in the level 2 and level 3. Skip connections allow the multi-layer features of the sketch to be effectively integrated so that the sketch simplification network can generate cleaner and clearer sketches.

Table 1. Architecture of the simplification network.

Level	Name	Type	Kernel	Stride	Size of Output
0	input	Input	-	-	256×256
1	down1	Down-convolution	3×3	2	128×128
2	down2	Down-convolution	3×3	2	64×64
	flat1	Flat-convolution	3×3	1	64×64
	flat2	Flat-convolution	3×3	1	64×64
3	down3	Down-convolution	3×3	2	32×32
	flat3	Flat-convolution	3×3	1	32×32
	flat4	Flat-convolution	3×3	1	32×32
4	down4	Down-convolution	3×3	2	16×16
	flat5	Flat-convolution	3×3	1	16×16
	flat6	Flat-convolution	3×3	1	16×16
	flat7	Flat-convolution	3×3	1	16×16
	flat8	Flat-convolution	3×3	1	16×16
	flat9	Flat-convolution	3×3	1	16×16
	flat10	Flat-convolution	3×3	1	16×16
	flat11	Flat-convolution	3×3	1	16×16
3	up1	Up-convolution	4×4	1/2	32×32
	flat12	Flat-convolution	3×3	1	32×32
	flat13	Flat-convolution	3×3	1	32×32
2	up2	Up-convolution	4×4	1/2	64×64
	flat14	Flat-convolution	3×3	1	64×64
	flat15	Flat-convolution	3×3	1	64×64
1	up3	Up-convolution	4×4	1/2	128×128
0	up4	Up-convolution	4×4	1/2	256×256

Discriminator Network. The goal of the discriminator network in our model is to distinguish whether the combination of the rough sketch x and the simplified sketch S (x) comes from real data or is the output of our simplification network. Since the purpose of our discriminator network is to assist the training of the simplification network, we use a relatively shallow CNN network as the discriminator network to avoid that the gradient of the parameters generated by the discriminator network tends to disappear when its performance is too strong compared to the simplification network.

As shown in Table 2, the discriminator network consists of 7 convolutional layers. The size of the input training sketch is 256×256, and the size of convolution kernels used in all convolution layers is 4×4. All convolutional layers use Leaky Rectified Linear Units (LReLU) as the activation function. The last layer of the discriminator network is a fully connected layer without activation function, which output the probability between 0 and 1.

Table 2. Architecture of the discriminator network.

Number	Name	Type	Kernel	Activate Function	Size of Output
0	input	Input	-	-	256×256
1	down1	Down-convolution	4×4	LReLu	128×128
2	down2	Down-convolution	4×4	LReLu	64×64
3	down3	Down-convolution	4×4	LReLu	32×32
4	down4	Down-convolution	4×4	LReLu	16×16
5	down5	Down-convolution	4×4	LReLu	8×8
6	down6	Down-convolution	4×4	LReLu	4×4
7	fc	Fully-connected	-	-	4×4

Self-Attention Mechanism. Given a real hand-drawn sketch, the detail features and structural features of lines are often long-distance related to each other. The convolutional layers in the simplification network can only handle the local information of the sketch pixels, and cannot handle the remote dependencies of the sketch pixels. In order to solve above problems, we introduce the self-attention module in the CGAN model based on the non-local model proposed by Wang et al. [17], so that the simplification network and the discriminator network can effectively modeling the relationship between long-distance pixel space of the sketch.

In the self-attention module, the image features output by the previous convolutional layer are first converted into two feature spaces f, g used to calculate the self-attention value. The self-attention value is then given by,

$$\beta_{j,i} = \frac{\exp(s_{ij})}{\sum_{i=1}^{N} \exp(s_{ij})}, \text{ where } s_{ij} = f(x_i)^T g(x_j). \tag{1}$$

where $\beta_{j,i}$ represents the relevance to which the model attends to the ith location when synthesizing the jth region. C represents the number of feature channels, and N represents the number of feature locations in the feature map output from the previous layer. The output of the attention layer is $o = (o_1, o_2, \ldots, o_j, \ldots, o_N) \in \mathbb{R}^{C \times N}$, where,

$$o_j = v(\sum_{i=1}^{N} \beta_{j,i} h(x_i)), h(x_i) = W_h x_i, v(x_i) = W_v x_i \tag{2}$$

where $W_g \in \mathbb{R}^{\bar{C} \times C}, W_f \in \mathbb{R}^{\bar{C} \times C}, W_h \in \mathbb{R}^{\bar{C} \times C}, W_v \in \mathbb{R}^{\bar{C} \times C}$ all represent the learned weight matrices. Zhang et al. [18] proved that no significant performance decrease when reducing the channel number of $\bar{C}$ to be C/k. For calculate efficiency, we choose k = 8 (i.e., $\bar{C} = C/8$) in all our experiments. The final self-attention value is given by,

$$y_i = \gamma o_i + x_i \tag{3}$$

where γ is a learnable scalar with an initial value of 0. Figure 3 shows the structure diagram of the self-attention module. By introducing the self-attention module in both the simplification network and the discriminator network, the simplification network better coordinates long-distance details and the discriminator network imposes more precise constraints on the global sketch structure.

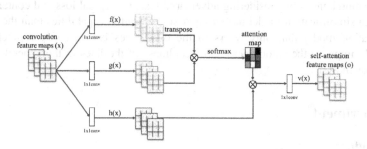

Fig. 3. Structure of the self-attention module.

3.3 Loss Function

The overall loss function includes three parts, which are adversarial loss L_{ADV}, perceptual loss L_{VGG} and content loss L_{MSE}.

$$L = \alpha L_{ADV} + \beta L_{VGG} + \gamma L_{MSE} \tag{4}$$

During the training of our model, the CGAN network updates the simplification network and the discriminator network by optimizing the adversarial loss L_{ADV}. We calculate it with the following formula:

$$L_{ADV} = E_{x,y}[\log D(x,y)] + E_{x,z}[1 - D(x, S(x,z))] \tag{5}$$

The perceptual loss uses the pre-trained VGG19 network to calculate the differences between high-dimensional features of the simplified sketch. It ensures the perception consistency of the simplified sketch generated by our model and the real sketch.

We defines the perceptual loss L_{VGG} based on the ReLU activation layer in the pre-trained VGG19 network proposed by Simonyan et al. [19]:

$$L_{VGG/i,j} = \frac{1}{W_{i,j}H_{i,j}} \sum_{a=1}^{W_{i,j}} \sum_{b=1}^{H_{i,j}} (\phi_{i,j}(y)_{a,b} - \phi_{i,j}(S(x,z))_{a,b})^2 \tag{6}$$

where $\phi_{i,j}$ represents the feature map output by the jth convolutional layer before the ith max pooling layer in the VGG19 network. $W_{i,j}$ and $H_{i,j}$ represent the dimensions of the corresponding feature maps in the VGG19 network respectively. As shown in the formula, the perceptual loss calculates the Euclidean distance between the feature maps of the generated simplified sketch and the real simplified sketch.

The content loss ensures that the sketch simplification model generates lines that are sufficiently clear at the pixel level when simplifying complex lines:

$$L_{MSE} = ||S(x,z) - y||^2 \tag{7}$$

By comprehensively considering adversarial loss, perceptual loss and content loss, our sketch simplification model has stronger simplification capabilities than the sketch simplification model that only refers to the differences between single pixels, and effectively improve the smoothness and cleanliness of the lines in the generated simplified sketches.

4   Experiment

4.1   Platform

Our experiments are implemented under the Tensorflow deep learning framework of the Linux operating system. The training was performed on a single GeForce GTX 1080 Ti GPU.

4.2   Dataset

We build a brand new sketch simplification dataset for our model training. Our dataset contains original sketches composed of 100 pairs of rough sketches and corresponding clean sketches. These sketches have different resolutions with an average of 580×620 pixels. The smallest sketch is 207×217 pixels and the largest is 1200×1445 pixels.

In order to build a variety of sketch object dataset, we collected 100 sketches of different types including anime portraits, animals, plants, daily necessaries, etc. on the Internet. On the basis of the existing sketch data, we invited 10 artists with drawing experience to draw a clean version of a rough sketch by the direct drawing approach, or a rough version of a clean sketch by the inverse drawing approach. The direct drawing approach means that the artist modifies the small and messy lines into clear and smooth lines based on the original rough sketch, and directly deletes lines that are too vague or meaningless. The inverse drawing approach refers to the approach of Simo-Serra et al. [4], that is, given a clean simplified sketch drawing, the artist is asked to make a rough version of that sketch. We use these approaches in the creation of the 100 pairs of training sketches, some examples can be seen in Fig. 4.

(a) Direct drawing approach

(b) Inverse drawing approach

Fig. 4. Some images in our sketch simplification dataset. The first row is the rough sketch, and the second row is the clean sketch.

Since the number of sketches in our dataset is relatively small, we further augment the number of sketches in the original dataset to five times by employing four augmentations including flipping, tone change, image slur, and noise. Figure 5 presents an example of these augmentations.

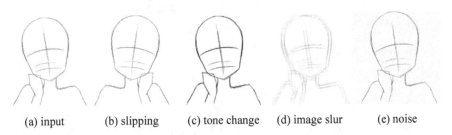

| (a) input | (b) slipping | (c) tone change | (d) image slur | (e) noise |

Fig. 5. We further augment the dataset by (b) slipping, (c) changing the tone of the input image, (d) slurring the input image, and (e) adding random noise.

4.3 Training Details

We set the size of the sketch in the dataset to 256×256 by scaling and cropping. For each sketch, we subtracted the mean gray value of sketches in the training dataset. We set the learning rate to 0.0002 and optimized it with the Adam optimizer with the momentum of 0.5. We trained our model for 150,000 iterations with a batch size of 1. For computing the loss function, we used the values of $\alpha = 1$, $\beta = 1$, and $\gamma = 100$.

We introduced the self-attention module in the convolutional layer of the simplification network whose output feature map size of 32×32. For the pre-trained VGG19 network, we selected the feature map output from the ReLU5_4 layer after a series of contrast experiments to calculate the perceptual loss.

5 Evaluation

5.1 Comparison with the State of the Art

We compare our model with the state-of-the-art sketch simplification models including Learning to Simplify (LtS) [4] and Mastering Sketching (MS) [5] in Fig. 6. It is easy to observe that for sketches with complicated structures and messy lines, as shown in Fig. 6(b), LtS tends to generate simplified sketches with a large number of fuzzy lines. Although the performance of MS has been improved on the basis of LtS, it still lead to the sticking of the sketch lines. As shown in Fig. 6(c), the lines of the hair and eyes are entangled, and the structure of them cannot be discerned accurately, the branches of flowers also appear fuzzy and deformed. These approaches ignored the high-level semantic features of the sketch and failed to retain different level of fine-grained lines in the sketch. As contrary, as shown in Fig. 6(d), our model generates simplified sketches with clear lines and distinguishable structures by making full use of the high-dimensional features of the sketch.

(a) input   (b) LtS

(c) MS   (d) Ours

Fig. 6. Comparation between the output of our model, LtS [4] and MS [5].

5.2 User Study

We also performed a user study to evaluate our model. We invited 20 participants including 10 men and 10 women at the age between 20–30 years old. Participants were asked to score the simplified sketches generated by different models between 1–10 in three aspects: image aesthetics, line simplicity, and content consistency, the higher score indicates better model performance.

We compare our model with the state-of-the-art sketch simplification models including Learning to Simplify (LtS) [4] and Mastering Sketching (MS) [5]. We selected 20 rough sketches as input of different sketch simplification models and got 60 simplified sketches. We then randomly presented these 60 sets of data to each participant, and counted scores of the simplified sketches in each set of data. Results are shown in Table 3. We can see that our model has higher scores than the state-of-the-art models in terms of image aesthetics, line simplicity and content consistency. This further illustrates that our model has better performance on simplifying sketches at the level of human perception.

Table 3. Results of our user study comparing our approach with the state-of-the-art sketch simplification models.

Model	Image aesthetics	Line simplicity	Content consistency	Average
LtS	8.18	7.52	8.18	7.96
MS	8.32	7.97	8.29	8.19
Ours	**8.99**	**8.61**	**9.08**	**8.89**

6 Conclusion

We proposed a sketch simplification model that simplifies rough sketches into clean and clear sketches, thus assisting the artist to efficiently carry out subsequent creation. To improve the quality of the simplified sketch, we introduced U-Net network, self-attention mechanism and perceptual loss into our model. Experiments show that our model achieves superior performance both qualitatively and quantitatively over the state-of-the-art alternative. The current conclusion is obtained from limited data experiments, a larger amount of data will be used for testing and evaluation in the future.

References

1. Baudel, T.: A mark-based interaction paradigm for freehand drawing. In: ACM Symposium on User Interface Software and Technology, pp. 185–192 (1994)
2. Bae, S.H., Balakrishnan, R., Singh, K.: Ilovesketch: as-natural-as-possible sketching system for creating 3d curve models. In: ACM Symposium on User Interface Software and Technology, pp. 151–160 (2008)
3. Grimm, C., Joshi, P.: Just drawit: A 3d sketching system. In: International Symposium on Sketch-Based Interfaces and Modeling, pp. 121–130 (2012)
4. Simo-Serra, E., Iizuka, S., Sasaki, K., et al.: Learning to simplify: fully convolutional networks for rough sketch cleanup. ACM Trans. Graph. 35(4) (2016)
5. Simo-Serra, E., Iizuka, S., Ishikawa, H.: Mastering sketching: adversarial augmentation for structured prediction. ACM Trans. Graph. 37(1) (2017)
6. Freeman, H.: Computer processing of line-drawing images. ACM Comput. Surv. 6(1), 57–97 (1974)
7. Zhang, T.Y., Suen, C.Y.: A fast parallel algorithm for thinning digital patterns. Commun. ACM 27(3), 236–239 (1984)
8. Janssen, R.D.T., Vossepoel, A.M.: Adaptive vectorization of line drawing images. Comput. Vis. Image Underst. 65(1), 38–56 (1997)
9. Bartolo, A., Camilleri, K.P., Fabri, S.G., et al.: Scribbles to vectors: preparation of scribble drawings for CAD interpretation. In: Proceedings of the 4th Eurographics workshop on Sketch-Based Interfaces and Modeling, pp. 123–130. ACM (2007)
10. Chen, J., Guennebaud, G., Barla, P., et al.: Non-oriented MLS gradient fields. Comput. Graph. Forum 32(8), 98–109 (2013)
11. Goodfellow, I.J., Pouget-Abadie, J., Mirza, M., et al.: Generative Adversarial Networks (2014)
12. Radford, A., Metz, L., Chintala, S.: Unsupervised representation learning with deep convolutional generative adversarial networks. In: International Conference on Learning Representations (2016)
13. Mirza, M., Osindero, S.: Conditional generative adversarial nets. In: Conference on Neural Image Processing Deep Learning Workshop (2014)
14. Isola, P., Zhu, J.Y., Zhou, T., et al.: Image-to-image translation with conditional adversarial networks. In: IEEE Conference on Computer Vision and Pattern Recognition (2017)
15. Hedau, V., Hoiem, D., Forsyth, D.: Recovering the spatial layout of cluttered rooms. In: IEEE International Conference on Computer Vision. IEEE (2009)

16. Ronneberger, O., Fischer, P., Brox, T.: U-Net: convolutional networks for biomedical image segmentation. Med. Image Comput. Comput. Assist. Interv. **9351**, 234–241 (2015)
17. Wang, X., Girshick, R., Gupta, A., et al.: Non-local neural networks. In: CVPR (2018)
18. Zhang, H., Goodfellow, I., Metaxas, D., et al.: Self-attention Generative Adversarial Networks. arXiv:1805.08318 (2018)
19. Simonyan, K., Zisserman, A.: Very Deep Convolutional Networks for Large-scale Image Recognition. arXiv:1409.1556 (2014)

Improved Method of Target Tracking Based on SiamRPN

Naiqian Tian, Rong Wang$^{(\boxtimes)}$, and Luping Zhao

Key Laboratory of Security Technology and Risk Assessment, People's Public
Security University of China, Beijing 102623, China
{13051128832, dbdxwangrong}@163.com

Abstract. An improved target tracking method based on SiamRPN is proposed, which solves the problem that SiamRPN method is prone to tracking drift
in complex scenes. First, in the feature extraction part, the deep reference network ResNet-50 is adopted to extract deep features, and a lightweight dual-
channel attention mechanism CBAM is introduced to refine feature extraction
along the channel and spatial dimensions. Secondly, SE module is introduced
into the bridge connection of ResNet-50 to weight the features, which can
achieve the purpose of improving the network's feature expression ability.
Finally, the region proposal network performs classification and regression
operations, thereby outputting tracking results. This paper conducts simulation
experiments on OTB2015 dataset and VOT2016 dataset. The results demonstrate that the tracking performance of the proposed method is improved compared to SiamRPN.

Keywords: Target tracking · Attention mechanism · Feature extraction

1 Introduction

As an important research hotspot of computer vision, target tracking has been universally applied in the field of public safety, providing important technical support for
preventing and fighting crime. Although there has been a lot of research, in the face of
complex scenes, such as rapid target movement and departure from the field of view,
the robustness of various tracking algorithms still has much room for improvement.

Target tracking aims to anticipate the movement trajectory of the target in the
subsequent frames according to the initial situation of the target in the video frame, so
as to understand and analyze its behavior. The target tracking methods include two
types: generative and discriminative. The generative method searches the target similar
area by generating the target appearance model, such as Meanshift algorithm [1].
However, this type of method is easy to ignore the background information of the
image and has poor robustness. In the discriminant method, the tracking problem is
transformed into a classification problem. Due to the combination of target information
and background information, the overall tracking performance is superior to the generative method, which can be divided into two categories including correlation filters
and deep learning. The tracking method based on the correlation filter effectively
improves the operation efficiency through Fourier transforms, such as KCF algorithm

© Springer Nature Singapore Pte Ltd. 2020
Y. Wang et al. (Eds.): IGTA 2020, CCIS 1314, pp. 140–153, 2020.
https://doi.org/10.1007/978-981-33-6033-4_11

[2] and ECO algorithm [3]. The recent application of deep functions in such methods has improved tracking accuracy. However, overall, these methods are slow [4]. Deep learning-based methods use deep convolutional networks to perform feature extraction and classification of targets, showing higher accuracy. Among them, the application of Siamese Network brings convenience to target tracking. Bertinetto et al. proposed SiamFC algorithm [5], which uses a fully convolutional Siamese Network for tracking, and the speed exceeds the real-time frame rate in the benchmark. Li Bo et al. introduced the RPN (Region Proposal Network) module based on SiamFC [5] and proposed SiamRPN algorithm [6], which increased the multi-scale detection capability of the model, resulting in a more accurate bounding box and demonstrating excellent tracking performance. However, in complex scenes such as background clutter and target deformation, the problem of target drift or failure is still prone to occur.

This paper proposes an improved target tracking method based on SiamRPN [6], which to a certain extent solves the problem that SiamRPN method is prone to tracking drift in complex scenarios. This method adopts ResNet-50 [7] as the reference network, introduces a lightweight dual-channel attention mechanism CBAM [8], infers the attention maps along two independent dimensions, and multiplies the input feature maps for adaptive feature refinement. The SE module [9] is further introduced to quantify the importance of feature maps and weighted to enhance the bridge connection of ResNet and improve the feature extraction of Siamese Network.

The experiments in this paper are conducted on OTB2015 dataset [10] and VOT2016 dataset [11]. Experimental results indicate that compared with SiamRPN, our method has higher success rate and precision on OTB2015 dataset, higher accuracy and EAO on the VOT2016 dataset, and the overall tracking performance has been improved.

The remainder of the paper is structured as follows: In the second part, the baseline method SiamRPN is introduced in detail. In the third part, we introduce our improved method. In the fourth part, we experiment with our method and analyze the results. In the fifth part, conclusions are presented and future research directions are discussed.

2 Baseline

2.1 SiamRPN

SiamRPN [6] introduces RPN (Region Proposal Network) into target tracking. Feature maps are extracted by template and detection branches of Siamese Network, in which the appearance information of the target on the template branch is transmitted to RPN features, which is used to classify the foreground and background information of a given picture. SiamRPN constructs the tracking problem into a local one-shot detection task, abandons traditional multi-scale testing and online fine-tuning, and implements offline end-to-end training. SiamRPN can run at a speed of 160 fps, proving its speed advantage.

The framework of SiamRPN is shown in Fig. 1. SiamRPN is composed of Siamese Network and RPN. The former is used for feature extraction, and the network structure and parameters of the upper and lower branches are identical. The latter is used to generate candidate regions, including two branches. One is for classification, which is used to score samples predicted as targets and backgrounds. The other is for regression, which is used to fine-tune the generated candidate regions and use cross-correlation to obtain the output of the two branches.

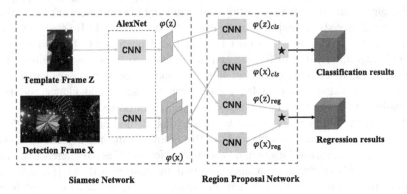

Fig. 1. The framework of SiamRPN.

SiamRPN applies the one-shot detection strategy for tracking tasks for the first time. Based on this, the tracking task can be expressed as:

$$\min_{w} \frac{1}{n} \sum_{i=1}^{n} L(\varsigma(\varphi(x_i; W); \varphi(z_i; W)), l_i) \tag{1}$$

Where n denotes the number of dataset, x represents the detection sample, l_i represents its corresponding label, z represents the template sample. The function ς and the function φ are used for feature extraction and region suggestion selection, respectively. In addition, minimizing the average loss L is the key to obtaining the parameter W. The value of the parameter W needs to be obtained by the following formula, where ϕ is the prediction function:

$$\min_{w} \frac{1}{n} \sum_{i=1}^{n} L(\phi(x_i; W), l_i) \tag{2}$$

2.2   Region Proposal Network

Region Proposal Network (RPN) first appeared in Faster R-CNN [12], makes the distinction between foreground and background information and the bounding box regression more accurate. In SiamRPN, RPN is divided into the classification branch and the regression branch. The classification branch is used to score the likelihood that the sample is predicted to be the target and background. Here, RPN convolves the template frame feature extracted from Siamese Network with the detection frame feature, which reduces the size of the feature map and generates new template feature map $[\varphi(z)]_{cls}$ and detection feature map $[\varphi(x)]_{cls}$. Moreover, $[\varphi(z)]_{cls}$ is used as the kernel to convolve $[\varphi(x)]_{cls}$, which generates the corresponding map. The regression branch performs similar operations as the classification branch and generates the position of each sample regression values, including dx, dy, dw, dh.

3 Proposed Method

SiamRPN [6] shows excellent tracking performance in general scenes. But with the addition of complex scenes, tracking drift or failure may occur. Especially when occlusion occurs and the target leaves the field of view, tracking is more difficult. In a complex background, it is impossible to accurately lock the tracking target only by the initial frame of the video. Besides, the target is displayed in the video with a large difference from the original image due to motion and other issues, which will also cause tracking drift and affect the tracking effect.

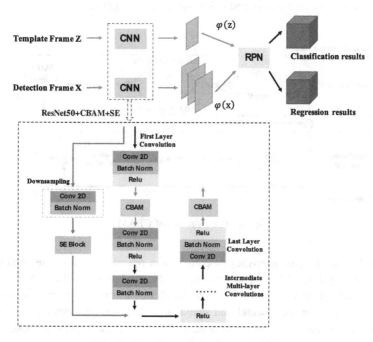

Fig. 2. The framework of our method.

In response to the above problems, we propose an improved tracking method based on SiamRPN, as shown in Fig. 2. To obtain more abundant deep features in the video frame, the deep reference network ResNet-50 [7] is adopted. Introduce a lightweight dual-channel attention mechanism CBAM [8] to enhance the attention to tracking targets along the channel and spatial dimensions. By learning the information to be emphasized or suppressed, the information flow in the network is effectively helped and the accuracy of ResNet-50 is greatly improved. In order to further enhance the reliability of the feature maps, the SE module [9] is introduced to quantify the importance of the features of different channels in the network bridge connection to obtain richer feature maps. Our method improves SiamRPN from the above point of view, and the tracking results of the algorithm have been improved in complex scenes such as rapid target movement and out of view.

3.1 Convolutional Block Attention Module

While the convolutional layer performs feature extraction, it also brings a lot of redundant calculations to the entire network. To this end, this paper introduces a lightweight dual-channel attention mechanism CBAM (Convolutional Block Attention Module) [8], which is dedicated to improving the attention of important features while suppressing unnecessary features and enhancing the representation ability of convolutional neural networks. The dual-channel method can effectively avoid tracking drift caused by the target leaving the field of view, and in most cases, the redundant overhead caused by parameters and calculations can be ignored.

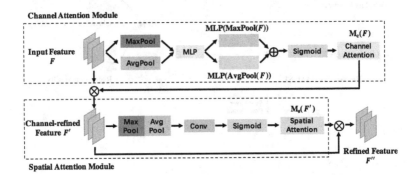

Fig. 3. The structure of CBAM.

According to the characteristics of convolution operation to fuse the channel and spatial information, CBAM emphasizes the attention points along the two dimensions of channel and space to express features. As shown in Fig. 3, given a feature map $F \in R^{C \times H \times W}$, passing the channel and spatial attention module, respectively infer the one-dimensional channel attention map $M_c \in R^{C \times 1 \times 1}$ and the two-dimensional spatial attention map $M_s \in R^{1 \times H \times W}$ along the two dimensions. The entire process is as follows:

$$F' = M_c(F) \otimes F \tag{3}$$

$$F'' = M_s(F') \otimes F' \tag{4}$$

Where $\otimes$ represents element-wise multiplication, F' is the multiplication result of the channel attention map and the feature map F, and F'' is the multiplication result of the spatial attention map and F', which is the final refined feature map.

3.2   Squeeze-and-Excitation Module

The convolutional layer captures feature maps in a comprehensive and equal way so that some features with interference options or meaninglessness are propagated through the network, which affects the accuracy of the network. SE module (Squeeze-and-Excitation Module) [9] is introduced to extract more representative feature maps and increase the tracking performance of the tracker in complicated scenes. As shown in Fig. 2, Se module is added after downsampling in the bridge connection of ResNet to ensure that the feature map can be weighted according to the importance of the content to be carried out when transferring in the network, and the calibration of feature validity is performed, which improves the quality of features obtained by the convolutional neural network and just adds only a small amount of additional calculation.

The SE module includes two operations: squeeze and excitation. Figure 4 shows the structure of the SE module. The feature X acquired by the convolutional layer generate feature descriptors [13] along the spatial dimension of each feature map through a squeeze operation. That is, the output vector of the input feature map after global pooling is reduced in size by the fully-connected layer, and ReLU activation is performed. All network layers can use information from the global acceptance domain according to the global embedding of feature responses in the channel dimension. The excitation operation uses the embedding generated during squeezing to obtain a set of feature map weights, that is, the output of the squeeze operation is upsampled through a fully connected layer followed by sigmoid activation, and finally generates a weighted feature map $\tilde{X}$.

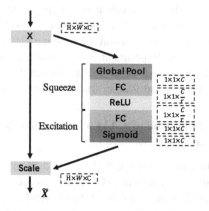

Fig. 4. The structure of SE module.

4   Experiment and Analysis

4.1   Datasets and Evaluation Criteria

Our method uses the COCO dataset [14] and VID2015 dataset [15] as the training set for the experiment, and the experimental testing and evaluation are performed on OTB2015 dataset and VOT2016 dataset.

OTB2015 dataset [10] contains 100 sequences, including various tracking types such as pedestrians, animals, and vehicles, covering 11 more complex scenes. They are labeled with 11 tags, where each sequence includes 3 to 7 tags. OTB2015 benchmark has two evaluation criteria: success and precision rate. The success rate is used to evaluate the average coincident area of the test position and the calibration position. The degree of compliance with the distance between the center position estimated by the tracking algorithm and the accurate position manually calibrated can be used to reflect the precision.

VOT2016 dataset [11] contains 60 sequences, each with 6 tags. VOT2016 benchmark has three evaluation criteria: accuracy, robustness, and expected average overlap (EAO). The accuracy can be used to represent the average overlap between the generation box and the calibration box in the case of successful tracking. The robustness reflects the failure rate in the tracking process. And the EAO evaluates the overall performance of the tracking algorithm.

4.2   Implementation Details

The method in this article uses Pytorch 0.4.1 as the development platform and is implemented in Python. The experiments are implemented on a computer with an Intel i5-8400, 16 GB RAM, Nvidia RTX2060 GPU. Set the total number of training rounds $epoch = 40$, $batch\_size = 8$, and the learning rate in the log space is reduced from 10^{-2} to 10^{-4}.

There are 4 sets of experiments in this paper: Experiment 1 is the comparison of CAM visualization results of the feature extraction networks. Experiment 2 is to compare the performance of our method and the baseline method SiamRPN in various scenarios of OTB2015 dataset. Experiment 3 is to compare the performance of our method and 6 other tracking methods on OTB2015 dataset. Experiment 4 is to compare the performance of our method and 9 other tracking methods on VOT2016 dataset.

4.3   Experiment 1: Comparison of CAM Visualization Results of the Feature Extraction Networks

The dual-channel attention mechanism CBAM [8] and SE module [9] are introduced into ResNet-50 as our improved network (Ours) and compared with the visualization results of CAM (Class Activation Mapping) [16] of AlexNet [17] and ResNet-50 [7]. The results are calculated from the final convolution output of the three networks. As shown in Fig. 5, it can be seen that the network in this article is significantly better than AlexNet and ResNet-50 for capturing target objects.

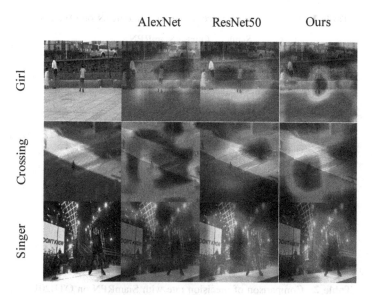

Fig. 5. CAM visualization results.

4.4 Experiment 2: Comparison with the Baseline Method SiamRPN on OTB2015

OTB2015 dataset [10] covers 11 complex scenes, which are illumination variation (IV), out-of-plane rotation (OPR), scale variation (SV), occlusion (OCC), deformation (DEF), motion blur (MB), fast motion (FM), in-plane rotation (IPR), out of view (OV), background clutter (BC), low resolution (LR). The tracking performance of our method and the baseline method SiamRPN is compared in the above scenarios. Table 1 and Table 2 present the detailed results.

From the Table 1 and Table 2, compared with SiamRPN, the success rate of our method is a litter lower in the four scenarios of OPR, DEF, IPR, and LR, and the precision rate is lower in the three scenarios of OPR, SV, and LR. But The success and precision rate in other scenarios have been improved. Because the method in this paper may lose part of the target's appearance information while suppressing the invalid information flow, the success rate of tracking in target deformation scenes decreases. And the feature representation extracted by the deep network is not sensitive to low resolution scenes, resulting in poor tracking effect in these scenes. Therefore, in the case of excluding target deformation and low resolution, our method can optimize the tracking effect, especially in the OV scene, our method has been greatly improved.

Table 1. Comparison of success rate with SiamRPN on OTB2015.

Scene	Ours	SiamRPN
IV	**0.668**	0.663
OPR	0.622	0.631
SV	**0.632**	0.628
OCC	**0.602**	0.597
DEF	0.608	0.628
MB	**0.638**	0.627
FM	**0.628**	0.606
IPR	0.631	0.636
OV	**0.620**	0.550
BC	**0.607**	0.601
LR	0.570	0.597
Overall	**0.642**	0.637

Table 2. Comparison of precision rate with SiamRPN on OTB2015.

Scene	Ours	SiamRPN
IV	0.873	0.873
OPR	0.852	0.855
SV	0.844	0.846
OCC	**0.800**	0.791
DEF	**0.838**	0.837
MB	**0.829**	0.821
FM	**0.818**	0.793
IPR	**0.866**	0.859
OV	**0.847**	0.728
BC	**0.820**	0.803
LR	0.805	0.870
Overall	**0.858**	0.851

4.5 Experiment 3: Comparison with Other Tracking Methods on OTB2015

On OTB2015 dataset, our method is compared with six representative tracking methods including SiamRPN [6], SiamMask [18], SiamFC [5], ECO-HC [3], Staple [19], CREST [20]. The experimental results of the above tracking methods on OTB2015 dataset are displayed in Fig. 6. We can see from the figure that our method has advantages in tracking precision compared to the other six representative tracking methods, surpassing the other six tracking methods.

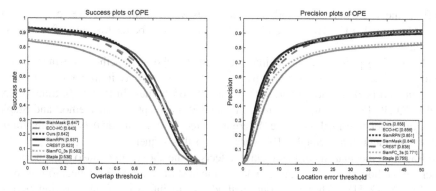

Fig. 6. Success plot and precision plot of OTB2015.

Fig. 7. Results of robustness analysis on partial sequences of OTB2015.

Three video sequences of complex scenes were selected from OTB2015 dataset to test the robustness of the above seven tracking methods. The test results are presented in Fig. 7.

Figure 7(a) presents the tracking effects of the above 7 methods on the Biker sequence. Here, the sequence mainly shows scenes such as fast motion, rotation, and out of view, with a total of 142 frames. In frame 72, the target partially leaves the line of sight, and the tracking drift of Staple [19] and CREST [20] occurs. In frame 125, the continuous drift of Staple [19] and CREST [20] occurs, and the error of the tracking frame of SiamFC [5] and ECO-HC [3] is large. Our method and SiamMask [18] work best in this sequence.

Figure 7(b) presents the tracking effects of the above 7 methods on the Football sequence. The sequence here mainly shows scenes such as occlusion, rotation, and background clutter, with a total of 362 frames. In frame 116, SiamRPN [6] and SiamFC [5] show tracking drift. In frame 289, occlusion occurs, SiamFC [5] and Staple [19] drift to the occlusion area, and other methods work well. Therefore, our method, SiamMask [18], ECO-HC [3], and CREST [20] have a better tracking effect in this sequence.

Figure 7(c) presents the tracking effects of the above 7 methods on the Singer2 sequence. The sequence here mainly shows scenes such as illumination variation, rotation, and background clutter, with a total of 366 frames. In frame 70, SiamFC [5] and ECO-HC [3] show tracking drift. In frame 300, except for our method and Staple [19], the other five methods have different degrees of tracking drift or tracking frame error. Our method and Staple [19] work best in this sequence.

Based on the analysis of the tracking effects of the above three video sequences, the tracking method proposed in this article has the best tracking performance, followed by the SiamMask [18]. Therefore, our method can maintain excellent tracking performance in the face of complex scenes such as illumination variation, out of view, and background clutter.

4.6   Experiment 4: Comparison with Other Tracking Methods on VOT2016

On VOT2016 dataset, our method is compared with nine representative tracking methods including SiamRPN [6], ECO-HC [3], CCOT [4], TCNN [21], MDNet_N [22], Staple [19], KCF [2], DAT [23], DeepSRDCF [24]. The detailed results is shown in Table 3. Our method ranks first in terms of accuracy and EAO. Figure 8 is the Accuracy-Robustness plot of the above methods, which can more intuitively see the advantages of our method in various aspects. Figure 9 shows the EAO ranking of the above methods.

Table 3. Comparison results of 10 tracking methods on VOT2016.

Tracker	Accuracy	Robustness	EAO
DeepSRDCF	0.529	0.326	0.276
DAT	0.469	0.480	0.217
KCF	0.491	0.569	0.192
Staple	0.547	0.378	0.295
MDNet_N	0.542	0.337	0.257
TCNN	0.555	0.268	0.324
CCOT	0.541	0.238	0.331
ECO-HC	0.542	0.303	0.322
SiamRPN	0.578	0.312	0.337
Ours	**0.606**	0.312	**0.346**

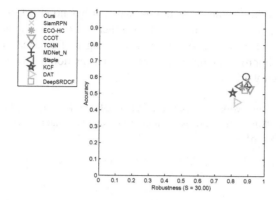

Fig. 8. Accuracy-robustness plot of VOT2016.

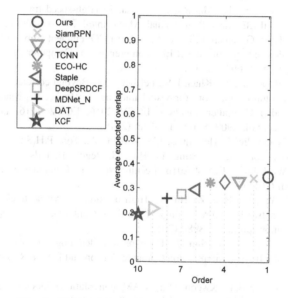

Fig. 9. The EAO scores in the VOT2016 challenge.

5 Conclusion

This paper proposes an improved target tracking method based on SiamRPN, which improves the feature extraction part. Our method uses the deep reference network ResNet-50 and introduces a lightweight dual-channel attention mechanism CBAM to heighten the extraction of useful features along the two dimensions of channel and space. At the same time, the SE module is introduced to enhance the network bridge connection and further improve the robustness of feature extraction. The evaluation results on OTB2015 dataset and VOT2016 dataset show that our method is more accurate and shows better tracking effects in complex scenes. Since the method in this paper only changes the feature extraction network, it has no effect on the flow of the original SiamRPN, so it is also applicable to other tracking methods based on deep learning. The method in this paper has poor tracking performance in low resolution scenes, so we will improve the deep network's representation of low resolution scenes as our next research work.

Acknowledgments. This work is supported by National Key Research and Development Project under Grant A19808. This work is supported by National Natural Science Foundation of China under Grant No. 61906199.

References

1. Cheng, Y.: Mean shift, mode seeking, and clustering. IEEE Trans. Pattern Anal. Mach. Intell. **17**, 790–799 (1995)
2. Henriques, J.F., Caseiro, R., Martins, P., Batista, J.: High-speed tracking with kernelized correlation filters. IEEE Trans. Pattern Anal. Mach. Intell. **37**, 583–596 (2015)
3. Danelljan, M., Bhat, G., Khan, F.S., Felsberg, M.: ECO: efficient convolution operators for tracking. In: Proceedings of the IEEE Conference on Computer Vision and Pattern Recognition, pp. 6931–6939 (2017)
4. Danelljan, M., Robinson, A., Khan, F.S., Felsberg, M.: Beyond correlation filters: learning continuous convolution operators for visual tracking. In: Leibe, B., Matas, J., Sebe, N., Welling, M. (eds.) Computer Vision – ECCV 2016. ECCV 2016. Lecture Notes in Computer Science, vol. 9909. Springer, Cham (2016)
5. Bertinetto, L., Valmadre, J., Henriques, J.F., Vedaldi, A., Torr, P.H.S.: Fully-convolutional siamese networks for object tracking. In: Hua, G., Jégou, H. (eds.) Computer Vision – ECCV 2016 Workshops. ECCV 2016. Lecture Notes in Computer Science, vol. 9914. Springer, Cham (2016)
6. Li, B., Yan, J., Wu, W., Zhu, Z., Hu, X.: High performance visual tracking with siamese region proposal network. In: Proceedings of the IEEE Conference on Computer Vision and Pattern Recognition, pp. 8971–8980 (2018)
7. He, K., Zhang, X., Ren, S., Sun, J.: Deep residual learning for image recognition. In: Proceedings of the IEEE Conference on Computer Vision and Pattern Recognition, pp. 770–778 (2016)
8. Woo, S., Park, J., Lee, J.Y., Kweon, I.S.: CBAM: convolutional block attention module. In: European Conference on Computer Vision, pp. 3–19 (2018)
9. Varshaneya, Balasubramanian, S., Gera, D.: RES-SE-NET: Boosting Performance of Resnets by Enhancing Bridge-connections (2019)
10. Wu, Y., Lim, J., Yang, M.H.: Object tracking benchmark. IEEE Trans. Pattern Anal. Mach. Intell. **37**, 1834–1848 (2015)
11. Kristan, M., Leonardis, A., Matas, J., Felsberg, M., Pflugfelder, R., Čehovin, L.: The visual object tracking Vot2016 challenge results. In: Hua, G., Jégou, H. (eds.) Computer Vision – ECCV 2016 Workshops. ECCV 2016. Lecture Notes in Computer Science, vol. 9914. Springer, Cham (2016)
12. Ren, S., He, K., Girshick, R., Sun, J.: Faster R-CNN: towards real-time object detection with region proposal networks. In: International Conference on Neural Information Processing Systems. IEEE Computer Society, Los Alamitos (2015)
13. Hu, J., Shen, L., Sun, G.: Squeeze-and-excitation networks. In: Proceedings of the IEEE Conference on Computer Vision and Pattern Recognition, pp. 7132–7141 (2018)

14. Lin, T., Maire, M., Belongie, S., Hays, J., Perona, P., Ramanan, D., Zitnick, C.L.: Microsoft COCO: common objects in context. In: European Conference on Computer Vision, pp. 740–755 (2014)
15. Russakovsky, O., et al.: ImageNet large scale visual recognition challenge. Int. J. Comput. Vision **115**(3), 211–252 (2015). https://doi.org/10.1007/s11263-015-0816-y
16. Zhou, B., Khosla, A., Lapedriza, A., Oliva, A., Torralba, A.: Learning deep features for discriminative localization. In: Proceedings of the IEEE Conference on Computer Vision and Pattern Recognition, pp. 2921–2929 (2016)
17. Krizhevsky, A., Sutskever, I., Hinton, G.E.: Imagenet classification with deep convolutional neural networks. In: Proceedings of the Advances in Neural Information Processing Systems, pp. 1097–1105 (2012)
18. Wang, Q., Zhang, L., Bertinetto, L., Hu, W., Torr, P.H.S.: Fast online object tracking and segmentation: a unifying approach. In: Proceedings of the IEEE Conference on Computer Vision and Pattern Recognition, pp. 1328–1338 (2018)
19. Bertinetto, L., Valmadre, J., Golodetz, S., Miksik, O., Torr, P.H.: Staple: complementary learners for real-time tracking. In: Proceedings of the IEEE Conference on Computer Vision and Pattern Recognition, pp. 1401–1409 (2016)
20. Song, Y., Ma, C., Gong, L., Zhang, J., Lau, R. W., Yang, M.: CREST: convolutional residual learning for visual tracking. In: 2017 IEEE International Conference on Computer Vision (ICCV), pp. 2574–2583 (2017)
21. Nam, H., Baek, M., Han, B.: Modeling and propagating CNNs in a tree structure for visual tracking. In: Proceedings of the IEEE Conference on Computer Vision and Pattern Recognition (2016)
22. Nam, H., Han, B.: Learning multi-domain convolutional neural networks for visual tracking. In: Proceedings of the IEEE Conference on Computer Vision and Pattern Recognition, pp. 4293–4302 (2016)
23. Possegger, H., Mauthner, T., Bischof, H.: In Defense of color-based model-free tracking. In: Proceedings of the IEEE Conference on Computer Vision and Pattern Recognition, pp. 2113–2120 (2015)
24. Danelljan, M., Hager, G., Khan, F. S., Felsberg, M.: Convolutional features for correlation filter based visual tracking. In: 2015 IEEE International Conference on Computer Vision (ICCV), pp. 621–629 (2015)

An Improved Target Tracking Method Based on DIMP

Luping Zhao, Rong Wang[⊠], and Naiqian Tian

Key Laboratory of Security Technology and Risk Assessment, People's Public Security University of China, Beijing 102623, China
908010028@qq.com, dbdxwangrong@163.com

Abstract. To address tracking failures of the DIMP method in scenarios of motion blur, illumination variation, and deformation, etc., this paper proposes an improved DIMP method, which uses the IResNet network with higher accuracy and better learning convergence for visual object tracking task. Firstly, we extract the depth features from a set of labeled sample images using the IResNet network, and then use a convolution block (Cls Feat) to classify the features. Secondly, the features are input into the model predictor to generate the weights to update the sample model. Finally, based on the weights, the convolution layer selects the region with the highest confidence score in the test frame as the target region. Experimental results on the benchmark OTB-2015 illustrate that our method, running at 53 fps, improves the precision and success rate by 1.8% and 1.1%, respectively, compared with DIMP. Therefore, our method effectively improves the tracking accuracy and robustness.

Keywords: DIMP · IResNet · Visual object tracking

1 Introduction

Visual object tracking refers to the continuous positioning of a given target in a video sequence. This technology has many application scenarios, such as intelligent security monitoring, human-machine interaction, and virtual reality. According to the appearance model generation, the target tracking method can be divided into a generative model and a discriminant model. Earlier target tracking algorithms focused on the study of generative models, which use a single mathematical model to describe the tracking target, such as the Meanshift algorithm [1] and the Camshift algorithm [2]. But such methods have significant limitations because they can not fully exploit or best utilize the background information of the image. The MOSSE [3] method uses a correlation filter for the first time in the object tracking domain, and then came some methods of KCF [4], DSST [5], ECO [6] and other correlation filter methods, which significantly improved the accuracy and speed of operation. Subsequently, deep learning began to be applied to the field of target tracking. And by its excellent feature modeling capabilities, deep learning became the mainstream research direction of target tracking, with representative methods such as SiamFC [7], SiamRPN [8], ATOM [9] and DIMP [10]. Martin Danelljan and others proposed the DIMP method in 2019, using a discriminative learning loss function. This method can make the most of the appearance

© Springer Nature Singapore Pte Ltd. 2020
Y. Wang et al. (Eds.): IGTA 2020, CCIS 1314, pp. 154–167, 2020.
https://doi.org/10.1007/978-981-33-6033-4_12

information in the image for the target model prediction and maximize the discriminative power of the predictive model. At the same time, they designed a model predictor to accelerate network convergence and perform model updates, improving the robustness of the tracking method. However, this method is prone to tracking drift when the tracking scenes are more complex, with challenges such as background clutters and occlusion.

This paper puts forward an advanced DIMP method that uses the IResNet [10] network with consistent improvements in accuracy and learning convergence for the tracking task, which somewhat improves the robustness of the DIMP method in the presence of challenges such as occlusion and background clutters. Firstly, a set of sample images with bounding box annotations is entered into the backbone network IResNet and an additional convolutional block Cls Feat to extract and classify the target features. Then the feature map is fed into a model predictor consisting of an initializer and a recurrent optimizer module [10]. The initializer uses the appearance information of objects to provide an initial prediction of the weights. And the optimizer processes the weights and outputs them by using both target and background information. Finally, according to the obtained weights, the convolutional layer extracts the target features from the test frame, classifies the features, and outputs the confidence score of the target.

2   Baseline

The DIMP method has two components: a target classification branch and a bounding box estimation branch. The function of the former is to make a distinction between the moving object and background. And the latter is used to improve the positioning accuracy of the prediction frame. Figure 1 illustrates the framework of this method.

The target classification branch takes a set of training set images with bounding box annotations as input samples, extracts the deep features using the backbone network ResNet-50 [11], and appends the convolutional block Cls Feat to classify the features. The features are then entered into the model predictor consisting of the initializer module and the recurrent optimizer module. The initializer effectively provides an initial estimate of the model weights f^0 using only the target appearance. The model is then processed by the recurrent optimizer, taking into account both the target sample and the background sample, and the model is updated. The model predictor outputs a weighted model f of the convolutional layer Conv. The convolutional layer performs a target classification operation on the features extracted from the test frame and outputs a confidence graph of the target features.

The bounding box estimation branch adopts the overlap maximization based architecture of the ATOM method. It adds the intersection over union (IoU) [12] branch to the target detection network for predicting the IoU between the target and a set of candidate prediction boxes. And it adjusts the output box by gradient-up-maximization IoU to obtain a more accurate target region.

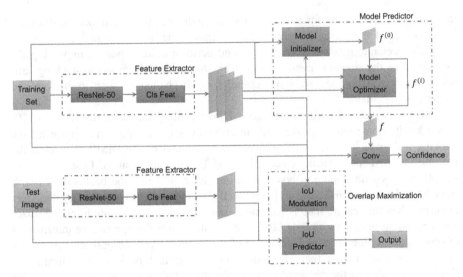

Fig. 1. An improved target tracking method based on DIMP

2.1 Discriminative Learning Loss

The model predictor employs a discriminative loss, which can make the best of information about the appearance of target and background and improve the predictive model's discriminatory power. The deep feature maps entered into the model predictor are recorded as $x_j \in X$, and the corresponding target center coordinate is recorded as $c_j \in R^2$. The model predictor consists of the training set $S_{train} = \{(x_j; c_j)\}_{j=1}^n$ of x_j and its output is the filter weights $f = D(S_{train})$ of the convolutional layer. According to the least-squares loss, we can get the loss function,

$$L(f) = \frac{1}{|S_{train}|} \sum_{(x,c) \in S_{train}} \|r(s,c)\|^2 + \|\lambda f\|^2 \tag{1}$$

The function $r(s, c)$ calculates the residual for each spatial location according to the ground-truth center coordinate c and the target confidence score $s = x * f$ [10]. Among them, λ is the regularization factor, and $*$ is the convolutional operation. Typically, $r(s, c) = s - y_c$, where y_c is the desired target fraction for each position, usually set as a Gaussian function centered at c [3].

Simple subtraction makes the model's regression-calibrated confidence score zero for negative samples, ignores the discriminatory ability of positive samples themselves, and makes the data unbalanced between target and background. DIMP proposes to use a hinge-like loss in r to alleviate this issue and optimize the classification results. The spatial weight function v_c is introduced, where the subscript c indicates the centrality dependence of the target. The negative fraction in the background is zeroed using $max(0, s)$. We can define the loss function as,

$$r(s, c) = v_c \cdot (m_c s + (1 - m_c) max(0, s) - y_c) \tag{2}$$

Here, $m_c \in [0, 1]$ is the mask of the target region. The parameters m_c, v_c, λ, y_c are all learned by the network model during the training. Without increasing loss, the model can reduce the propensity for negative samples in the model training process.

2.2 Model Predictor

The model predictor includes an initializer module and a recurrent optimizer module. The initializer further reduces iteration times of model predictor. It contains a convolutional layer and a precise ROI pooling [13]. The ROI pooling extracts features from the target region, processes them into feature maps of the same dimension as a target model, and averages these feature maps to obtain an initial weight model $f^{(0)}$. Finally, the weights are fed into the recurrent optimizer for processing to obtain the final discriminative model. The recurrent optimizer predicts filter weights f by minimizing the error in Eq. (1). From Eqs. (1) and (2), we can get the expression for the gradient of the loss ∇L concerning the filter f,

$$f^{(i+1)} = f^{(i)} - \alpha \nabla L\left(f^{(i)}\right) \tag{3}$$

Here, α is the learning rate, and the DIMP method calculates α based on the steepest descent methodology [14, 15]. This method can accelerate the convergence speed of gradient descent and predict a strong discriminant filter with only a few iterations. Loss is estimated by using the quadratic function at the current estimate $f^{(i)}$,

$$L(f) \approx \tilde{L}(f) = \frac{1}{2}\left(f - f^{(i)}\right)^T Q^{(i)}\left(f - f^{(i)}\right) + \left(f - f^{(i)}\right)^T \nabla L\left(f^{(i)}\right) + L\left(f^{(i)}\right) \tag{4}$$

In order to minimize $L(f)$ in the gradient direction, we need to find a suitable α for the fastest descent. Substitute formula (3) for $L(f) \approx 0$, find the derivative on both sides of the formula,

$$\frac{d}{d\alpha} \tilde{L}\left(f^{(i)} - \alpha \nabla L\left(f^{(i)}\right)\right) = 0 \tag{5}$$

$$\alpha = \frac{\nabla L(f^{(i)})^T \nabla L(f^{(i)})}{\nabla L(f^{(i)})^T Q^{(i)} \nabla L(f^{(i)})} \tag{6}$$

$Q^{(i)} = \left(J^{(i)}\right)^T J^{(i)}$ is a positive definite square, which determines the quadratic model and α in Eq. (4), where $J^{(i)}$ is the Jacobian determinant of the residuals at $f^{(i)}$. The updated α and $L(f)$ are substituted into Eq. (3) to obtain the output weight model f for the model predictor.

3 Proposed Method

The DIMP method uses ResNet as its backbone network for feature extraction. The Down-sampling layer and the last ReLU of every set of ResBlocks in ResNet will negatively affect the propagation of the information flow and generate significant information losses. It makes the tracking method incapable of making the best of the image information under the scenarios of occlusion, background clutters, etc., which may lead to tracking drift and tracking failure.

Accounting for these above shortcomings, we propose an improved target tracking method, the overall framework is similar to the DIMP method, as shown in Fig. 2. Changing the arrangement of modules in the original ResBlock, dividing each stage into three types: Start ResBlock, Middle ResBlock, and End ResBlock. This approach provides a better way of information flow through the network. Simultaneously, the improvement to the way of projected shortcuts can reduce information loss, gain greater accuracy, and get better learning convergence.

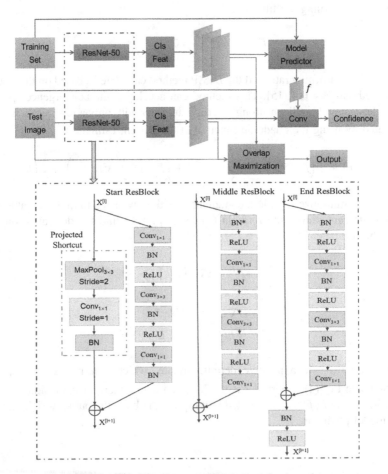

Fig. 2. Diagram of our methodological framework.

3.1   Information Flow Through the Network

The main structure of ResNet is ResBlock. Each ResBlock contains three convolution layers and three ReLU layers. But the last ReLU layer will zero the negative signal, which will harm the information dissemination.

The solution of IResNet is to divide the network structure into different stages. As shown in Fig. 3 and take ResNet-50 as an example, it is divided into six parts, which are four main stages, a start-stage, and an end-stage. Each stage contains multiple ResBlocks. Four main stages have [3, 4, 6] ResBlocks, respectively. Each stage consists of one Start ResBlock, multiple Middle ResBlocks, and one End ResBlock, and the number of Middle ResBlocks corresponding to each stage is [1, 2, 4]. This structure only recombines different components of ResBlock. It provides a better channel for the information flow to propagate through the network layer, without increasing the parameters and complexity of the network.

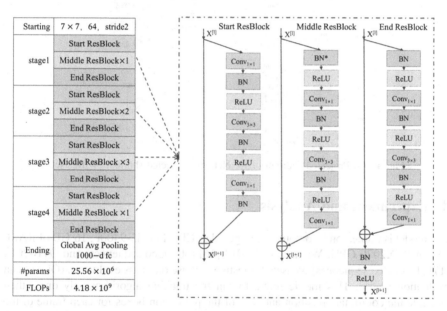

Fig. 3. Proposed ResBlocks (* the first BN in the first middle resblock is eliminated in each stage [10]).

3.2   Projected Shortcut

In Stage ResBlock, the feature map needs to be down-sampled by projection shortcuts because the dimensions of the input feature map and the output feature map do not match. ResNet applies a 1×1 Conv with stride 2 to unify the features' spatial sizes, followed by normalization of the channels with a BN layer. Since a 1×1 Conv with stride 2 can lose 75% important information and introduce noise, negatively affecting the main information flow, IResNet proposes a way to improve the down-sampling operations.

As shown in Fig. 4, for the spatial projection, IResNet uses a 3 × 3 max pooling with stride 2; for the channel projection, a 1 × 1 Conv with stride 1 is used, followed by a BN layer. This approach allows for integrated consideration of all the information from the feature maps, selecting the element with the highest degree of activation in the next step, and reducing information loss.

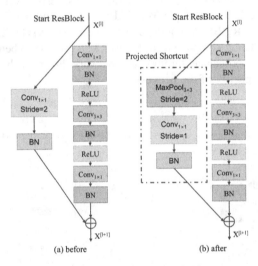

Fig. 4. (a) Projected shortcut of ResNet. (b) Projected shortcut of IResNet.

4 Experiment and Analysis

The model is trained on a computer configured as CPU Intel i5-9400F, 16 GB of RAM, and an RTX2070 GPU. We use OTB-2013 [16] (50 video sequences) and OTB-2015 [17] (100 video sequences) datasets for testing and use one-pass evaluation (OPE) as an evaluation metric. This metric refers to run the tracking algorithm only once, then generate the coordinate position and size of the prediction boxes for each frame of the image, and finally evaluate the tracking effect using the Precision plot and Success plot.

Four sets of experiments are mainly completed: experiment 1 is the selection of model training rounds for epoch; experiment 2 compares the effects of our approach with baseline DIMP; experiment 3 is the robustness analysis of our method and some mainstream tracking methods; experiment 4 is the applicability study of our method for other algorithms.

4.1 Experiment 1: Selection of Model Training Rounds for Epoch

Epoch is the process of performing a complete training of a model using all the data of the training dataset, and it defines the times the training dataset will work. The iterations times of weights in the neural network increases with the increase of epoch, and the curve changes from underfitting to overfitting. In the interest of fitting convergence, it will be obliged to train all the data repeatedly. Therefore, choosing the right epoch is critical to the performance of the training model. To obtain the best performance model, we use the Lasot [18], Got-10 k [19], and COCO2014 [20] datasets to train our model. We set $epoch = 50, batch\_size = 4$, initial learning rate $\alpha = 2e - 4$, and use Adam [21] with a learning rate decay of 0.2 every 15th epoch [10]. The improved models trained under different epochs are tested on the OTB-2015 dataset, as shown in Fig. 5.

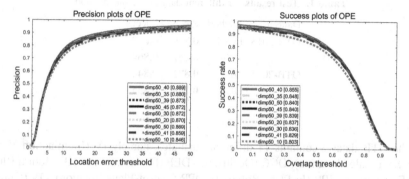

Fig. 5. Comparisons of models trained for different epochs.

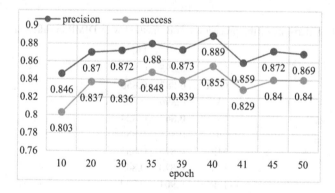

Fig. 6. The tracking effect of our method fluctuates with changes in *epoch*.

From the experiment we can get that with the gradual increase of epoch, the tracking effect of our approach presents a trend of fluctuation rising first and then fluctuation falling. The best results of overall precision and success rates appear at $epoch = 40$, as demonstrated in Fig. 6. Therefore, we choose the model generated at $epoch = 40$ as the final model.

4.2   Experiment 2: Comparison Between Our Method and DIMP

The comparison of our approach with DIMP under OTB-2013 and OTB-2015 datasets is recorded in Table 1. On the OTB-2013 dataset, the precision rate of our method is 89.7%, while the success rate is 86.4%. Compared with the baseline DIMP, we obtain improvements by 2.7% and 1.4%. On the OTB-2015 dataset, our method's precision rate is 88.9% and the success rate is 85.5%, which is 1.8%, and 1.1% better than the baseline, respectively. On a single RTX2070 GPU, our method runs at around 53 fps.

Table 1. Test results for different datasets of our method.

Dataset	Method	Precise	Success
OTB-2013	DIMP	**0.870**	0.850
	OURS	**0.897**	0.864
OTB-2015	DIMP	**0.871**	0.844
	OURS	**0.889**	0.855

The OTB dataset contains 11 different attributes, namely IV (Illumination Variation), SV (Scale Variation), OCC (Occlusion), DEF (Deformation), MB (Motion Blur), FM (Fast Motion), IPR (In-Plane Rotation), OPR (Out-of-Plane Rotation), OV (Out-of-View), BC (Background Clutters) and LR (Low Resolution) [17]. Tables 2 and 3 summarizes the performance of this method and DIMP under 11 different attributes.

Table 2. Comparisons of the precision of our method with DIMP in different scenarios.

Scene	OURS	DIMP
IV	**0.895**	0.885
OPR	**0.887**	0.856
SV	**0.890**	0.863
OCC	**0.862**	0.824
DEF	0.886	**0.889**
MB	**0.878**	0.853
FM	**0.873**	0.845
IPR	**0.901**	0.892
OV	**0.836**	0.781
BC	**0.836**	0.786
LR	0.897	**0.903**
Overall	**0.889**	0.871

From Tables 2 and 3, and we know that, compared with DIMP, our method has a lower precision rate and success rate under both deformation and low resolution scenarios. Images in low-resolution scenes have poor clarity and contain more noisy information. Since the improved method retains more information, it introduces more

Table 3. Comparisons of the success of our method with DIMP in different scenarios.

Scene	OURS	DIMP
IV	0.869	0.872
OPR	**0.847**	0.821
SV	**0.870**	0.845
OCC	**0.831**	0.797
DEF	0.815	0.828
MB	**0.878**	0.845
FM	**0.847**	0.819
IPR	**0.857**	0.856
OV	**0.783**	0.747
BC	**0.808**	0.774
LR	0.827	0.836
Overall	**0.855**	0.844

noise than the original method, so its performance is degraded. Besides, its success rate is relatively weak in the scene with significant illumination variation. But on the whole, the performance of this method is improved under all other scenarios.

4.3 Experiment 3: Robustness Analysis of Our Methodology and Some Mainstream Tracking Methods

The experiment is conducted under the OTB-2015 dataset. Figure 7 illustrates our approach's evaluation results with the baseline DIMP and four mainstream trackers ATOM, ECO, SiamRPN, and Siamfc3s. We can see that the proposed method performs well in precision and success rates, and is superior to other methods mentioned above.

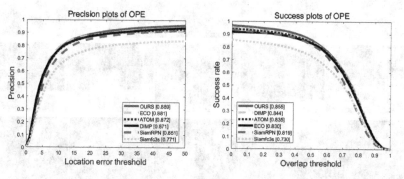

Fig. 7. Comparisons of ours approach with mainstream trackers on OTB-2015.

In this experiment, three typical video sequences Bird1, Jump, and Ironman in the OTB-2015 dataset, are used to analyze our method's robustness. The tracking results are shown in Fig. 8.

The comparison result on the Bird1 sequence is shown in Fig. 8a. The attributes included in the sequence are DEF, FM, and OV. A large occlusion area occurred during the 115th to 184th frames, resulting in tracking failure of SiamRPN, ECO, and Siamfc3s. At frame 260, as the tracking target showed a large shape change, DIMP and ATOM showed a decrease in tracking accuracy. And at frame 290, ATOM completely lost the target. On the Bird1 sequence, only our tracking method can achieve full tracking and perform well.

The comparison result on the Jump sequence is shown in Fig. 8a. The attributes included in the sequence are SV, OCC, DEF, MB, FM, IPR, and OPR. At frame 20, the five methods other than the one we proposed failed to deal better with the out-of-plane rotation and rapid motion of the tracking object, their tracking frame differing significantly from the real target. By frame 90, the ATOM, SiamRPN, ECO, and Siamfc3s experienced tracking drift due to continuous scale variation and motion blur of the tracking target. On the Jump sequence, our method and DIMP can better reduce the impact of scale variation and tracking target rotation on track.

The comparison result on the Ironman sequence is shown in Fig. 8c. This video sequence has more complex and challenging scenes to track, containing ten properties beyond DEF. After experiencing challenges such as constant background clutters, fast motion, and illumination variation, the DIMP, SiamRPN, ECO, and Siamfc3s all lost their targets. At frame 165, the tracking target underwent in-plane rotation, and the tracker ATOM showed significant deviation. Our method and ATOM were able to do the basic tracking on the Ironman sequence, with the best performance being the proposed method.

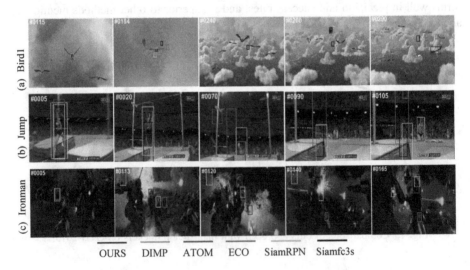

Fig. 8. Robustness analysis on three sequences from OTB-2015.

From the analysis above, it follows that compared to other mainstream approaches, our approach performs best in tracking the three video sequences selected. Our method exhibits strong robustness both in the face of external challenges such as occlusion and background clutters, as well as the internal challenges of tracking the target's rotation, motion blur, and so on.

4.4 Experiment 4: The Applicability of Our Method for Other Algorithms

To verify the applicability of IResNet to other tracking methods, we propose an improved ATOM method, which replace the backbone of the ATOM tracking method by IResNet. The method used in Experiment 1 is used to select the number of training rounds, which is finally selected $epoch = 40$. The improved ATOM method and the baseline ATOM method are tested on the OTB-2015 dataset. Our method The performance of the two methods under the 11 different attributes labeled by OTB-2015 is shown in Tables 4 and 5.

As the tables show, the accuracy of the improved ATOM method is 88.4%, which is 1.2% better than the baseline method, and the success rate is 84.4%, which is 0.9% better than the baseline method. Our approach is slightly less precise in the two scenarios of illumination variation and motion blur. It has a lower success rate in the two scenes of motion blur and fast motion. Therefore, IResNet can be added to other tracking methods to improve tracking effectiveness.

Table 4. Comparison of the precision of our method with ATOM in different scenarios.

Scene	OURS	ATOM
IV	0.876	**0.879**
OPR	**0.871**	0.854
SV	**0.891**	0.888
OCC	**0.832**	0.830
DEF	**0.841**	0.839
MB	0.853	**0.858**
FM	**0.866**	0.864
IPR	**0.896**	0.866
OV	**0.856**	0.836
BC	**0.842**	0.811
LR	**0.995**	0.989
Overall	**0.884**	0.872

Table 5. Comparison of the success of our method with ATOM in different scenarios.

Scene	OURS	ATOM
IV	**0.866**	0.852
OPR	**0.821**	0.799
SV	**0.867**	0.857
OCC	**0.808**	0.801
DEF	**0.780**	0.778
MB	0.826	**0.842**
FM	0.827	**0.839**
IPR	**0.855**	0.822
OV	**0.807**	0.776
BC	**0.820**	0.775
LR	**0.927**	0.918
Overall	**0.844**	0.835

5   Conclusion

In our work, we propose our architecture based on the DIMP tracking method. It replaces the original backbone with a better performing IResNet network, which improves the tracking effect of the baseline method in multiple scenarios and effectively improves the anti-interference ability of target tracking. The improved method in this paper is equally applicable to other tracking methods. From the test results on the OTB-2015 dataset, we can know that our method's precision and precision improve in 8 scenarios, but it performs poorly in scenes with low resolution and deformation. The next research work can consider introducing an attention mechanism to enhance the tracking effect.

Acknowledgments. This work is supported by National Key Research and Development Project under Grant A19808. This work is supported by National Natural Science Foundation of China under Grant No. 61906199.

References

1. Cheng, Y.: Mean shift, mode seeking, and clustering. In: IEEE Transactions on Pattern Analysis and Machine Intelligence, pp. 790–799 (1995)
2. Black, J., Ellis, T., Rosin, P.: Multi view image surveillance and tracking. In: 2002 Workshop on Motion and Video Computing, Proceedings, pp. 169–174 (2002)
3. Bolme, D.S., Beveridge, J.R., Draper B.A., et al.: Visual object tracking using adaptive correlation filters. In: CVPR, pp. 2544–2550 (2010)
4. Henriques, J.F., Caseiro, R., Martins, P., et al.: High-speed tracking with kernelized correlation filters. In: IEEE Transactions on Pattern Analysis and Machine Intelligence, pp. 583–596 (2015). arXiv:1404.7584

5. Danelljan, M., Häger, G., Khan, F.S., et al.: Accurate scale estimation for robust visual tracking. In: BMVC (2014)
6. Danelljan, M., Bhat, G., Khan, F.S., et al.: ECO: efficient convolution operators for tracking. In: CVPR, pp. 6931–6939 (2017). arXiv:1611.09224
7. Bertinetto, L., Valmadre, J., Henriques, J.F., Vedaldi, A., Torr, P.H.S.: Fully-convolutional siamese networks for object tracking. In: Hua, G., Jégou, H. (eds.) ECCV 2016. LNCS, vol. 9914, pp. 850–865. Springer, Cham (2016). https://doi.org/10.1007/978-3-319-48881-3_56
8. Li, B., Yan, J., Wu, W., et al.: High performance visual tracking with siamese region proposal network. In: CVPR, pp. 8971–8980 (2018)
9. Danelljan, M., Bhat, G., Khan, F.S.: ATOM: accurate tracking by overlap maximization. In: CVPR, pp. 4660–4669 (2019). arXiv:1811.07628
10. Bhat, G., Danelljan, M., Van Gool, L.: Learning discriminative model prediction for tracking. In: ICCV, pp. 6182–6191 (2019). arXiv:1904.07220
11. Duta, I.C., Liu, L., Zhu, F., et al.: Improved residual networks for image and video recognition. In: CVPR (2020). arXiv:2004.04989
12. He, K., Zhang, X., Ren, S., Sun, J.: Deep residual learning for image recognition. In: CVPR, pp. 770–778 (2016). arXiv:1512.03385
13. Jiang, B., Luo, R., Mao, J., Xiao, T., Jiang, Y.: Acquisition of localization confidence for accurate object detection. In: Ferrari, V., Hebert, M., Sminchisescu, C., Weiss, Y. (eds.) Computer Vision – ECCV 2018. LNCS, vol. 11218, pp. 816–832. Springer, Cham (2018). https://doi.org/10.1007/978-3-030-01264-9_48
14. Nocedal, J., Wright, S.J.: Numerical Optimization, 2nd edn. Springer, New York (2006). https://doi.org/10.1007/978-0-387-40065-5
15. Shewchuk, J.R.: An introduction to the conjugate gradient method without the agonizing pain. In: Technical report. Carnegie Mellon University (1994)
16. Wu, Y., Lim, J., Yang, M.H.: Online object tracking: a benchmark. In: CVPR, pp. 2411–2418 (2013)
17. Wu, Y., Lim, J., Yang, M.H.: Object tracking benchmark. IEEE Trans. Pattern Anal. Mach. Intell. 37(9), 1834–1848 (2015)
18. Fan, H., Lin, L., Yang, F., et al.: LaSOT: a high-quality benchmark for large-scale single object tracking. In: CVPR (2019). arXiv:1809.07845
19. Huang, L., Zhao, X., Huang, K.: Got-10k: a large highdiversity benchmark for generic object tracking in the wild. In: CVPR (2018). arXiv:1810.11981
20. Lin, T.-Y., et al.: Microsoft COCO: common objects in context. In: Fleet, D., Pajdla, T., Schiele, B., Tuytelaars, T. (eds.) ECCV 2014. LNCS, vol. 8693, pp. 740–755. Springer, Cham (2014). https://doi.org/10.1007/978-3-319-10602-1_48
21. Kingma, D.P., Ba. J.: Adam: a method for stochastic optimization. In: ICLR, pp. 1–15 (2014). arXiv:1412.6980

Infrared Small Target Recognition with Improved Particle Filtering Based on Feature Fusion

Qian Feng$^{(\boxtimes)}$, Dongjing Cao, Shulong Bao, and Lu Liu

Beijing Institute of Space Mechanics and Electricity, Beijing 100094, China
fengqian_cast@foxmail.com

Abstract. Aiming at the problem of tracking weak targets in different scenarios, an improved particle tracking method is proposed. This paper firstly uses background prediction and extracts the gray and motion features of weak targets to suppress the background of the image, then uses the krill herd algorithm to update the particle weights, and finally uses three different types of scenes to verify the tracking effect of the algorithm. It is experimentally demonstrated that compared with the traditional algorithm, this algorithm enhances the tracking ability of small targets in different infrared scenes.

Keywords: Infrared small target · Feature fusion · Krill herd optimization · Particle filter

1 Introduction

Target recognition is one of the important research topics in computer vision, image processing and machine learning. In recent years, with the rapid development of infrared imaging technology in the field of remote sensing, infrared dim and small moving target detection has also become a research hotspot in image processing technology [1–5].

It is not easy to obtain shapes, textures and other features of the dim and small target. As a result, methods like moment invariant based on the geometric feature of the target, template matching based on the texture feature, mean shift based on the grayscale, probability and density of the target and other traditional extended target tracking methods are hard to guarantee the detection and tracking performance of weak and small targets [6]. Only the information of the target on the sequence image can be used to track and identify while detecting through the dynamic characteristics of the inter-frame correlation. Particle filtering has developed into a critical research direction of small target tracking algorithms for the better performance in the non-linear and non-Gaussian scenes [7, 8]. However, traditional particle filtering is prone to tracking missing under a single feature of the target under strong noise interference. Therefore, extracting multiple features of the target and integrating the particle filter helps to improve the stability of detection. For the detection of dim and small targets in the complex background of infrared sequence images, this paper first uses the multi-frame difference method to suppress the background and uses the grayscale and dynamic

© Springer Nature Singapore Pte Ltd. 2020
Y. Wang et al. (Eds.): IGTA 2020, CCIS 1314, pp. 168–180, 2020.
https://doi.org/10.1007/978-981-33-6033-4_13

feature of small target to delete the false points. At the same time, considering the large amount of calculation and the problem of sample depletion in resampling of particle filter, the strategy of krill herd optimization algorithm to update particle weights is added to track the dim and small targets in infrared sequence images.

2 Algorithm Framework

In this paper, the algorithm is divided into four modules: preprocessing, gray feature extraction, motion feature extraction and improved particle filter algorithm optimized by krill herd. The framework of the entire algorithm is as follows (Fig. 1):

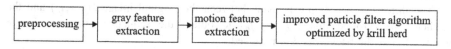

Fig. 1. Algorithm framework

The detail of the proposed algorithm is given in Sects. 2.1–2.4 below.

2.1 Preprocessing

It is no surprise that background suppression is an essential step in preprocessing to improve the detection accuracy of infrared dim targets and accurately separate the background, noise and target points. Common backgrounds can be divided into stationary backgrounds and non-stationary backgrounds. The pixels in the stationary background have strong spatial correlation. The false points are mainly composed of the noise of the detector, which will cause small changes in the background. They can be suppressed by spatial filtering. Pixels in a non-stationary background have a strong time-domain correlation so that the frame difference can be used before threshold segmentation to highlight the target and suppress the background.

Infrared images containing small targets are consist of targets, noise and background. The specific model is as follows:

$$f(x,y) = f_T(x,y) + f_B(x,y) + n(x,y) \tag{1}$$

Where $f(x,y)$ is the infrared image, $f_T(x,y)$, $f_B(x,y)$, $n(x,y)$ represent the target, background and noise [9].

Because of the long distance, the target usually appears like a dot, and the background is a large area of gentle variation with strong pixel correlation. The infrared radiation intensity of the target is generally higher than that of the surrounding background. Therefore, small targets and noises appear as isolated bright spots in infrared images, and the gray value is much larger than the pixels in their neighborhood.

Considering the non-stationary background of common infrared remote sensing images and the long distance of imaging, this article focuses on small targets moving at a constant speed under non-stationary backgrounds. This paper adopts the first few frames as the prediction background, and then segments the images to obtain suspicious targets.

Since the target occupies very few pixels in a single-frame image, the image of the previous k frames can be taken and averaged as the predicted background image. If the original image is $f(x, y, t_n)$ and the predicted background image is $f'(x, y, t_n)$, then the predicted background image is:

$$f'(x, y, t_n) = \frac{1}{k} \sum_{n=1}^{k} f(x, y, t_n) \tag{2}$$

The results show that k taking 10 can better balance the relationship between the amount of calculation and robustness.

The difference processing is made between the original image $f(x, y, t_n)$ and the predicted background image $f'(x, y, t_n)$ to obtain the sequence difference image $g(x, y, t_n)$ after the background is removed:

$$g(x, y, t_n) = |f(x, y, t_n) - f'(x, y, t_n)| \tag{3}$$

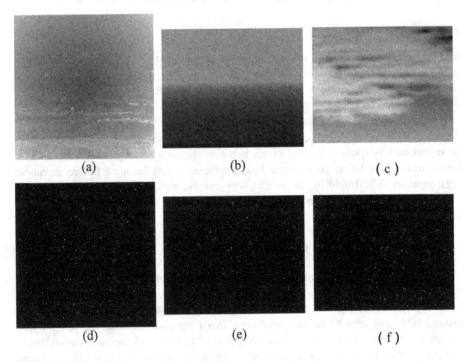

(a)   (b)   (c)

(d)   (e)   (f)

Fig. 2. Preprocessing results.

The results of the preprocessing experiment are as follows:

Among them, (a), (b) and (c) in Fig. 2 are the three kinds of backgrounds in the video sequence, and (d), (e) and (f) are the background suppression results calculated by Eqs. (2) and (3). As can be seen from the figure, the algorithm can suppress the background well and highlight the target point using the gray and motion features of the targets.

2.2 Gray Feature Extraction

After predicting the background image, the obtained image can be subtracted with value T to obtain a binary image $F(x, y, t_n)$, in which the non-zero gray value are the set of suspected target points:

$$F(x, y, t_n) = \begin{cases} 0, & g(x, y, t_n) < T \\ g(x, y, t_n), & g(x, y, t_n) \geq T \end{cases} \tag{4}$$

Where $F(x, y, t_n)$ is the binary map of the suspect target point set, $g(x, y, t_n)$ is the sequence difference image after background removal, and T is the segmentation threshold selected according to experience.

According to the 4 neighborhoods traversal image $F(x, y, t_n)$, we can find all the suspected target points with connected 4 domains. Cause the infrared weak small target spots are usually slightly larger than the noise points, we will arrange the connected domain according to the size and select top k largest connected domains, which will be used as candidate target points, and the others are eliminated as noise. At this point, we have obtained the grayscale features of the suspected target points with position information.

2.3 Motion Feature Extraction

According to the positional correlation of the movement of the small targets between frames, the distance of the candidate target points between frames can be used as the basis for discrimination. In this paper, the shortest Euclidean distance is used as the trajectory of the target points and the initial target position is determined from the candidate target points.

$$D_k^n = \sqrt{\left|c\left(x_k^n\right) - c\left(x_k^{n-1}\right)\right|^2 + \left|c\left(y_k^n\right) - c\left(y_k^{n-1}\right)\right|^2} \tag{5}$$

Where D_k^n represents the distance of the k_{th} candidate target point between the n_{th} frame and the $n-1_{th}$ frame, and $c\left(x_k^n\right)$ represents the centroid x coordinate of the k_{th} candidate target point in the n_{th} frame, $c\left(x_k^{n-1}\right)$ represents the centroid x coordinate of the $k-1_{th}$ candidate target point in the $n-1_{th}$ frame, $c\left(y_k^n\right)$ represents the centroid y coordinate of the k_{th} candidate target point in frame $n-1$, $c\left(y_k^{n-1}\right)$ represents the x coordinate of the centroid of the k_{th} candidate target point in the $n-1_{th}$ frame. According to the experimental results of this paper, k takes 4.

Infrared dim targets move slowly on the imaging surface, and sometimes the distance between frames is less than one pixel. As a sequence of the random and messy appearance and no correlation between frames of noise, it is helpful for us to use motion distance of candidate target points between two adjacent frames as a criterion to effectively eliminate noise points.

2.4 Improved Particle Filter Algorithm Optimized by Krill Herd

Assuming that the target is moving at a constant speed, the state variable of the target is defined as $S = \{x, y, v_x, v_y\}$, where (x, y) is the position of the center of the target, (v_x, v_y) is the horizontal and vertical speed of the target. We define the state transition matrix F and random noise W_k as:

$$F = \begin{bmatrix} 1 & 0 & 1 & 0 \\ 0 & 1 & 0 & 1 \\ 0 & 0 & 1 & 0 \\ 0 & 0 & 0 & 1 \end{bmatrix}, \ W_k = \begin{bmatrix} S_p * rand \\ S_p * rand \\ S_v * rand \\ S_v * rand \end{bmatrix} \tag{6}$$

Where S_p represents the status constant variable in the horizontal direction, and S_v represents the status constant variable in the vertical direction. According to the experiment, S_p takes 25 and S_v takes 5.

The steps of the improved particle filter algorithm optimized by krill herd are as follows (Fig. 3):

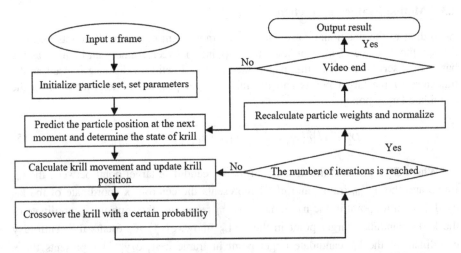

Fig. 3. The steps of the improved particle filter algorithm optimized by krill herd.

Step 1: Initialization

Establish the target motion model:

$$X_k = FX_{k-1} + W_k \tag{7}$$

Where F is the state transition matrix, W_k is the random noise.

Sampling independent and identically distributed particles $\left\{x_k^i, \frac{1}{N}\right\}_{i=1}^N$ according to

$$x_k^i \sim q\left(x_k^i | x_{k-1}^i, z_k\right) \tag{8}$$

Set iterations I to 1, set krill group size N, maximum induction speed N^{max}, maximum foraging speed V_f, maximum random diffusion speed D^{max}, time interval Δt, maximum number of iterations I_{max}, the inertia weight of induced motion w_n and inertial weight w_f of foraging behavior.

Step 2: Prediction

Calculate the weight of the new particle set $\left\{x_k^i, w_k^i\right\}_{i=1}^N$ according to the current observed value z_k:

$$w_k^i = w_{k-1}^i \frac{p\left(z_k | x_k^i\right) p\left(x_k^i | x_{k-1}^i\right)}{p\left(x_k^i | x_{k-1}^i, z_k\right)} \tag{9}$$

Use the state function of Eq. (7) to calculate the state value x_k^i of the particle at the next moment, then use Eq. (9) to calculate the weight of each particle. The particle state value x_k^i is used as the position of krill X_i.

Step 3: Update

$$N_i^{new} = N^{max}\alpha_i + w_n N_i^{old} \tag{10}$$

$$\alpha_i = \alpha_i^{local} + \alpha_i^{target} \tag{11}$$

Where N indicates the induced motion speed, max indicates the maximum speed, α_i indicates the individual induced direction, α_i^{local} indicates the neighboring individual induced direction, α_i^{target} indicates the optimal individual induced direction, $w_n \in [0, 1]$ indicates the inertial weight of the original motion.

$$F_i^{new} = V_f\beta_i + w_f F_i^{old} \tag{12}$$

$$\beta_i = \beta_i^{food} + \beta_i^{best} \tag{13}$$

Where F_i^{new} is the speed vector of the foraging behavior of the i_{th} krill, V_f is the speed of foraging behavior, β_i^{food} is the effect of food on the i_{th} krill, and β_i^{best} is the optimal fitness of the i_{th} krill, $w_f \in [0, 1]$ is the inertia weight of foraging behavior, and F_i^{old} is the speed vector of the last foraging behavior.

$$D_i^{new} = D^{max}\left(1 - \frac{I}{I_{max}}\right)\delta \qquad (14)$$

Where D^{max} is the maximum speed of random diffusion, I_{max} is the maximum number of iterations, and δ is a directional vector with each variable obeying $[-1, 1]$ uniform distribution.

$$X_i^l = X_i^{l-1} + \Delta t\left(N_i^{new} + F_i^{new} + D_i^{new}\right) \qquad (15)$$

Where Δt is the time interval.

Use Eqs. (10) (12) (14) to calculate the induced foraging and diffusion movement of each krill individual, and finally calculate the position change of each krill in each iteration according to Eq. (15).

Step 4: Cross operation

$$C_r = 0.2\hat{K}_{i,best} \qquad (16)$$

Where C_r is the crossover probability [10–12], $\hat{K}_{i,best}$ represents the best fitness of the i_{th} krill.

Calculate the crossover probability according to Eq. (16) and perform crossover operation on individual krill groups.

Step 5: Determine the Termination Conditions
Whether the maximum number of iterations I_{max} is reached, otherwise step 3.

Step 6: Weight Normalization

$$w_k^i = \frac{w_k^i}{\sum_{i=1}^{N_s} w_k^i} \qquad (17)$$

$$\hat{x}_k = \sum_{i=1}^{N} w_k^i x_k^i \qquad (18)$$

Recalculate the weight of each particle, update the particle weight according to Eq. (9), normalize the particles according to Eq. (17), and finally output the estimated state value of each particle according to Eq. (18).

Step 7: Iteration
Repeat steps 2 to 6 until the prediction is completed.

3   Simulation Results and Analysis

Three sets of video sequences with different backgrounds were selected for the experiment to confirm the performance of the method in this paper. The first test video came from the detection and tracking data set of small and weak aircrafts with infrared images under the ground/air background of Hui Bingwei et al. [13], the simulation experiment was performed on the hardware platform of AMD Ryzen 5 4600H CPU 3.00 GHz and memory 16 GB, and the development environment is MatlabR2017b (Table 1).

Table 1. Properties of experimental scenes.

Video sequence	Backgrounds	Frame length	Image size	Target size
1	Ground	256	256×256	2×2
2	Sea and sky	280	280×228	2×2
3	Clouds	245	250×200	2×2

The experiment compares the classical particle filtering algorithm and the improved method in this paper from the tracking time per frame and error under three common backgrounds. Define the tracking error as the Euclidean distance between the target centroid in the tracking algorithm and the calibrated target centroid:

$$error = \sqrt{(x - x^*)^2 + (y - y^*)^2} \tag{19}$$

The number of particles in the algorithm is 300.

Video sequence 1 is the background with ground. (a) and (b) in Fig. 4 are the tracking effects of the basic particle filtering algorithm and the improved particle filtering algorithm on the video, respectively, from the 30th, 120th, 180th, and 220th frames of the video. Figure 4(c) is the comparison of the single frame operation time of the two algorithms, and Fig. 4(d) is the comparison of the tracking errors of the two algorithms.

It can be seen from Fig. 4 (a1) (b1) that the basic particle filtering algorithm is disturbed and the centroid shifted during the first few frames of the video, and the particle filter algorithm optimized by krill herd can track the target more stably and has better robustness. It can be seen from Fig. 4(c) that adding particle optimization does not greatly extend the calculation time. The calculation time of the two algorithms is similar, and the basic particle filtering algorithm is slightly faster. It can be seen from Fig. 4(d) that the error of the optimized particle filter algorithm is smaller than that of the basic particle filter algorithm in the middle of the video.

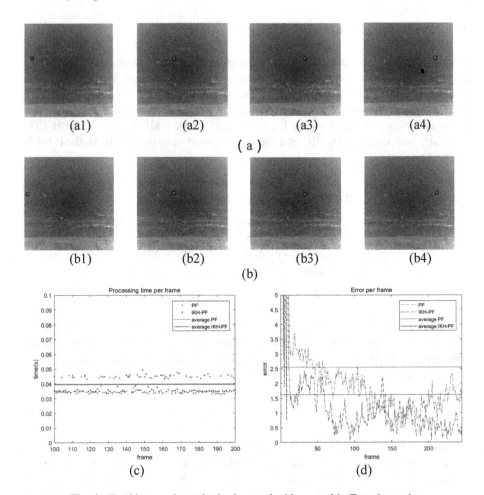

Fig. 4. Tracking results under background with ground in Experiment 1.

Video sequence 2 is the sea-sky background. (a) and (b) in Fig. 5 are the tracking results of the basic particle filtering algorithm and the improved particle filtering algorithm on the video, respectively from the 30th, 120th, 180th, and 220th frames of the video. Figure 5(c) is the comparison of the single frame operation time of the two algorithms, and Fig. 5(d) is the comparison of the tracking errors of the two algorithms.

As can be seen from Fig. 5(a1) (b1), the gray level of the background is similar in the video, so the target characteristics are more obvious, and both algorithms have good performance. Figure 5(c) displays that the calculation time of the two algorithms is similar, and the basic particle filtering algorithm is slightly faster. What's more, Fig. 4 (d) demonstrates that the error of the two algorithms are about the same in the front and the end part of the video, and optimized particle filter algorithm is smaller than that of the basic particle filter algorithm in the middle of the video.

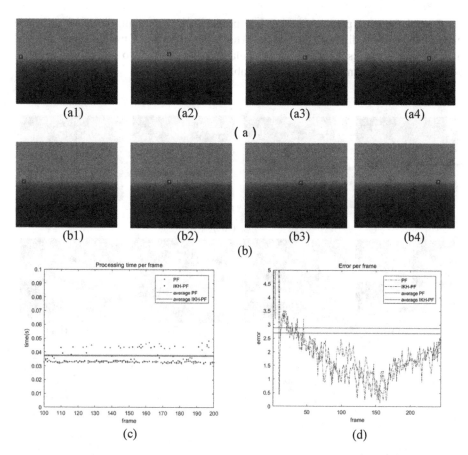

Fig. 5. Tracking results under background with sea and sky in Experiment 2.

Video sequence 3 is a background with clouds. (a) and (b) in Fig. 6 are the tracking results of basic particle filtering algorithm and improved particle filtering algorithm on the video, respectively from the 40th, 80th, 160th, and 200th frames of the video. Figure 6(c) is the comparison of the single frame operation time of the two algorithms, and Fig. 6(d) is the comparison of the tracking errors of the two algorithms.

It can be seen from Fig. 6(a1) (b1) that in backgrounds where the grayscale is close to the target, the elementary particle filtering algorithm suffers some interference and there are cases of loss. The particle filter algorithm optimized by the krill herd can track the target more stable, and the robustness is better. In addition, Fig. 6(c) illustrates that the calculation time of the two algorithms is similar, and the basic particle filter algorithm is slightly faster. Furthermore, the particle filter algorithm with krill herd optimization is highly robust under the background with large interference in Fig. 6(d).

Considering that particle filtering is a probabilistic algorithm, the results of each experiment will be different, so we have run each experiment twelve times and the average of the results is taken. The calculation data of the two algorithms in three backgrounds is shown in the following Tables 2 and 3:

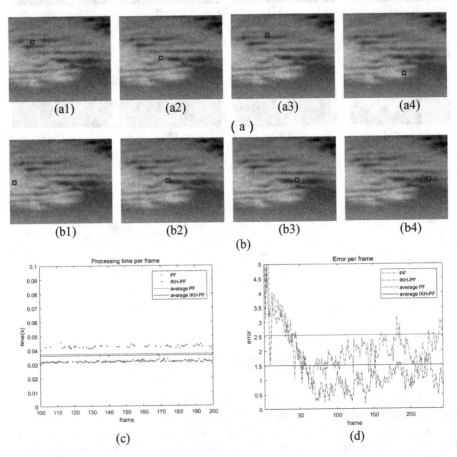

Fig. 6. Tracking results under background with clouds in Experiment 3.

Table 2. Processing time of algorithm

Video sequence	Algorithm	Average processing time(s)	Total processing time(s)	Average error
1	PF	0.0374	9.8401	2.5438
	IKH-PF	0.0398	10.1765	1.5007
2	PF	0.0374	9.0376	3.8052
	IKH-PF	0.0378	9.4603	2.6754
3	PF	0.0362	8.7630	2.5502
	IKH-PF	0.0377	9.6131	1.6163

Table 3. Rate of algorithm

Video sequence	Algorithm	Missing detection rate (%)	False alarm rate (%)	Accuracy rate (%)
1	PF	2.33	3.67	89.33
	IKH-PF	1.67	1.33	92.67
2	PF	4.00	3.33	93.33
	IKH-PF	2.67	2.00	95.00
3	PF	2.67	4.67	87.67
	IKH-PF	1.33	2.00	90.33

4 Conclusions

The above experiments demonstrate that the preprocessing algorithm using grayscale and motion features proposed in this paper can effectively suppress background in complex environments and highlight the small dim target. Then the algorithm uses the particle filter algorithm based on krill herd optimization to track the target. As we can see, although part of the operation speed is sacrificed, the robustness and accuracy of tracking are effectively improved. As a result, the detection of infrared dim targets can be accurately achieved.

References

1. Deng, H., Sun, X., Liu, M., et al.: Entropy-based window selection for detecting dim and small infrared targets. Pattern Recogn. **61**, 66–77 (2017)
2. Song, M., Wang, S., Lü, T., et al.: A method for infrared dim small target detection in complex scenes of sky and ground. Infrared Technol. **40**(10), 996–1001 (2018). (in Chinese)
3. Gao, C., Meng, D., Yang, Y., et al.: Infrared patch-image model for small target detection in a single image. IEEE Trans. Image Process. **22**(12), 4996–5009 (2013)
4. Gao, C., Zhang, T., Li, Q.: Small infrared target detection using sparse ring representation. IEEE Aerosp. Electron. Syst. Mag. **27**(3), 21–30 (2012)
5. Shi, Y., Wei, Y., Yao, H., et al.: High-boost-based multiscale local contrast measure for infrared small target detection. IEEE Geosci. Remote Sens. Lett. **15**(1), 33–37 (2018)
6. Zhang, S., Huang, X., Wang, M.: Algorithm of infrared background suppression and small target detection. J. Image Graph. **21**(8), 1039–1047 (2016). (in Chinese)
7. Han, H., Ding, Y.S., Hao, K.R., et al.: Particle filter for state estimation of jump Markov nonlinear system with application to multi-targets tracking. Int. J. Syst. Sci. **44**(7), 1333–1343 (2013)
8. Han, H., Ding, Y.S., Hao, K.R., et al.: An evolutionary particle filter with the immune genetic algorithm for intelligent video target tracking. Comput. Math Appl. **62**(7), 2685–2695 (2011)
9. Hao, Chen, Caiwen, Ma., Yuecheng, Chen, et al.: Multi dim feature detection and tracking algorithm based on multi feature fusion in complex background. Acta Photonica Sinica **38**(9), 2444–2448 (2009). (in Chinese)

10. Zhang, Z., Jiang, Q., Xu, L.: Airborne single observer passive location based on krill herd immune particle filter. Fire Control Command Control **40**(4), 92 (2015). (in Chinese). https://doi.org/10.3969/j.issn.1002-0640.2015.04.023
11. Guo, W., Gao, Y., Liu, P.: An improved krill herd algorithm with adaptive inertia weight. J. Taiyuan Univ. Technol. **47**(5), 651 (2016). (in Chinese). https://doi.org/10.16355/jcnkiissn1007-9432tyut,2016.05.017
12. Liu, Z., Lu, H., Ren, J.: Co-evolutionary gravitational krill herd algorithm. Adv. Eng. Sci. **50**(6), 217 (2018). https://doi.org/10.15961/j.jsuese.201800012
13. Hui, B.W., Song, Z.Y., Fan, H.Q., et al.: Weak and small aircraft target detection and tracking data set in infrared sequence images. [DB/OL]. Sci. Data Bank (2019). https://doi.org/10.11922/csdata.2019.0074.zh. Accessed 28 Oct 2019

Target Recognition Framework and Learning Mode Based on Parallel Images

Zihui Yin[1], Rong Meng[1], Zhilong Zhao[1], He Yin[1], Zhedong Hu[2], and Yongjie Zhai[2(✉)]

[1] Ltd. Maintenance Branch, State Grid Hebei Electric Power Co., Shijiazhuang 050070, China
yinzihui530@sohu.com, mrr311@163.com, 401448030@qq.com, 245802737@qq.com
[2] Department of Automation, North China Electric Power University, Baoding 071003, China
{201602030206, zhaiyongjie}@ncepu.edu.cn

Abstract. In the application of deep learning algorithms based on large-scale data sets, some problems, such as insufficient samples, imperfect sample quality, and high cost of building large data sets, emerge and restrict algorithm performance. In this paper, a target recognition framework and learning mode based on parallel images are proposed, and the application verification is carried out by taking the insulator target recognition in the transmission line as an example. This paper uses the artificial image generation technology to establish the insulator data set NCEPU-J, and then proposes the target recognition framework PITR and three learning modes, namely OriPITR, TrsPITR, and MutiPITR. The insulator strings with the piece number of 7, 11 and 14 are verified, and the recognition accuracy is significantly improved. The results show that the target recognition framework and learning mode based on parallel images are feasible and effective.

Keywords: Insulator recognition · Parallel image · PITR (Target recognition based on parallel Image)

1 Introduction

As the key component for electrical insulation, wire and cable connection, and physic support in transmission lines, insulators are of great importance to the whole system and are the vital contents of transmission line inspection. In the insulator recognition

---

This work was supported in part by Research on Intelligent Video Analysis Technology and System in Unattended Substation Based on Artificial Intelligence under grant number kj2019-036, by the National Natural Science Foundation of China (NSFC) under grant number 61871182, 61773160, by Beijing Natural Science Foundation under grant number 4192055, by the Natural Science Foundation of Hebei Province of China under grant number F2020502009, by the Fundamental Research Funds for the Central Universities under grant number 2018MS095, 2020YJ006, and by the Open Project Program of the National Laboratory of Pattern Recognition (NLPR) under grant number 201900051.

Y. Wang et al. (Eds.): IGTA 2020, CCIS 1314, pp. 181–192, 2020.
https://doi.org/10.1007/978-981-33-6033-4_14

process, there exist some problems such as insufficient samples, imperfect sample quality, and high cost of constructing large data sets [1]. Traditional image processing technologies mainly focus on feature extraction and recognition of image edge, corner point, line and contour of insulator samples. The main techniques used in image processing include image matching, feature description, morphology and image segmentation. Among them, image matching methods use image matching algorithm combined with insulator features to carry out feature index and realize insulator recognition [2]. Feature description methods mainly make use of target color [3, 4], texture [5], contour [6] and other features of the insulator, and set the threshold value. It combines the image processing algorithm to realize insulator recognition. Morphological methods and image segmentation algorithms are based on morphological algorithm combined with image segmentation algorithm [7, 8]. Yang et al. used LBP (Local Binary Patterns) and HOG (Histogram of Oriented Gradients) algorithms for feature extraction, and then adopted AdaBoost classifier for target recognition [9]. Zhai et al. extracted the image features using the LSD (Line Segment Detector) algorithm, and then realized the recognition of insulator string using cluster analysis and Ada-Boost classifier [10]. Zhang et al. used SSDA (Sequential Similarity Detection Algorithm) algorithm for image processing of test samples and training samples, and used gray invariance and rotation invariance of Harris operator for corner detection to recognize the insulators [11]. With the large-scale application of artificial intelligence in the field of image processing, it provides the direction for us to conduct insulator recognition.

With the development of image processing technology based on machine learning and deep learning, the automatic extraction of features and training and learning of samples can be realized for insulator recognition s. Zhou established an insulator identification and detection model by using the Faster R-CNN network model [12]. Liu et al. established a transmission line target recognition and detection model using the improved SSD method [13]. By comparing the structures of R-CNN, Fast R-CNN and Faster R-CNN, Huang et al. further studied the specific application of Faster R-CNN in the insulator recognition and detection [14]. Although these researches made some improvement in the algorithm for the insulator recognition, but the whole frame combining algorithm and data was not considered. Therefore, for some specific goals such as insulators, the improvement in model to enhance the recognition accuracy, expand the sample set, make advantage of deep learning algorithm under the condition of insufficient samples and low sample quality is still a focus of research.

Wang and his team proposed the parallel vision theory [15, 16] and made some breakthroughs in the application of artificial intelligence scenarios [17, 18], which strongly confirmed the great application value and prospect of artificial samples in the field of deep learning. By introducing the concept of parallel image [19], this paper studies the problem of small sample recognition for specific targets from the perspective of combining the algorithm with data. The target recognition framework and various learning modes are then established according to the principle of parallel images. With the research on insulator as a specific object, the feasibility and validity of target recognition framework and learning mode based on parallel images are verified. It is expected to provide new models and methods to broaden the research directions of low-sample learning and zero-sample learning (Fig. 1).

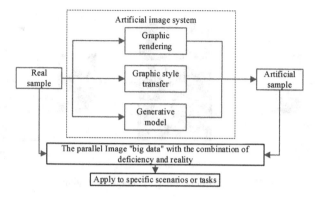

Fig. 1. Flow chart of parallel images

2 Establishment of Artificial Image Data Set

2.1 Artificial Image Generation Method

The insulator parameters of the model are determined according to the national insulator standard (GB/T 1386), and the model meshing is carried out by using 3DS MAX. The meshed models with different piece numbers are shown in Fig. 2 which are exported into OBJ format. Input the OBJ files of different insulator models into KeyShot software to render.

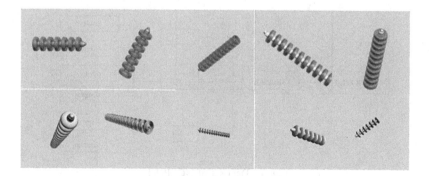

Fig. 2. Meshed insulator models

In the rendering of insulators, depending on the real-time rendering function of KeyShot software, different materials (glass, ceramics, etc.), different number of pieces (7, 11, and 14), and different backgrounds (forest, road, city, building, seaside) are rendered and animated. Through the interception of each frame of animation by KeyShot software, a total of more than 50,000 artificial images are produced (Fig. 3).

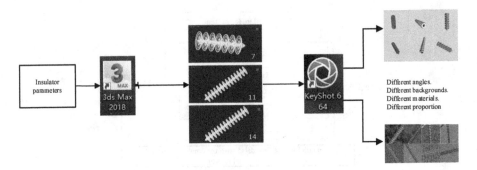

Fig. 3. Artificial image generation system

2.2 Data Set Structure

Based on the sample data establishment method of parallel image principle and after screening more than 50,000 artificial images, an insulator data set NCEPU-J is established. The NCEPU-J data set in Fig. 4 includes insulator sample data of different materials with different pieces in different environments. There are 10,000 actual samples and 12,500 artificial samples. The data set is stored in the classification folder for insulator classification identification with different pieces. J7, J11, and J14 are the sub-folders with a total of 22,500 images. For each category, there are 7,500 images 2500 with a single background and 5000 with a complex background. The 22500 images are named from J1 to J22500.

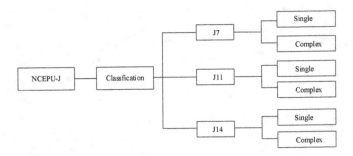

Fig. 4. Data set of NCEPU-J

3   Target Recognition Framework Based on Parallel Images

Based on deep learning and artificial sample generation technology, this paper proposes a target recognition framework based on parallel image (PITR), as shown in Fig. 5. On the basis of fusing large-scale specific target data set of artificial images and real images, ITR framework is equipped with various neural network models such as CNN, RNN, GNN and Faster R-CNN. It contains the evaluation correction module and a variety of learning modes are derived based on it, which provide solutions for transfer learning, small sample learning and multi-task learning.

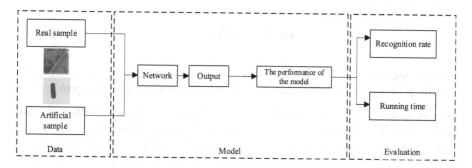

Fig. 5. Framework of PITR

Three learning modes are designed based on the PITR framework, namely OriPITR (see in Fig. 6); TrsPITR (see in Fig. 7), MutiPITR (see in Fig. 8).

OriPITR. OriPITR mode incorporates the idea of parallel images, and applies it to the data input layer. In the early stage of input, the two samples with different properties are randomly scrambled, and then the mixed sample set is input into the neural network model, so as to enrich the sample data set to solve the problem of neural network model overfitting or identification accuracy decline due to the small number of specific target samples. This pattern mainly addresses the single-task problem. The network module shown in Fig. 6 is one of the neural networks which can be either a classification model or a detection model.

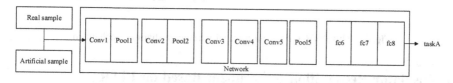

Fig. 6. OriPITR mode

TrsPITR. TrsPITR mode reflects the idea of parallel images from another perspective. First of all, traditional image transformation processing (including image rotation, mirroring, compression, resolution reduction, etc.) is carried out on actual samples. After increasing the number of actual images, they enter into the data input layer and the network module is trained. At the same time, the artificial samples generated by the artificial image generation system are sent into the data input layer. On the premise of maintaining the robustness of network modules, the number of artificial samples will be increased continuously until an index is found to make the comprehensive performance of network modules with the best rapidity and robustness, and then the network module

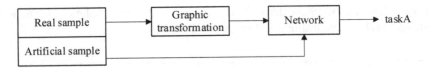

Fig. 7. TrsPITR mode

training stops. If the module is a classification model, it is aimed at the classification problem. If the module is a detection model, it is aimed at the detection problem.

MutiPITR. MutiPITR mode combines the parallel the image idea with multi-task learning mode to solve the multi-task learning problem with few samples of specific targets. In this mode, part of it adopts OriPITR, and the other part is the traditional deep learning training mode. Specifically, the actual samples are used to train Network A while the actual samples and artificial samples are combined to train the Network B, and parameters are shared between these two modules. The parameter sharing mechanism is shown in Fig. 8(b). The processing of tasks, task A and task B, is realized through the parameter sharing of the convolutional layer and pooling layer within the two network modules. Task A and Task B can be both time detection tasks and classification tasks.

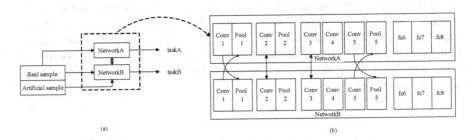

Fig. 8. MutiPITR mode and internal parameter sharing mechanism

4   Insulator Recognition Experiment Based on PITR Framework

4.1   Experiment

In order to realize the classification of different number insulators, the target recognition framework PITR based on parallel image is adopted in this paper. The data from the NCEPU-J data set were used to classify the insulator strings with 7, 11 and 14 pieces. The sample increase experiment was set up, and the sample number is shown in Table 1 to observe the insulator recognition effect in the following two situations:

a) The number of total samples increases, and the proportion of training samples in total samples increases.

b) The ratio of the actual samples to the artificial samples remains unchanged, the number of total samples increases, and the proportion of the training sample increases.

Table 1. Insulator sample setting

Total sample number	Actual sample number	Artificial sample number	Ratio of actual sample to artificial sample
3000 * 3	2000 * 3	1000 * 3	2:1
4000 * 3	3000 * 3	1000 * 3	3:1
5000 * 3	4000 * 3	1000 * 3	4:1
6000 * 3	4000 * 3	2000 * 3	2:1
6000 * 3	5000 * 3	1000 * 3	5:1
7500 * 3	5000 * 3	2500 * 3	2:1

By the obtained recognition results, the feasibility of the three learning modes proposed in this paper, OriPITR, TrsPITR and MutiPITR, is analyzed.

First, the total number of samples was increased from 3000 * 3 to 7500 * 3 (six cases in total, shown in Table 1), and the proportion of training samples in the total number of samples increased from 50% to 90% to observe and obtain its accuracy and loss curves under different conditions. At the same time, when the ratio of the actual samples to artificial samples is 2:1 and the total samples are 3000 * 3, 6000 * 3 and 7500 * 3, the proportion of training samples in the total samples was increased from 50% to 90% to obtain the test accuracy and loss curve. The effects of different expansion ratio and training ratio on network performance were analyzed by comparing the six cases. The number of positive samples is 22,500, all of which are images with insulators, while the negative samples are images without insulators. The proportion of positive and negative samples in any group is 1:1. The samples were selected after random shuffles. Figure 9 is the picture of some insulators.

Fig. 9. Images of insulators

The insulator recognition algorithm is adopted, and the specific process is as follows:

(1) Load the NCEPU-J data set and mark all insulator samples in the data set to generate corresponding labels.
(2) Calculate the number of insulator images in each category.
(3) The proportion factor of training sample size to total sample size is set, and the data set is divided into training data set and test data set.
(4) Set the image pixel and convert all samples into a unified pixel value of 512 * 512 * 3.
(5) Define a convolutional neural network. The training option is stochastic gradient descent with momentum (SGDM), with an initial learning rate of 0.01. The output is of category 3, and the loss is cross entropy loss.
(6) Preserve the trained neural network.
(7) The test images are classified and the average accuracy is calculated. The average accuracy is the percentage of the correctly classified images verified by the classification network.
(8) Select the insulator string with the piece number of 7, 11 and 14 for verification and display the test results (Fig. 10).

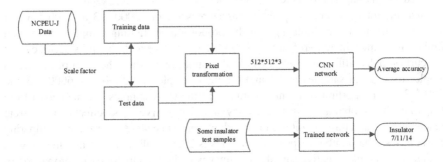

Fig. 10. Flow chart of classification algorithm

4.2   Structure of Sample Classification Network

In this paper, the convolutional neural network (CNN) is adopted as the classification network structure. The input layer is an insulator image of 512 * 512 * 3, and then a convolutional layer is connected. The filter size is 3 * 3, the number of filters is 8, and the default step is 1 to ensure that the space output size is the same as the input size. It is followed by a normalization layer, Relu nonlinear activation function, and a maximum pooled subsampling layer. The above sampling layer from convolutional layer to maximum pooling is regarded as a small module. After this module, a module with the same structure and the filter number of 16 is accessed, then a set of modules consisting of a filter 3 * 3 is connected, consisting of a convolutional layer with the number of 32, a mass normalization layer and a Relu nonlinear activation function layer. At last, the full connection layer, SoftMax layer and classification layer are put in. The output of the full connection layer is 3. The specific structure is demonstrated in Fig. 11.

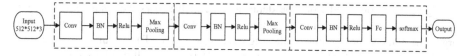

Fig. 11. Structure of classification network

4.3 Results and Discussion

After the classification experiment, with the increasing proportion of training samples in the total sample number, the results of insulator classification accuracy with pieces of 7, 11 and 14 are listed in Table 2.

Table 2. Results of insulator classification experiment accuracy under different sample numbers and training ratios

Total number of samples	Proportion of the total number of training samples				
	50%	60%	70%	80%	90%
3000 * 3	65.4%	69.2%	74.3%	78.2%	76.5%
4000 * 3	67.3%	71.2%	75.3%	78.6%	77.1%
5000 * 3	69.2%	79.3%	78.2%	80.3%	77.3%
6000 * 3	70.4%	78.5%	80.2%	85.3%	82.6%
6000 * 3	72.6%	77.3%	81.4%	88.6%	84.5%
7500 * 3	74.3%	79.6%	83.5%	90.1%	88.3%

Figure 12 is obtained from the data in Table 2. It shows that when the total sample number is the same, the accuracy with the training sample proportion of 80% is higher. This shows that there is a game process between the test sample and the training sample on the network model. It is necessary to explore a balanced range to achieve better training results. In addition, it can also be seen that the training accuracy is improved with the increase of the total sample number when the proportion of training samples remains unchanged. It confirms the prior knowledge that the deep learning model is generally based on the training of large data sets. On the other hand, it also indicates that the OriPITR model proposed in this paper based on the principle of parallel images has a good effect.

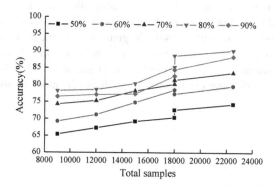

Fig. 12. Accuracy of different training sample numbers

Figure 13 is obtained from the data in Table 2. It shows the changes of training accuracy with the increase of total sample number at different training proportions. The larger the total sample number, the higher the accuracy. The accuracy of the total sample number of 18,000 (2:1) exceeds that of 18,000 (5:1). It can be seen that when the total sample number is the same, when the real sample number is approaching the artificial sample number, the accuracy decreases. It indicates that the number of real samples is still the main part in the network training process, and artificial samples will damage the robustness of the network, but accelerate the speed of network training to some extent. The rationality of TrsPITR model is verified from the side.

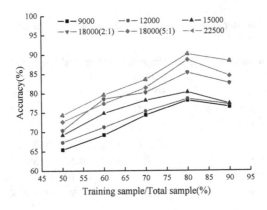

Fig. 13. Accuracy of different total sample numbers

Figure 14 is obtained from the data in Tables 1 and 2. It shows the change of the accuracy of the network model with the increase of training samples when the ratio of real samples and artificial samples remains the same (2:1). It can be clearly seen that when the total sample number increases, the model accuracy improves; the training sample proportion increases, so does the model accuracy. It is consistent with the analysis results on Figs. 12 and 13. In addition, the line chart with a total sample number of 9,000 and a total sample number of 22,500 are relatively parallel, indicating that the network model trained under the two sample numbers has a certain degree of similarity. It provides some ideas for target recognition based on the principle of parallel image; that is to train the network model on the large data set of artificial image expansion produced by the artificial sample generation framework, which offers the possibility of knowledge migration applied on small data. In this case, the training parameters of the two network models can be exchanged to provide reference for multi-task learning, thus confirming the feasibility of MutiPITR mode to some extent.

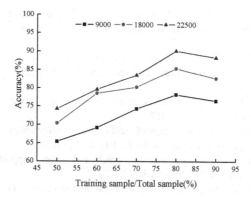

Fig. 14. Accuracy under the same ratio of actual sample number to artificial sample number

5 Conclusion

(1) 3DS MAX was used to mesh the insulator and generate artificial sample images. By using KeyShot software to intercept each frame of the image, a total of more than 50,000 artificial images were produced. After image screening and cleaning, the insulator image data set NCEPU-J was established.

(2) A target recognition framework PITR based on parallel image theory was proposed, and three learning modes based on PITR framework, OriPITR, TrsPITR and MutiPITR, were addressed.

(3) Three kinds of insulators with different number of pieces were identified by experiment, and the feasibility and effectiveness of the three learning modes were verified by designing insulator recognition experiments with different sample numbers and training ratios. Further research and implementation of the three learning modes are the key jobs for the next step.

(4) With the help of the recognition method of insulators, this paper not only provides a new model and a new method for broadening the field of deep learning and computer vision, but also provides a reference for speeding up the construction of "three types and two networks" and improving the intelligence level of the power industry.

References

1. Tong, W., Yuan, J., Li, B.: Application of image processing in patrol inspection of overhead transmission line by helicopter. Power Syst. Technol. **34**(12), 204–208 (2010)
2. Zhang, G., Liu, Z., Han, Y., et al.: A fast matching method of fault detection for rod insulators of high-speed railways. J. China Railway Soc. **35**(5), 27–33 (2013)
3. Zhai, Y., Wang, D., Zhang, M., et al.: Fault detection of insulator based on saliency and adaptive morphology. Multimedia Tools Appl. **76**(9), 1–14 (2017)
4. Zhang, M.: Recognition and fault detection of the key components in transmission lines based on aerial images, North China Electric Power University (2018)

5. Wang, W., Wang, Y., Han, J., et al.: Recognition and drop-off detection of insulator based on aerial image. In: International Symposium on Computational Intelligence and IEEE, pp. 162–167 (2017)
6. Li, B.F., Wu, D.L., Cong, Y.: A method of insulator detection from video sequence. In: International Symposium on Information Science and Engineering, Shanghai, China, pp. 386–389 (2012)
7. Wang, D.: Research on insulator detection and location methods in aerial images, North China Electric Power University (2017)
8. Ge, Y., Li, B., Liang, S.: Application of mathematical morphology to edge detection of insulator image. High Volt. Apparatus 48(1), 101–109 (2012)
9. Yang, H.: Detection of catenary insulator crack and locate supporter based on image processing, Southwest Jiaotong University (2017)
10. Zhai, Y., Wang, D., Zhang, M.: Recognition of insulator string based on cluster analysis and AdaBoost algorithm. Sens. World 22(09), 7–11 (2016)
11. Zhang, G., Liu, Z.: Fault detection of catenary insulator damage/foreign material based on corner matching and spectral clustering. Chin. J. Sci. Instrum. 35(06), 1370–1377 (2014)
12. Xin, Z.: Research on fault detection method of power insulator based on deep learning, Shaanxi University of Science and Technology (2019)
13. Liu, Y., Wu, T., Jia, X., et al.: A multi-scale target detection method for transmission lines. Instrumenttion 26(01), 15–18 (2019)
14. Huang, W., Zhang, F., Li, P., et al.: Exploration and application on faster R-CNN based insulator recognition. Southern Power Syst. Technol. 12(09), 22–27 (2018)
15. Wang, K.-F., Gou, C., Wang, F.-Y.: Parallel vision: an ACP-based approach to intelligent vision computing. Acta Automatica Sinica 42(10), 1490–1500 (2016)
16. Li, L., Lin, Y.-L., Cao, D.-P., et al.: Parallel learning-a new framework for machine learning. Acta Automatica Sinica 2017(01) (2017)
17. Tian, Y.L., Xuan, L., Feng, W.K., et al.: Training and testing object detectors with virtual images. IEEE/CAA J. Automatica Sinica 5(02), 154–161 (2018)
18. Yue, F.W., Chao, G., Rehg, J.M., et al.: Parallel vision for perception and understanding of complex scenes methods, framework, and perspectives. Artif. Intell. Rev. 48(3), 299–329 (2017)
19. Wang, K., Lu, Y., Wang, Y., et al.: Parallel imaging: a new theoretical framework for image generation. PR&AI 30(07), 577–587 (2017)

Crowd Anomaly Scattering Detection Based on Information Entropy

Fujian Feng[1(✉)], Shuang Liu[1], Yongzhen Pan[2], Qian Zhang[1], and Lin Wang[1]

[1] College of Data Science and Information Engineering,
Guizhou Minzu University, Guiyang 550025, China
fujian_feng@gzmu.edu.cn
[2] College of Software Engineering, South China University of Technology,
Guangzhou 510006, China

Abstract. Crowd anomaly scattering detection is an important indicator in crowd anomalous events and plays an important role in scene event determination. At present, the special detection methods of abnormal scattering behavior cannot fully mine and make use of the specific information, and the accuracy of these methods is not sensitive to the rapid capture of abnormal scattering behavior. In this paper, we propose a crowd anomaly scattering detection method based on motion velocity and information entropy. In addition, in order to accurately determine the degree of crowd anomalous scattering, we propose a new evaluation method for crowd anomalous scattering. Finally, the performance of the proposed method is verified through UMN dataset. The experimental results show that the proposed method can accurately detect the abnormal scattering behavior in the medium-low density scenarios.

Keywords: Crowd anomaly scattering · Motion velocity · Information entropy · Anomaly detection · Crowd anomaly evaluation

1 Introduction

With the rapid development of intelligent security system [1], intelligent community [2], and intelligent city [3], the analysis of crowd abnormal behavior has become a hot spot of crowd events research [4]. At present, intelligent monitoring has been increasingly applied in public places such as stations, hospitals, campuses and shopping malls. The scattered panic is one of the most common abnormal phenomena in the crowd, but there are relatively few methods to detect the abnormal scattering. It is very important to make full use of the existing intelligent monitoring and other resources, to monitor the overall abnormalities of the crowd by abnormal scattering detection, and to protect the safety of people's life and property in time.

In the past decade, some research achievements have been made in the analysis of crowd abnormal behavior. For example, Boominathan et al. [5] designed a bilinear neural network to combine the deep and shallow information of the crowd. It did well in crowd counting under large scale variations and severe occlusion. Zhang et al. [6] proposed a Multi-column Convolutional Neural Network (MCNN) to map the image to

© Springer Nature Singapore Pte Ltd. 2020
Y. Wang et al. (Eds.): IGTA 2020, CCIS 1314, pp. 193–205, 2020.
https://doi.org/10.1007/978-981-33-6033-4_15

its crowd density map, which overcame the impact of arbitrary perspective and size variations. Cao et al. [7] used a encoder-decoder network, called Scale Aggregation Network (SANet), for extracting multi-scale features. Then it could generate high-resolution density maps. Li et al. [8] proposed an easy-trained end-to-end network called CSRNet to perform accurate count estimation under Congested Scene. In terms of abnormal behavior detecting and analysis, Singh et al. [9] proposed ScatterNet Hybrid Deep Learning (SHDL) to estimate the human pose. The orientations between the limbs of the estimated pose are next used to identify the violent individuals. Rabiee et al. [10] combined the simplified Histogram of Oriented Tracklets (sHOT) with Dense Optical Flow (DOF) to form a novel abnormal behavior descriptor. It is shown to be very effective in detecting crowd abnormal behavior in crowded scenes. Wang et al. [11] proposed a framework for fall detection based on automatic feature learning methods. Arroyo et al. [12] built an expert video-surveillance system, which focus on the real-time detection of potentially suspicious behaviors in shopping malls. Arivazhagan et al. [13] used the dense trajectory features, Short term biometric features and texture features to judge the loitering behavior in crowd. This method could effectively distinguish the offensive loiter and innocuous loiter. To sum up, although there are some achievements in the research of crowd density anomaly analysis, abnormal trajectory detecting and tracking, abnormal behavior detecting and analysis, the works in crowd anomaly scattering are few.

The crowd anomaly evaluation method is the key of correctly detecting abnormal situations. So far, scholars at home and abroad have proposed some methods for abnormal behavior detection. For example, Marsden et al. [14] proposed an anomaly detection method based on mean motion speed and crowd density, which used Gaussian Mixture Model (GMM) and Support Vector Machine (SVM) to evaluate the crowd anomaly. Bera et al. [15] presented a real-time anomaly detection algorithm based on trajectory level behavior learning. This algorithm takes the distance between the local pedestrians' trajectory features and the global trajectory features as the anomaly metric, and it is suitable to detect the local anomaly in medium or low-density scenes. Zhao et al. [16] used a method based on the evolution of spatial position relationship (ESPR) features to detect crowd spread behavior and crowd gather behavior. Hao et al. [17] extracted the Gabor features from videos to represent the crowd motion patterns and used an enhanced gray level co-occurrence matrix model to evaluate the abnormal behaviors. Although above methods have been put into practice as evaluation criteria, there are few achievements on the evaluation methods of the crowd anomaly scattering.

In this paper, we propose a new approach to detect the crowd anomaly scattering. This method selects Yolo v3 and Deep-sort to construct the multi-target detection and tracking model. Based on the model, the speed and direction information are taken as the input values of the abnormal scattering detection model. The evaluation index value of the chaotic degree of crowd is calculated by the abnormal scattering detection model, and the evaluation index value of the chaos degree of the crowd is judged. If the evaluation index value of the chaos of the crowd exceeds the threshold value, the abnormal warning of the crowd will be made. In order to improve the accuracy of scattering detection, this paper proposes a new evaluation index of the degree of crowd anomaly scattering based on the velocity of motion and information entropy. Finally, experiments are carried out on UMN datasets to verify the effectiveness of our method.

2 Multi-target Detection and Tracking

In this section, we will introduce the related work about multi-target detection and tracking. Multi-target detection and tracking is defined as a comprehensive technique which integrate object detection with multiple object tracking (MOT) to locate a category of objects while maintaining a unique identity for each of the targets over the time [18]. According to the number of tracking cameras, we can divide the multi-target detection and tracking methods into approaches worked on single camera and approaches worked on multi-camera. Our own approach focuses on multi-target detection and tracking based on one camera. In order to construct the efficient and reliable pipeline (see Fig. 1), we choose Yolo v3 as the detector and Deep-sort as the tracker. The detailed description of Yolo v3 and Deep-sort will be given in the next two paragraphs.

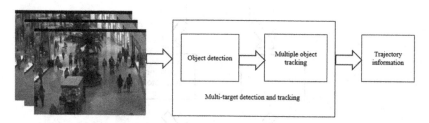

Fig. 1. Multi-target detection and tracking pipeline.

Yolo v3 algorithm is the most popular one-stage object detection model in target detection field. Unlike previous methods that regard object detection as a classification problem, Yolo's innovation is to use the border as a regression task and solve it with a single neural network. The whole flow path of Yolo v3 is as follows (Shown in Fig. 2). At the beginning, Yolo system splits the given video stream into some single images. Then the single images are resized to 416 * 416 * 3 and transported into Yolo v3's convolution neural network (CNN) as input data. After the process of CNN, we get the bounding boxes and the class probabilities of the objects. It is worth mentioning that Yolo v3 is one of the few object detection methods which have both high speed and great accuracy at the same time.

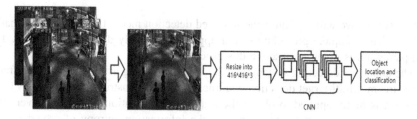

Fig. 2. The work flow of Yolo v3 algorithm.

In the past several years, online multiple object tracking attracts lots of attention in multi-target tracking field. Deep-sort is a Simple Online and Realtime Tracking (SORT) algorithm improvement that reach a high performance. The work flow of Deep-sort is as follows (Shown in Fig. 3). Firstly, it takes the output of object detection model (the location of pedestrians) as input. When the input of a new frame comes, a standard Kalman filter with constant velocity motion and linear observation model is used to predict the possible location of every pedestrian. Then an appearance descriptor which is a pre-trained CNN extracts the appearance feature of previous tracks' bounding boxes and current frame's bounding boxes. Finally, Deep-sort utilizes both the location and appearance information to associate current positions and previous frames.

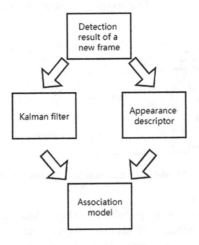

Fig. 3. The work flow of deep-sort.

Multi-target detection and tracking is the foundation of our scattered detection method of crowd abnormality. Through processing the trajectory information provided by multi-target detection and tracking model, velocity and direction of pedestrians can be calculated and passed to the abnormality detection model as input.

3   Scattered Detection Method of Crowd Abnormality

In this section, we will introduce the scattered detection method of crowd abnormality in detail. The complete procedure of our approach is clearly displayed in Fig. 4. The next paragraph is given to describe this procedure.

The abnormal detecting system gets the video streaming as input. For every frame in the stream, Multi-target detection and tracking model constructed with Yolo v3 and Deep-sort is firstly applied to obtain the trajectory information of crowd. Leveraging the trails, the movement speed factor and the information entropy of crowd can be calculated. The evaluation index of crowd chaos takes both the movement speed factor and information entropy into consideration. Then a judgement is used: comparing the index we just got to the threshold we set. If it exceeds the threshold, Abnormal warning will be triggered. If not, continue to the next frame.

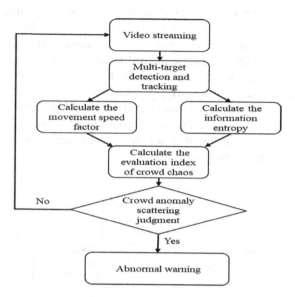

Fig. 4. Procedure of scattered detection method of crowd abnormality

3.1 Movement Speed Factor of the Crowd

Movement speed is a significant feature of crowd, that frequently used to reflect the energy contained in crowd. Only when the scalar and directional differences of velocity are large can the crowd disorder be reflected, it is recognized that due to the high speed, the amount of crowd will reach a high level. Therefore, in this paper, we introduce a measure called movement speed factor.

We will illustrate the research ideas of movement speed factor in this paragraph. To begin with, we denote the number of people as N and the speed of pedestrian j as v_j, the average speed of crowd in time t can be mathematically described as:

$$\bar{v}_t = \frac{\sum_{j=1}^{N} v_j}{N} \tag{1}$$

In order to normalize the speed information as well as effectively distinguish running behavior from walking behavior, a sigmoid function is employed in calculating the movement speed factor in time t:

$$V_t = Sigmoid(\gamma \bar{v}_t) \tag{2}$$

γ is the compressibility factor which satisfies the equation:

$$\gamma * k = 0.5 \tag{3}$$

k is the boundary value between running speed and walking speed that we set and its value can be adjusted. In most cases, k is always set to be 5 km/h.

This paragraph will introduce the process of crowd movement speed factor (Shown in Fig. 5). Since the Multi-target detection and tracking model has got the trajectory information of every pedestrian, we can take the length of trajectory vector in a unit time to represent the speed. Then the average speed of crowd can be gained by formula (1). Through setting the boundary value k, the compressibility factor γ can be got by formula (3). Finally, we use the formula (2) to compute the crowd movement factor of time t. Finishing the introduction of crowd movement speed factor, next section will focus on the information entropy of crowd.

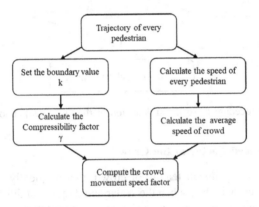

Fig. 5. The process of crowd movement speed factor.

3.2   The Information Entropy of the Crowd

In thermodynamics, entropy is a physical quantity to measure the disorder degree of molecular motion. Besides, entropy can also work on interpreting the chaotic state of other things. In order to solve the problem of quantitative measurement of information, C. E. Hannon borrowed the concept of entropy from thermodynamics and applied it to the problem of information quantization, which is called information entropy, which is used to eliminate random uncertainty. In this section, we will introduce how we apply the information entropy to our method.

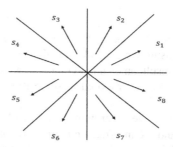

Fig. 6. The vectors are classified into 8 types.

In this paragraph, we will introduce the calculation formula of information entropy. Firstly, d_i represents the motion vector of pedestrian I and the vectors are classified into 8 types $s_j (j = 1, \ldots 8)$ according to their direction (Shown in Fig. 6). When given a vector, we use $II\{d_i \in s_j\}$ to judge whether the vector d_i belongs to type s_j. If d_i is part of s_j, $II\{d_i \in s_j\} = 1$. If not, $II\{d_i \in s_j\} = 0$. Based on the settings above, P_j, which represents the ratio of the number of vectors belonging to s_j to the total quantity of vectors n, can be mathematically described as:

$$P_j = \frac{\sum_{i=0}^{n} II\{d_i \in s_j\}}{n}, \quad (j = 1, \ldots 8) \tag{4}$$

Then, the information entropy of crowd in time t is computed as follow:

$$W_t = -\sum_{j=1}^{8} P_j \cdot \log P_j \tag{5}$$

This paragraph will be described the process of using information entropy to obtain the trajectory information of each pedestrian. We use the start position and end position of a pedestrian in a unit time to construct the motion vector. After that, we divide the vectors into 8 types as the description in the last paragraph. The next step is to calculate the P_j by formula (4). Finally, we can get the information entropy by formula (5). So far, we have obtained the movement speed factor and the information entropy of crowd. In next section, we will illustrate how to utilize such information to evaluate the state of crowd.

3.3   Evaluation Method

The Crowd anomaly scattering is a kind of crowd abnormal events that contains several abnormal characteristics. If using the general anomaly detection methods, it is likely to have a miscalculation. As a result, an evaluation method which focus on the features of crowd anomaly scattering must be proposed to solve it. Next, we will introduce our evaluation method by details.

In the sections above, we have given introduction and detailed description to the movement speed factor V_t and the information entropy W_t of the crowd. Here, we will introduce the evaluation index of crowd chaos. The evaluation index of crowd chaos L_t is used to display the disorder degree of the crowd in time t, and it can be calculated as:

$$L_t = \alpha W_t + (1 - \alpha) V_t \tag{6}$$

where α is an adjustable parameter which is usually set to be 0.5. If more consideration is be taken on information entropy, it could be set higher. Conversely, it would be set lower when the movement speed factor is the focus point. Then, we compare L_t with the threshold S:

$$Abnormaly = \begin{cases} 1, & L_t \geq S \\ 0, & L_t < S \end{cases} \tag{7}$$

where 1 denotes crowd anomaly scattering, o denotes that there is no abnormality in the crowd, threshold S of a scene is an empirical value which can be got by doing some experiments in this scene. Through our experiments, the threshold S is found to be a number up and down at 0.7. If L_t exceeds the threshold, it will be considered that the crowd anomaly scattering happens.

The process of evaluation method will be introduced in this paragraph. Through the processing flow of the previous sections, we have got the movement speed factor and the information entropy of the crowd. The next step is to setting the adjustable parameter and the value of threshold. The adjustable parameter is a variable that changes according to the personal understanding on the abnormal situations. Threshold is an empirical value that should be learnt from the experiments in the specific scenario. Then the evaluation index of crowd chaos can be computed by formula (6). The final step is to make the comparison of formula (7) and trigger the abnormal warning if the evaluation index of crowd chaos exceeds the threshold.

4 Experiments and Analysis Results

In this section, we will introduce the experiments to support that our crowd anomaly scattering detection method can work in different real scenes.

4.1 Experiments Setup

Experiment assesses the performance of our crowd anomaly scattering detection method on UMN dataset. The UMN dataset consists of several scenarios in which some crowd anomaly scattering events happen. Three scenes in this dataset are chosen to verify our method:

Scene A: Some people walk on the lawn, and finally scatter.
Scene B: Some people walk in the corridor, and finally scatter.
Scene C: Some people walk in the square, and finally scatter.

Besides the difference in where the scattering event happens, there are some other differences between three scenes. In Scene A, the speed and motion direction of crowd are maintained at normal level. In Scene B, most person stay still in the corridor. In Scene C, people walk round and round in the square. Due to these differences, the boundary value k, adjustable parameter α and threshold S are set according to specific situation as follows (Table 1):

Table 1. The settings in three scenes

Scene	k	α	S
A	5	0.50	0.65
B	5	0.50	0.65
C	5	0.30	0.70

The experimental computer is equipped with one Intel(R) Core (TM) CPU i7-10710u 1.10 GHz, 16 GB memory, one MX 250 GPU and Windows 10 OS.

4.2 Experimental Analysis Based on Three Scenarios

The experiment results will be displayed in the form of line charts. And the evaluation is carried out according to the following two metrics:

- Accuracy rate: The abnormal frames that our method finds divide all the abnormal frames.
- Misjudgment rate: The number of normal frames that are judged as abnormal frames divide all the frames judged to be anomalous.

Three-line charts are used to record the results of three scenes separately. Figure 7 shows the experiment results of Scene A, B and C. The solid line in each chart represents how the evaluation index of crowd chaos changes with frames. The straight dotted line is the threshold line. All the frames which stride over the threshold line are detected as abnormal frames. Figure 8 shows the normal and abnormal images of three scenes.

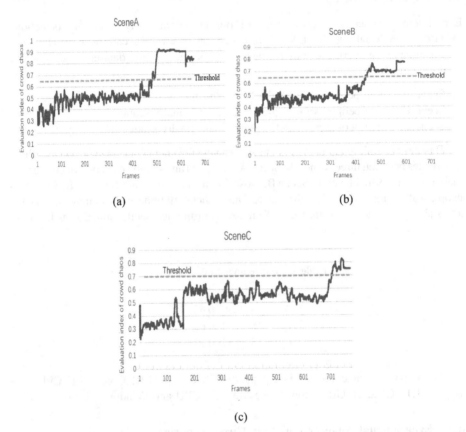

Fig. 7. Experimental results of three different scenarios with UMN dataset. (a) Experiment results of Scene A, (b) experiment results of Scene B, (c) experiment results of Scene C

The accuracy rate and misjudgment rate of three scenes are shown in Table 2. Three-line charts show us that our method performs well in distinguishing the normal and abnormal situations. In some scenarios like Scene A and Scene B, the evaluation index of crowd chaos stays gentle before entering the abnormal situation. In some other scenarios like Scene C, the evaluation index of crowd chaos vibrates rapidly owing to the disorder or high walking speed of crowd. Even so, these features can be caught and parameters can be set better to get good results like in Table 2.

The data in Table 2 shows our method reaches a high level in crowd anomaly scattering detection. A good work in anomaly detection should not only play well in detecting the abnormal situations, but also have a good performance in avoiding misjudgments. In all three scenes, our method holds high accuracy rate and low misjudgment rate. For example, if using the common anomaly detection methods to deal with Scene 3, some parts where the people walk around may be detected to be abnormal. But our approach could adjust the parameters to decrease the misjudgment rate to a very low level.

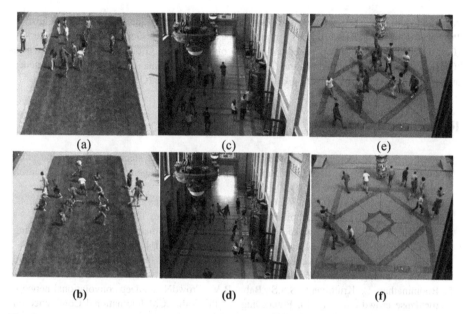

(a)   (c)   (e)

(b)   (d)   (f)

Fig. 8. The normal and abnormal images of three scenes. (a) The normal image of Scene A, (b) the abnormal image of Scene A, (c) the normal image of Scene B, (d) the abnormal image of Scene B, (e) the normal image of Scene C, (f) the abnormal image of Scene C

Table 2. The accuracy rate and miscarriage rate of three scenes

Scene	Accuracy rate	Misjudgment rate
A	97.26%	0.01%
B	100%	0.05%
C	94.37%	0%

5   Conclusion

This paper proposed a crowd anomaly scattering detection method based on motion velocity and information entropy. The velocity and direction information of pedestrians are obtained by the Yolo v3 target detection and Deep-sort multi-target tracking. The velocity information is normalized by sigmoid function, and the walking and running states are effectively distinguished. Therefore, the detection model of abnormal scattering detection is constructed, and information entropy is introduced to measure the degree of crowd disorder. The evaluation index of abnormal scattering detection is proposed by combining motion speed and information entropy. Experiments were performed on three scenes of UMN dataset, where our approach achieved very persuasive results. In the future, the key work is to explore how to detect the abnormal scattering detection in high-density situation.

Acknowledgments. This work was supported by Science Technique Department of Guizhou Province (QKHJC [2019]1164), Scientific Research Project of Guizhou Minzu University (GZMU[2019]QN03), Major Research Project of Innovation Group of Guizhou Province (QJHKY [2018]018), National Natural Science Foundation of China (61802082), Youth Science and Technology Talents Cultivating Object of Guizhou Province (QJHKY[2018]141).

References

1. Kumar, S.A., Kumaresan, A.: Towards building intelligent systems to enhance the child safety and security. In: 2017 International Conference on Intelligent Computing and Control, pp. 1–5. IEEE, Coimbatore (2017)
2. Anthopoulos, L., Janssen, M., Weerakkody, V.A.: Smart Cities and Smart Spaces: Concepts, Methodologies, Tools, and Applications. IGI Global, Hershey (2019)
3. Komninos, N., Bratsas, C., Kakderi, C., Tsarchopoulos, P.: Smart city ontologies: improving the effectiveness of smart city applications. J. Smart Cities **1**(1), 31–46 (2016)
4. Feng, Y., Yuan, Y., Lu, X.: Learning deep event models for crowd anomaly detection. Neurocomputing **219**, 548–556 (2017)
5. Boominathan, L., Kruthiventi, S.S.S., Babu, R.V.: CrowdNet: a deep convolutional network for dense crowd counting. In: Proceedings of the 24th ACM International Conference on Multimedia, pp. 640–644. ACMMM, Amsterdam (2016)
6. Zhang, Y., Zhou, D., Chen, S., et al.: Single-image crowd counting via multi-column convolutional neural network. In: Proceedings of the IEEE Conference on Computer Vision and Pattern Recognition, pp. 589–597. IEEE, Las Vegas (2016)
7. Cao, X., Wang, Z., Zhao, Y., et al.: Scale aggregation network for accurate and efficient crowd counting. In: Proceedings of the European Conference on Computer Vision (ECCV), pp. 734–750. ECCV, Munich (2018)
8. Li, Y., Zhang, X., Chen, D.: CSRNet: dilated convolutional neural networks for understanding the highly congested scenes. In: Proceedings of the IEEE Conference on Computer Vision and Pattern Recognition, pp. 1091–1100. IEEE, Salt Lake City (2018)
9. Singh, A., Patil, D., Omkar, S.N.: Eye in the sky: real-time drone surveillance system (DSS) for violent individuals identification using ScatterNet hybrid deep learning network. In: Proceedings of the IEEE Conference on Computer Vision and Pattern Recognition Workshops, pp. 1629–1637. IEEE, Salt Lake (2018)
10. Rabiee, H., Mousavi, H., Nabi, M., Ravanbakhsh, M.: Detection and localization of crowd behavior using a novel tracklet-based model. Int. J. Mach. Learn. Cybernet. **9**(12), 1999–2010 (2017)
11. Wang, S., Chen, L., Zhou, Z., Sun, X., Dong, J.: Human fall detection in surveillance video based on PCANet. Multimedia Tools Appl. **75**(19), 11603–11613 (2015)

12. Arroyo, R., Yebes, J.J., Bergasa, L.M., et al.: Expert video-surveillance system for real-time detection of suspicious behaviors in shopping malls. Expert Syst. Appl. **42**(21), 7991–8005 (2015)
13. Arivazhagan, S.: Versatile loitering detection based on non-verbal cues using dense trajectory descriptors. Multimedia Tools Appl. **78**(8), 10933–10963 (2019)
14. Marsden, M., McGuinness, K., Little, S., et al.: Holistic features for real-time crowd behaviour anomaly detection. In: 2016 IEEE International Conference on Image Processing (ICIP), pp. 918–92. IEEE, Phoenix (2016)
15. Bera, A., Kim, S., Manocha, D.: Realtime anomaly detection using trajectory-level crowd behavior learning. In: Proceedings of the IEEE Conference on Computer Vision and Pattern Recognition Workshops, pp. 50–57. IEEE, Boston (2016)
16. Zhao, C., Zhang, Z., Hu, J., et al.: Crowd anomaly event detection in surveillance video based on the evolution of the spatial position relationship feature. In: 2018 IEEE 14th International Conference on Control and Automation (ICCA), pp. 247–252. IEEE, Anchorage (2018)
17. Hao, Y., Xu, Z.J., Liu, Y., et al.: Effective crowd anomaly detection through spatio-temporal texture analysis. Int. J. Autom. Comput. **16**(1), 27–39 (2019)
18. Wojke, N., Bewley, A., Paulus, D.: Simple online and realtime tracking with a deep association metric. In: 2017 IEEE international conference on image processing (ICIP), pp. 3645–3649. IEEE, Beijing (2017)

Computer Graphics

View Consistent 3D Face Reconstruction Using Siamese Encoder-Decoders

Penglei Ji[1], Ming Zeng[2], and Xinguo Liu[1(✉)]

[1] State Key Lab of CAD and CG, Zhejiang University, Hangzhou 310000, China
jpl@zju.edu.cn, xgliu@cad.zju.edu.cn
[2] School of Informatics, Xiamen University, Xiamen 361000, China
zengming@xmu.edu.cn

Abstract. This paper addresses the problem of reconstructing the high-quality 3D face shape and high-fidelity texture from a single image. Traditional methods produce a 3D facial shape consistent with only the viewpoint of the input image. The shape and texture of un-exposed regions are estimated from a classic linear 3D Morphable Model (3DMM), leading to unsatisfactory results. Instead, we propose a novel method which is capable of inferring view-consistent shape and texture in unseen regions. Our method is based on a multi-task encoder-decoder framework which simultaneously learns a coarse 3DMM shape, a texture map and a normal map. To build the underlying nonlinear mapping between the input image and the whole shape/texture, we introduce siamese networks to learn a complete texture map and normal map by utilizing the multi-view images of the same person. Experiments show that our method produces more detailed results compared with the previous methods and can generate consistent and complete facial shape and texture.

Keywords: 3D face reconstruction · View consistent · Siamese network

1 Introduction

The goal of the monocular 3D face reconstruction is to build the face shape and texture from a single image. Traditional optimization based methods [18,20] estimate the parameters of a low-dimension parametric face model, *e.g* the 3DMM [1] using the color, intensity or shading information. With the rapid development of the neural networks, some approaches [19,37,48] regress the low-dimensional shape and texture parameters directly using the convolutional neural networks. However, due to the 3DMM is built from a limited number of face scans using the PCA method, it is insufficient to represent highly nonlinear fine-grained shape details, only leading to coarse geometry and approximate texture. To reconstruct the high-fidelity faces, several improved methods [34,35,38–40] have been proposed to synthesize more face details into the shape and generate the photo-realistic textures. However, these methods can only produce good

© Springer Nature Singapore Pte Ltd. 2020
Y. Wang et al. (Eds.): IGTA 2020, CCIS 1314, pp. 209–223, 2020.
https://doi.org/10.1007/978-981-33-6033-4_16

Fig. 1. The first and third rows show the original images, the second and forth rows show the corresponding rendered faces using the predicted camera parameters, shape and texture map. Our method is robust to deal with the shadow (e), eye glasses (f), side view (g) and occlusion (h).

results from the same view direction as the original input, can not predict consistent shape details and textures for different views of the same person.

In this paper, we design a multi-task encoder-decoder to solve this problem. Considering the relationships between the shape, texture and normal, we predict the camera parameters, the face shape, the texture map and the normal map using a unified encoder to learn the consistency between them. The main idea of the encoder-decoder is to simulate the image generation procedure and then minimize the difference between the input and the rendered image. A rendering layer is implemented to generate a rendered image using the predicted camera parameters, shape and texture. The normal map is finally fused into the coarse shape for more details. Some reconstruction results are shown in Fig. 1.

To make the predicted 3D shape to be consistent, even with extreme poses or occlusion, we parameterize the coarse face shape as 3DMM shape parameters and train a network to regress them directly in the same way as the pervious approaches [37,48].

For the texture map, we use a non-linear decoder, which consists of several de-convolution layers to regress it from an encoded latent vector. Compared to the linear 3DMM texture model, our texture decoder is more powerful and effective. The architecture of the normal map decoder is same as the texture. The regressed normal map is finally taken to synthesize details into the coarse shape. However, the decoded texture map or normal map may not be consistent and complete for the multi-view images of a same person. To overcome this problem, we introduce the siamese network [4] into our network. The basic idea is to input a pair of images of a same person and then minimize the difference between the two decoded outputs. This design ties the information from multi views together and is useful to generate consist texture and normal maps. Wu *et al.* [42] utilizes multiple views to recover the 3D facial geometry. Our siamese networks also utilize multi-view images to recover the consistent texture and normal maps.

The main contributions of our work are summarized as follows:

1. We design a multi-task encoder-decoder architecture to regress the camera projection parameters, the 3DMM shape parameters, a texture map and a normal map from a single image.
2. We introduce siamese network to constrain the decoded texture maps and the normal maps of the multi-view images of the same person to be similar. This design is beneficial to generate more consistent and complete texture maps and normal maps, especially for extreme poses or occlusion.
3. We fuse the regressed normal information into the coarse shape obtained from the linear 3DMM model. Our method is able to generate more precise face details and more consistent texture maps than previous work.

2 Related Work

2.1 3D Morphable Model

Blanz and Vetter [1] propose a statistical 3D morphable model to describe the facial shape and texture. 3DMM is built from a set of scanned high-resolution face models via PCA. Basel Face Model (BFM) [26] is an implementation of the 3DMM and is constructed from 100 male and 100 female scanned faces. Booth *et al.* [2,3] build a more powerful 3DMM to improve the texture performance. Based on the 3DMM, 3D face reconstruction can be regarded as a parameter estimation problem. Some researchers solve a non-linear optimization problem to fit the projection of the 3D model to the 2D image [22,30,31,36]. Jiang *et al.* [18] and Roth *et al.* [32] both use a coarse-to-fine strategy to reconstruct 3DMM faces based on the photometric consistency and the normal information.

2.2 Learning-Based Methods

Learning-based methods have been widely used in the facial landmark detection. Cao *et al.* [7] introduce the cascaded regression into the face alignment. Dollar *et al.* [9] regard the 2D facial landmark detection as a regression problem. Yi *et al.* [45] adopt the Convolutional Neural Networks (CNN) to detect the facial landmarks using a coarse-to-fine strategy. Wu *et al.* [43] borrow the idea from human pose prediction and regress the landmark heat-maps using a stacked hourglass network.

CNNs are also widely used to reconstruct 3D face [19,37,48]. Zhu *et al.* [48] propose a novel face profiling method to synthesize a large number of extreme face poses without manually labeling. Genova *et al.* [12] import the facial recognition network into the loss functions, and propose the batch distribution loss, loop-back loss and multi-view identity loss to solve the problem. Guler *et al.* [14] propose a quantized regression architecture to fuse the semantic segmentation information and the dense regression networks. Bulat and Tzimiropoulos [5] combine a novel landmark localization architecture and a residual block to solve the 2D face alignment.

Except the 3DMM representation, some novel shape parameterization methods are proposed. Jackson *et al.* [17] represent the 3D face as a volumetric grid, and then regress the volumetric directly without predicting the vertices coordinates. However, the output models do not have a consistent representation and the computational complexity is relatively high. Feng *et al.* [10] use the position map to record the 3D face shape in texture space, and train a CNN to regress the position map directly. However, this method ignores the spatial information and produces distinct stripes.

2.3 High-Fidelity Face Reconstruction

To generate a high-fidelity face shape and texture, Tewari *et al.* [34] regress the shape, expression, reflectance and illumination parameters based on the classic 3DMM. They introduce a linear correction model to enhance the representation ability of the 3DMM for more details. Tran *et al.* [40] propose a nonlinear 3DMM to represent the shape and texture via an encoder-decoder architecture. However, the shape and texture of the invisible regions are not consistent or complete. Tewari *et al.* [35] use a deep encoder to extract semantic latent vector, and then use a model-based decoder to reconstruct the 3D face. In [23], a self-supervised bootstrapping process is adopted to update the synthetic training corpus iteratively.

Richardson *et al.* [29] introduce an end-to-end CNN framework, which consist of a CoarseNet and a FineNet. They train the FineNet using the Shape-from-Shading (SfS) [46] information. Guo *et al.* [15] also propose a coarse-to-fine learning framework with three convolutional networks to add face details by fusing SfS. Tran *et al.* [38] use a convolutional encoder-decoder to translate an image into a bump map, which can be used to add more facial details into the coarse shape.

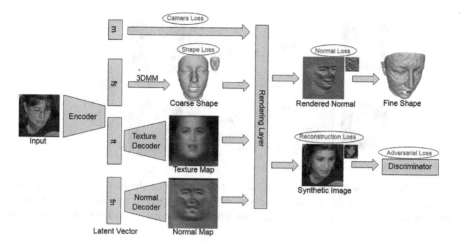

Fig. 2. The architecture of the proposed multi-task encoder-decoder.

Some researchers design some special hardwares, *e.g* the polarized gradient illumination to capture the high- and mid-frequency shape and reflectance as the training data. Huynh *et al.* [16] employ the diffusely-lit facial texture maps to synthesize the facial shape in a learning-based manner. Yamaguchi *et al.* [44] design a neural network to infer high-resolution skin reflectance maps and high- and mid- frequency displacement maps. Trigeorgis *et al.* [41] propose a data-driven method to estimate the face normal from a single image. The fully convolutional network recovers more accurate normals and shape than the SfS due to more accurate normal maps for training. Gecer *et al.* [11] utilize the GANs to fit models and generate high quality textures.

In our method, we extend the classic linear 3DMM parameter regression architecture into a multi-task encoder-decoder to regress the camera projection parameters, the 3DMM shape parameters, a non-linear texture map and a normal map. The main training goal is to minimize the difference between the input image and the rendered image. To generate more consistent and complete texture maps and normal maps for the invisible areas, we introduce the siamese network to solve this problem. The regressed normal maps are then used to refine the coarse shape for more details.

3   The Proposed Method

We design a multi-task siamese encoder-decoder to learn the mapping functions from a single image to camera projection parameters, 3DMM shape parameters, a texture map and a normal map.

The overview of our method is shown in Fig. 2. Firstly, the encoder extract a fixed-dimension feature vector, including the camera parameters m, the 3DMM shape parameters f_s, the texture map latent vector f_t and the normal map latent

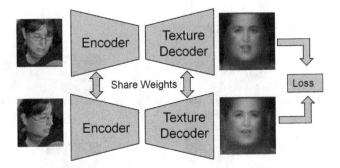

Fig. 3. The siamese texture network takes two images of the same person as input, and generate two texture maps. The siamese loss is utilized to minimize the difference between two output texture maps. It is beneficial to obtain a consistent and complete texture map.

vector f_n. The coarse face shape is obtained according to the 3DMM parametric model. Then, a texture decoder and a normal decoder are designed to convert the latent vector f_t and f_n into a texture map and normal map. The rendering layer generates a rendered image according to the camera parameters, shape and texture. The entire training procedure aims to minimize the difference between the input and the synthetic image. We implement a mesh refinement procedure using the predicted normal map to refine the coarse shape. An additional discriminator is designed to evaluate the photo-realistic quality of the rendered images.

For the shape prediction, we regress the parameters of the 3DMM shape model directly. For the texture map and normal map, we adopt the similar architecture with some de-convolution operations to regress them.

Also, based on the consideration that the multi-view images of a person should have the same texture map and normal map. Hence, we introduce the siamese network to decode the texture map and the normal map, in order to obtain a consistent and complete texture and normal for extreme poses or occlusion.

3.1   Siamese Network for Consistent Texture and Normal Map

For the multi-view images of a person, the predicted texture map should be similar, even under extreme poses or occlusions. Based on this consideration, we introduce the siamese network to decode the texture map. The goal is to minimize the difference between the two siamese networks outputs of two views. The architecture of the texture siamese network is shown in Fig. 3. The input is a pair of images of the same person, and the output is two texture maps. The two encoder-decoders share the weights and are implemented using a same network in practise.

The basic idea for normal map is same as that for the texture map. Recovery the normal from a single image is an under-constrained problem. Based on the assumption that the surface depth is inversely related to the image intensity, we can obtain the highly approximate depth and normal information. In the 3D face reconstruction application, we mainly focus on the perceptual salient features, e.g significantly change of the pixel intensity, which is the reason why the image intensity approximation is sufficient in our application.

We use a relatively simple method [21] to generate the ground-truth normal map. Firstly, the luminance of each pixel $L = 0.213R + 0.715G + 0.072B$ is computed as an indictor of the shape. Then, we use a bilateral filter to generate the filtered value $p(x, y)$, which is stored in the log space of the original luminance space. Then, the recovered depth is calculated as following:

$$d(x, y) = (\frac{\sigma^n p(x, y)}{1 - p(x, y)})^{1/n} \tag{1}$$

where n is the exponent value, and σ is the log average luminance of the pixel values.

The normal map can be obtained as the cross product of the depth map gradients.

$$n = [1, 0, \nabla_x d]^T \times [0, 1, \nabla_y d]^T \tag{2}$$

It should be noticed that the decoded normal maps are represented in a unified world coordinate system and they should be converted to a specific view coordinate system according to the camera rotation matrix. The two normal siamese networks share the weights and are used to guarantee the normal maps of the same person are similar. This is helpful to generate the consistent and complete normal map, especially for the extreme poses or occlusion.

To fuse the normal map with the coarse shape, the low-frequency component is from the face shape, while the high-frequency component comes from the normal map. With the recovered normal map and the coarse shape, we use the method [25] to refine the facial shape for more details. The full shape refinement is defined as the following optimization problem:

$$argmin_p\{\lambda E(p) + (1 - \lambda)E(n)\} \tag{3}$$

where the p represents all the vertices, the position error $E(p)$ is defined as the sum of the squared distances between the optimized vertices and original vertices, the normal error $E(n)$ is defined as the sum of squared projections of tangents to the optimized surface into the corrected normal field.

In our method, the recovered normal maps are used to refine the shape as a post-processing procedure. As an alternative, the shape refinement procedure can be implemented in the rendering layer, and the entire network can be trained in an end-to-end manner. However, the shape refinement procedure using the predicted normal map is time-consuming, and it takes about 10 ms for a mesh with about 50000 vertices, which damages the performance of the training procedure.

3.2   Network Architecture

Our multi-task encoder-decoder is consists of an encoder, a siamese normal decoder, a siamese texture decoder and a discriminator. An additional rendering layer is adopted to generate the rendered image and normal.

The encoder extracts a fixed dimension representation of the entire image based on CNN. In our experiment, the input face image is resized to 96×96, and the output latent vector $x \in \mathbb{R}^{599}$ consists of the camera projection parameters $m \in \mathbb{R}^{12}$, the shape coefficients $f_s \in \mathbb{R}^{228}$ which includes the geometry coefficient $\alpha_{geo} \in \mathbb{R}^{199}$ and the expression coefficient $\alpha_{exp} \in \mathbb{R}^{29}$, the texture latent vector $f_t \in \mathbb{R}^{199}$ and the normal latent vector $f_n \in \mathbb{R}^{160}$.

Compared to the non-linear shape decoder [40] which uses two fully connected layers to regress more than 50,000 point coordinates, the direct regression strategy is not easy to train and is able to obtain a consistent shape. Therefore we still choose the linear 3DMM shape as our coarse shape. The input is the regressed 3DMM shape parameters f_s, and the output facial shape is calculated as:

$$S = \bar{S} + A_{geo}\alpha_{geo} + A_{exp}\alpha_{exp} \tag{4}$$

where S is a 3D face model with N vertices, $\bar{S}$ is the mean shape, A_{shp} and A_{exp} are the geometry and the expression principal component basis respectively.

Due to the limited size of the training data, the linear 3DMM texture can not describe the great changes of the facial appearance, $e.g$ the different races and the different ages. We design a nonlinear texture decoder to regress a $128 \times 128 \times 3$ texture map using several de-convolution operations from the encoded latent space f_t.

The architecture of the normal decoder is same as the texture decoder, which is used to regress a $128 * 128 * 3$ normal map stored in texture space. The purpose of the normal decoder is to predict a normal map, which is used to refine the shape for more details. The goal of the siamese network is to generate more consistent texture maps and normal maps.

The Generative Adversarial Nets (GANs) [13] have been widely used to synthesize the photo-realistic images [28,33] or change the facial expression of a single image [8,27,47]. The input of our discriminator network is the original input image, either the rendered result using the predicted camera projection, shape and texture map. Here, the discriminator is consists of four 3×3 convolution layers with filter numbers are 32, 64, 128 and 1.

For the rendering layer, we adopt the weak perspective projection to render the synthetic image. The rendering layer takes the predicted camera projection parameters, the 3D shape and a 128×128 texture map as the input, and generate a 96×96 rendered image. The texture coordinate (u, v) of each vertex v_i is used to obtain the texture value to render the output. We also feed the camera parameters, the 3D shape and the 128×128 normal map into the rendering layer to generate the 96×96 rendered normal, which is trained by the ground-truth normal maps generated using SfS.

3.3 Loss Functions

The total loss function is defined as:

$$L = \lambda_m L_m + \lambda_{rec} L_{rec} + \lambda_{adv} L_{adv} + \lambda_{shp} L_{shp} + \lambda_{norm} L_{norm} + \lambda_s L_s \quad (5)$$

where all these λ parameters are used to balance these loss terms.

Camera Loss. $L_m = ||m - \hat{m}||_2$, where m and $\hat{m}$ are the estimated and the ground-truth camera projection parameters, respectively. The camera parameter is represented as a 3×4 projection matrix.

Reconstruction Loss. $L_{rec} = ||I - \hat{I}||_1$ asks the rendered image I to be similar to the input image $\hat{I}$. The rendered image I is the output of the rendering layer using the predicted camera parameters, shape and texture map.

Adversarial Loss. We also introduce the adversarial loss L_{adv} to improve the photo-realistic quality of the rendered images.

Shape Loss. $L_{shp} = \sum_{i=1}^{N} ||V_i - \hat{V}_i||_2$ is defined as the vertex differences between the recovered shape and the ground truth, where N is the vertex number of the shape, V_i and $\hat{V}_i$ are the ith vertex in the predicted and ground-truth shape. The model shape is consists of the geometry and the expression.

Normal Loss. $L_{norm} = ||N_r - \hat{N}_r||_1$, where N_r is the rendering result using the camera projection parameters m and the predicted normal map N, and $\hat{N}_r$ is the ground truth normal map obtained using the SfS method. It should be noticed that the $\hat{N}_r$ is represented in the view coordinate system and should be converted to the unified world coordinate system according to the camera parameters.

Siamese Loss. $L_s = \lambda_t L_t + \lambda_n L_n$ consists of the texture siamese loss L_t and the normal siamese loss L_n. The texture siamese loss $L_t = ||T_1 - T_2||_1$, where T_1 and T_2 are two output texture maps of two views of the same person. The normal siamese loss $L_n = ||N_1 - N_2||_1$, where N_1 and N_2 are two decoded normal maps represented in texture space.

4 Experiments

4.1 Implementation Details

We use two datasets in our experiments. **300W-LP** is an extension of 300 W database using the face profiling technique [48]. The generated multi-view images are used to train the siamese networks. The annotations of 300 W are generated using the multi-feature fitting approach and landmark constraints. The database contains $122, 450$ face images with the assistance of the flip operation. **AFLW2000-3D** contains the fitted 3D faces of the first 2000 samples in the AFLW database [24], which only contains the visible 2D facial landmarks. To evaluate the 3D facial alignment accuracy, the fitted 3D faces and 68 landmarks are reconstructed by [48].

We select 90% samples from the 300WLP [48] as our training data. We firstly crop the input images according to the 68 ground-truth landmarks and

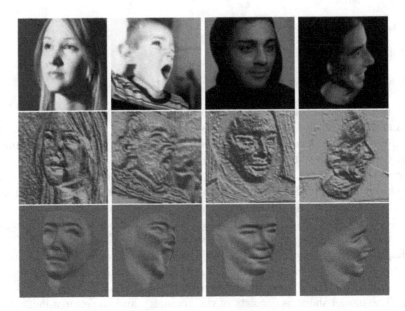

Fig. 4. The decoded normal maps from a single image. The first row displays the input images, the second row gives the corresponding ground-truth normal map represented in the unified world coordinate system, and the last row shows the rendered normal in the current views using the regressed normal map, the camera parameters and the shape.

resize them to 96 × 96. The annotated 3DMM parameters are used to generate the ground-truth shape models. We use the SfS method mentioned above to generate the ground-truth normal maps. The 3DMM parametric model we used is generated from the Basel Face Model [26] and the FaceWarehouse [6].

We implement the network using Tensorflow and use the Adam optimizer to train our network with an initial learning rate 0.005 and decays half after 10 epoches. The total training epoch number is set to 30, and the batch size is set to 64.

In Fig. 1, some rendered images predicted from a single image are shown to valid the representation ability of our method. The experimental results show that our method is able to predict a good camera parameters, a shape and a texture map from a single image. We also valid the effectiveness of our method to deal with the shadow (e), eye glasses (f), side view (g) and occlusion (h).

We also visualize the regressed normal in Fig. 4. The first row shows the input face images. The second row gives the corresponding ground-truth normal map represented in the unified world coordinate system. The last row shows the rendered normal image from the decoded normal map based on the predicted camera parameters and coarse shape.

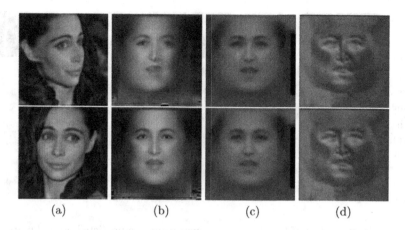

(a)   (b)   (c)   (d)

Fig. 5. The effect of the siamese network for the texture and normal map. (a) two views, (b) the predicted texture without the siamese loss, (c) and (d) are the predicted texture and normal maps with siamese loss, respectively.

4.2 Analysis on the Siamese Network

To evaluate the effectiveness of the siamese network, we remove the texture loss in our training procedure as a contrast experiment. As shown in Fig. 5, the two generated texture maps (c) with the siamese loss are more consistent and complete than results (b) obtained without siamese loss. Column (d) shows the generated normal maps using siamese loss.

4.3 Analysis on the Normal Refinement

We compare the results obtained using the normal refinement technique with the original predict output 3D shape. The comparison results are shown in Fig. 6. It is obvious that our mesh refinement procedure is able to generate more face details from a single image. The wrinkles or hairlines are recovered well with the assistance of the normal refinement technique. Also, due to recovered normal maps being more consistent and complete, our method is able to add face detail into the entire face area, even though some areas are invisible.

4.4 Comparison with the State-of-the-art Methods

We compare our method with several 3D face reconstruction methods, including 3DMM-CNN [37], Extreme-3DMM-CNN (Extreme-CNN) [38], Position Regression Network(PRNet) [10], Volumetric Regression Network (VRN) [17]. The 3DMM-CNN regresses the 3DMM parameters of the shape and texture directly. The Extreme-CNN extends the 3DMM-CNN to predict a bump map to add more facial details. The PRNet proposes a novel position map representation for the 3D facial shape. The VRN performs direct regression of a representation of the 3D facial geometry.

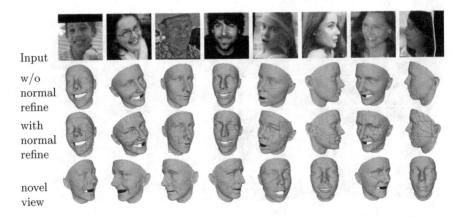

Input

w/o
normal
refine

with
normal
refine

novel
view

Fig. 6. The visual effect of the shape refinement using the predicted normal map. The refinement procedure adds some details into the coarse shape. The first row shows the input images, the second row shows the results obtained by the 3DMM parameters without the normal refinement. The third rows shows the refined shape with the predicted normal maps. The last row shows some invisible regions of the recovered high-quality 3D faces.

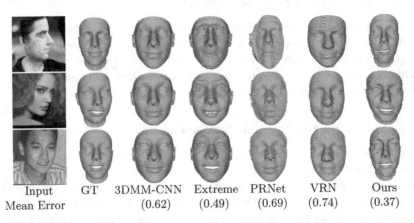

| Input | GT | 3DMM-CNN | Extreme | PRNet | VRN | Ours |
| Mean Error | | (0.62) | (0.49) | (0.69) | (0.74) | (0.37) |

Fig. 7. The comparison of the recovered shape with several state-of-the-art methods. All the reconstructed geometries are aligned with the ground-truth using normalization and rigid iterative closest point. The mean distance of all vertices between the estimated and the ground-truth shapes are calculated as the alignment error. Our method produces the lowest alignment error. For the mean alignment error, lower is better.

The comparison results are shown in Fig. 7. The 3DMM-CNN lacks several details and can only obtain a coarse shape due to the limitation of the parametrical model from PCA. The generated shape of PRNet has obvious stripe due to the limitation of the texture space parameterization. The VRN relies highly on the resolution of the facial volumetric geometry.

We also calculate the mean alignment error between the ground-truth and the reconstructed shape. All the reconstructed shapes need to be normalized to have the same scale as the ground-truth and then aligned using the rigid iterative closest point algorithm for the error calculation. The mean vertex distance of our method is 0.37, which is the lower than 3DMM-CNN (0.62), Extreme-CNN (0.49), PRNet (0.69), VRN (0.74). The experiments show that our method is able to recover more precise shape than other methods.

5 Conclusion

This paper addresses the problem of reconstructing the high-quality 3D face geometry and high-fidelity texture. In this paper, we propose a multi-task encoder-decoder for 3d face reconstruction from a single image. Under the multi-task framework, the correlations among camera parameters, the texture, and the normal map are naturally encoded in the input image. To generate view-consistent high-fidelity results, both of texture and geometry details, we introduce the siamese network for the texture and normal map decoder to utilize the multi-view images of the same person. An additional shape refinement technique is utilized to optimize the coarse shape for more details using the predicted normal maps. Experiments show that, compared with previous methods, our method is capable of inferring consistent details and texture for unexposed regions, with extremely large poses or occlusions.

Acknowledgments. This work was supported by the National Natural Science Foundation of China under Grants Nos. 61872317.

References

1. Blanz, V., Vetter, T.: A morphable model for the synthesis of 3D faces. In: Proceedings of the 26th Annual Conference on Computer Graphics and Interactive Techniques (1999)
2. Booth, J., Antonakos, E., Ploumpis, S., Trigeorgis, G., Panagakis, Y., Zafeiriou, S.: 3D face morphable models "in-the-wild". In: CVPR (2017)
3. Booth, J., Roussos, A., Zafeiriou, S., Ponniahy, A., Dunaway, D.: A 3D morphable model learnt from 10,000 faces. In: CVPR (2016)
4. Bromley, J., et al.: Signature verification using a siamese time delay neural network. In: NIPS (1993)
5. Bulat, A., Tzimiropoulos, G.: How far are we from solving the 2D and 3D face alignment problem? (and a dataset of 230,000 3d facial landmarks). In: ICCV (2017)
6. Cao, C., Weng, Y., Zhou, S., Tong, Y., Zhou, K.: Facewarehouse: a 3D facial expression database for visual computing. IEEE Trans. Visual. Comput. Graph. **20**(3), 413–425 (2014)
7. Cao, X., Wei, Y., Fang, W., Jian, S.: Face alignment by explicit shape regression. In: CVPR (2012)

8. Choi, Y., Choi, M., Kim, M., Ha, J.W., Kim, S., Choo, J.: Stargan: Unified generative adversarial networks for multi-domain image-to-image translation. In: CVPR
   (2018)
9. Dollar, P., Welinder, P., Perona, P.: Cascaded pose regression. In: CVPR (2010)
10. Feng, Y., Wu, F., Shao, X., Wang, Y., Zhou, X.: Joint 3d face reconstruction and
    dense alignment with position map regression network. In: ECCV (2018)
11. Gecer, B., Ploumpis, S., Kotsia, I., Zafeiriou, S.: Ganfit: Generative adversarial
    network fitting for high fidelity 3D face reconstruction. In: CVPR (2019)
12. Genova, K., Cole, F., Maschinot, A., Sarna, A., Freeman, W.T.: Unsupervised
    training for 3D morphable model regression. In: CVPR (2018)
13. Goodfellow, I.J., et al.: Generative adversarial nets. In: NIPS (2014)
14. Guler, R., Trigeorgis, G., Antonakos, E., Snape, P., Zafeiriou, S., Kokkinos, I.:
    Densereg: fully convolutional dense shape regression in-the-wild. In: CVPR (2017)
15. Guo, Y., Zhang, J., Cai, J., Jiang, B., Zheng, J.: Cnn-based real-time dense face
    reconstruction with inverse-rendered photo-realistic face images. In: TPAMI (2018)
16. Huynh, L., et al.: Mesoscopic facial geometry inference using deep neural networks.
    In: CVPR (2018)
17. Jackson, A.S., Bulat, A., Argyriou, V., Tzimiropoulos, G.: Large pose 3D face
    reconstruction from a single image via direct volumetric CNN regression. In: ICCV
    (2017)
18. Jiang, L., Zhang, J., Deng, B., Li, H., Liu, L.: 3D face reconstruction with geometry
    details from a single image. IEEE Trans. Image Process. **27**, 4756–4770 (2018)
19. Jourabloo, A., Liu, X.: Large-pose face alignment via CNN-based dense 3D model
    fitting. In: CVPR (2016)
20. Kemelmacher-Shlizerman, I., Basri, R.: 3D face reconstruction from a single image
    using a single reference face shape. TPAMI **33**(2), 394–405 (2010)
21. Khan, E.A., Reinhard, E., Fleming, R.W., Blthoff, H.H.: Image-based material
    editing. ACM Trans. Graph. **25**(3), 654 (2006)
22. Kim, H., et al.: Deep video portraits. ACM Trans. Graph. **37**(4), 1–14 (2018)
23. Kim, H., Zollöfer, M., Tewari, A., Thies, J., Richardt, C., Theobalt, C.: Inverse-
    facenet: Deep single-shot inverse face rendering from a single image. In: CVPR
    (2018)
24. Kostinger, M., Wohlhart, P., Roth, P.M., Bischof, H.: Annotated facial landmarks
    in the wild: a large-scale, real-world database for facial landmark localization. In:
    ICCVW (2012)
25. Nehab, D., Rusinkiewicz, S., Davis, J., Ramamoorthi, R.: Efficiently combining
    positions and normals for precise 3D geometry. ACM Trans. Graph. **24**(3), 536–
    543 (2005)
26. Paysan, P., Knothe, R., Amberg, B., Romdhani, S., Vetter, T.: A 3d face model for
    pose and illumination invariant face recognition. In: IEEE International Conference
    on Advanced Video and Signal Based Surveillance (2009)
27. Pumarola, A., Agudo, A., Martinez, A., Sanfeliu, A., Moreno-Noguer, F.: Ganima-
    tion: anatomically-aware facial animation from a single image. In: ECCV (2018)
28. Radford, A., Metz, L., Chintala, S.: Unsupervised representation learning with
    deep convolutional generative adversarial networks. In: ICLR (2016)
29. Richardson, E., Sela, M., Or-El, R., Kimmel, R.: Learning detailed face reconstruc-
    tion from a single image. In: CVPR (2016)
30. Romdhani, S., Vetter, T.: Efficient, robust and accurate fitting of a 3D morphable
    model. In: ICCV (2003)
31. Romdhani, S., Vetter, T.: Estimating 3d shape and texture using pixel intensity,
    edges, specular highlights, texture constraints and a prior. In: CVPR (2005)

32. Roth, J., Tong, Y., Liu, X.: Adaptive 3D face reconstruction from unconstrained photo collections. In: CVPR (2016)
33. Shrivastava, A., Pfister, T., Tuzel, O., Susskind, J., Wang, W., Webb, R.: Learning from simulated and unsupervised images through adversarial training. In: CVPR (2017)
34. Tewari, A., Zollhöfer, M., Garrido, P., Bernard, F., Kim, H., Pérez, P., Theobalt, C.: Self-supervised multi-level face model learning for monocular reconstruction at over 250 hz. In: CVPR (2018)
35. Tewari, A., Zollöfer, M., Kim, H., Garrido, P., Bernard, F., Perez, P., Christian, T.: MoFA: model-based deep convolutional face autoencoder for unsupervised monocular reconstruction. In: ICCV (2017)
36. Thies, J., Zollhöfer, M., Stamminger, M., Theobalt, C., Nießner, M.: Demo of face2face: real-time face capture and reenactment of RGB videos. In: CVPR (2016)
37. Tran, A.T., Hassner, T., Masi, I., Medioni, G.: Regressing robust and discriminative 3D morphable models with a very deep neural network. In: CVPR (2017)
38. Tran, A.T., Hassner, T., Masi, I., Paz, E., Medioni, G.: Extreme 3D face reconstruction: seeing through occlusions. In: CVPR (2018)
39. Tran, L., Liu, F., Liu, X.: Towards high-fidelity nonlinear 3D face morphable model. In: CVPR (2019)
40. Tran, L., Liu, X.: Nonlinear 3D face morphable model. In: CVPR (2018)
41. Trigeorgis, G., Snape, P., Kokkinos, I., Zafeiriou, S.: Face normals "in-the-wild" using fully convolutional networks. In: CVPR (2017)
42. Wu, F., Bao, L., Chen, Y., Ling, Y., Song, Y., Li, S., Ngan, K.N., Liu, W.: Mvf-net: multi-view 3d face morphable model regression. In: CVPR (2019)
43. Wu, W., Qian, C., Yang, S., Wang, Q., Cai, Y., Zhou, Q.: Look at boundary: a boundary-aware face alignment algorithm. In: CVPR (2018)
44. Yamaguchi, S., et al.: High-fidelity facial reflectance and geometry inference from an unconstrained image. ACM Trans. Graph. 37(4), 1–14 (2018)
45. Yi, S., Wang, X., Tang, X.: Deep convolutional network cascade for facial point detection. In: CVPR (2013)
46. Zhang, R., Tsai, P.S., Cryer, J.E., Shah, M.: Shape from shading: a survey. TPAMI 21(8), 690–706 (2002)
47. Zhu, J.Y., Park, T., Isola, P., Efros, A.A.: Unpaired image-to-image translation using cycle-consistent adversarial networkss. In: ICCV (2017)
48. Zhu, X., Lei, Z., Liu, X., Shi, H., Li, S.: Face alignment across large poses: a 3D solution. In: CVPR (2016)

An Angle-Based Smoothing Method
for Triangular and Tetrahedral Meshes

Yufei Guo[1,4(✉)], Lei Wang[2], Kang Zhao[3], and Yongqing Hai[4]

[1] Lab, The Second Academy of China Aerospace Science
and Industry Corporation, Beijing, China
[2] China Highway Engineering Consultants Corporation, Beijing, China
[3] Research and Development Centre, Sichuan Aerospace Chuannan Initiating
Explosive Technology Limited, Luzhou, China
[4] Department of Mechanics and Engineering Science, College of Engineering,
Peking University, Beijing, China
yfguo@pku.edu.cn

Abstract. In the computer graphics and finite element simulations, both minimum angle and maximum angle are widely used to evaluate the mesh quality. To increase the minimum angle and decrease the maximum angle, here we present an angle-based smoothing method. The method tends to relocate the free node to make the pairs of adjacent angles incident to it equal. In some examples and special cases, our method works better for the purpose of increasing the minimum angle and decreasing the maximum angle than other smoothing methods.

Keywords: Mesh smoothing · Laplacian smoothing · Angle-based smoothing · Triangular mesh · Tetrahedral mesh

1 Introduction

Mesh quality plays an essential role in computer graphics and finite element applications since it affects the simulation accuracy and efficiency. Topological optimization and geometrical optimization are two main approaches to improve mesh quality. Topological optimization improves the mesh via inserting or deleting nodes as well as reconnecting nodes [1–9]. Geometrical optimization is also called smoothing since it improves the mesh by simply moving the positions of nodes without changing their topologies. Smoothing can play an important role in mesh optimization.

There are several kinds of mesh smoothing methods, such as Laplacian smoothing, variations of Laplacian smoothing, optimization-based smoothing, and their combinations.

1.1 Laplacian Smoothing

Laplacian smoothing [10, 11] is the most commonly used method for mesh smoothing. It improves mesh by iteratively moving every free node to the centroid of the polygon formed by its adjacent nodes. This method is very efficient since the centroid can be

© Springer Nature Singapore Pte Ltd. 2020
Y. Wang et al. (Eds.): IGTA 2020, CCIS 1314, pp. 224–234, 2020.
https://doi.org/10.1007/978-981-33-6033-4_17

calculated straightforward. But its optimization performance is limited, and it may result in some worse-quality elements or even invalid elements.

1.2 Variations of Laplacian Smoothing

To overcome the disadvantages of the Laplacian smoothing, some modified methods have been presented. One common modification of Laplacian smoothing is constrained Laplacian smoothing, which is also called smart Laplacian smoothing [12]. This modification's basic idea is that node movement takes place only when the mesh quality is improved. Blacker and Stephenson [13] presented a length-weighted Laplacian method. This modification tends to keep elements along the exterior of the region perpendicular to the boundary and converges rapidly. There are also some methods based on element geometric transformation. D. Vartziotis *et al.* [14–16] developed a novel smoothing method named GETMe (geometric element transformation method). This method moves all nodes of one element at the same time to improve the element shape. Inspired by GETME, Shuli. Sun *et al.* [17] presented another method based on element geometric transformation too. This method involves two main stages. The first stage is the geometric deformation of all elements through a stretching-shrinking operation. The second stage is to determine each node position by a weighted average strategy. Abdalla G. M. *et al.* [18] proposed a simple push-pull optimization (PPO). Different from Laplacian smoothing, this method does not move the node but the polygon. The method involves three steps based on Delaunay triangulation, and its dual, the Voronoi diagram. Zhou *et al.* [19] presented an angle-based method for two-dimensional mesh smoothing. This method is better at increasing the minimum angle and decreasing the maximum angle compared with Laplacian smoothing. However, it obtains the final position of a free node in an iterative manner, which is inefficient. And the method can not be applied to there-dimensional mesh smoothing. Lin *et al.* [20] presented a vertex-ball spring smoothing method, which can avoid the occurrence of invalid elements effectively and efficiently.

1.3 Optimization-Based Smoothing

Optimization-based smoothing methods use mesh quality measures to define a cost function, then change the mesh optimization problem as a minimizing cost function problem [21]. The common quality measures include minimum/maximum angle [22–24], aspect ratio [25], distortion metrics [11, 26, 27], and so on. In practice, these measures are always combined to improve the mesh. There is a stand-alone library consisting of state-of-the-art algorithms for mesh quality improvement, called Mesquite [28]. Optimization-based smoothing can obtain higher quality meshes than Laplacian smoothing. However, its computational cost is 30–40 times more than Laplacian smoothing [19].

1.4 Combinations of Laplacian Smoothing and Optimization-Based Smoothing

This is an approach to develop hybrid methods combining the methods above. Combinations can always get a better-quality mesh than a single one. Lori A. [29] proposed three smoothing methods combining a smart variant of Laplacian smoothing with an optimization-based smoothing. Chen *et al.* [30] proposed a method for quadratic mixed surface meshed combining Laplacian smoothing with an optimization-based smoothing. David A. [10] presented a method that constrains Laplacian smoothing on Delaunay triangulations.

1.5 Our Contributions

Like the Variations of Laplacian smoothing method, our goal is to devise a method which can obtain a better mesh with a bigger minimum angle and a smaller maximum angle than Laplacian smoothing. We studied the Laplacian smoothing in a new view and presented an angle-based smoothing method correspondingly. The method is slightly less efficient than Laplacian smoothing. But this method can guarantee a better mesh quality.

The rest of the paper is organized as follows. In Sect. 2, we briefly review the Laplacian smoothing and smart Laplacian smoothing. In Sect. 3, we understand the Laplacian smoothing in a new view and give the idea of our method. In Sect. 4, we introduce our angle-based smoothing method for triangular meshes and tetrahedral meshes respectively. Some smoothing examples are present in Sect. 5 to compare the Laplacian smoothing method and our method, including their variations. In the last section, we summarize our method.

2 Related Work

2.1 Laplacian Smoothing

Laplacian smoothing method improves mesh by iteratively moving every free node to the centroid of the polygon formed by its adjacent nodes. Every step is shown in Fig. 1.

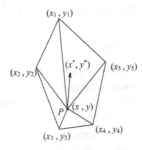

Fig. 1. Laplacian smoothing.

Supposing (x, y) is the initial coordinate of node P, and (x_i, y_i) is the coordinate of i-th neighboring node of P, according to Laplacian smoothing, the new coordinate of P is expressed as follows:

$$(x*, y*) = \frac{1}{n} \sum_{i=1}^{n} (x_i, y_i) \tag{1}$$

where n is the number of neighboring nodes of P.

2.2   Smart Laplacian Smoothing

This is a simple variation of Laplacian smoothing. Some quality measures $f(x, y)$ are calculated before and after the calculation of the new coordinate of P. Only if the quality of the local mesh is improved, the method relocates the node P to the new coordinate. The steps of the method are as follow.

1) Compute $f(x, y)$.
2) Compute (x^*, y^*) according to Formula 1.
3) Compute $f(x^*, y^*)$.
4) If $f(x^*, y^*) > f(x, y)$, set $(x, y) = (x^*, y^*)$.

3   Another Way to Understand Laplacian Smoothing

In Laplacian smoothing, if we do not focus on adjacent nodes of P but on edges formed by its adjacent nodes, we can find another way to compute the coordinate (x^*, y^*). Setting (x_j, y_j) and (x_k, y_k) coordinates of the two nodes from one edge, we can obtain a coordinate (x_i^*, y_i^*) which forms an equilateral triangle with the two coordinates, as shown in Fig. 2. The coordinate is computed as:

$$(x_i*, y_i*) = \frac{1}{2}\left(x_j + x_k, y_j + y_k\right) + \frac{\sqrt{3}}{2}\left(y_k - y_j, x_j - x_k\right) \tag{2}$$

And the following equation is set up:

$$(x*, y*) = \frac{1}{n}\sum_{i=1}^{n}(x_i*, y_i*) = \frac{1}{n}\sum_{i=1}^{n}(x_i, y_i) \tag{3}$$

It is to say that Laplacian smoothing tends to relocate the node P to make its adjacent triangles equilateral. Inspired by this idea, we present a method that tends to

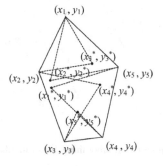

Fig. 2. Another way to understand Laplacian smoothing.

relocate the node P to make the pairs of adjacent angles incident to it equal. This leads to increase the minimum angle and decrease the maximum angle of the mesh.

4 Angle-Based Smoothing Method

This section will describe our method for triangular and tetrahedral meshes respectively. The idea of our method is to compute the intersecting points of the angular bisectors or dihedral angular bisectors and then obtain the new coordinate of P through an average strategy of intersecting points.

4.1 Angle-Based Smoothing for Triangular Meshes

In contrast to Laplacian smoothing computing the coordinate to obtain an equilateral triangle of the edge formed by the adjacent nodes, our angle-based smoothing compute the intersecting point of the angular bisectors at both ends of the edge, as shown in Fig. 3. With the purpose to make each pair adjacent angle equal, our method is very conducive to improve the minimum angle and maximum angle of the mesh. Supposing v_{i-1} and v_{i+1} are two vectors from adjacent node P_i (x_i, y_i), the angular bisectors through P_i can be written as:

$$(x - x_i, y - y_i)\left(\frac{v_{i-1}}{||v_{i-1}||} - \frac{v_{i+1}}{||v_{i+1}||}\right) = 0, \text{ if } \left(\frac{v_{i-1}}{||v_{i-1}||} - \frac{v_{i+1}}{||v_{i+1}||}\right) \neq 0 \quad (4)$$

$$(x - x_i, y - y_i)\begin{pmatrix} 0 & 1 \\ -1 & 0 \end{pmatrix}\left(\frac{v_{i-1}}{||v_{i-1}||}\right) = 0, \text{ if } \left(\frac{v_{i-1}}{||v_{i-1}||} - \frac{v_{i+1}}{||v_{i+1}||}\right) = 0 \quad (5)$$

Then, the computation of the intersecting point of the angular bisectors at both ends of an edge can be seen as the solution of a system of linear equations of two unknowns. Supposing $(x_i{}^*, y_i{}^*)$ is the i th interesting point's coordinate, the new coordinate of P can be computed as:

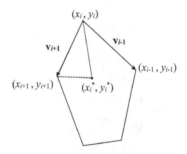

Fig. 3. Angle-based smoothing for triangular meshes.

$$(x*, y*) = \frac{1}{n} \sum_{i=1}^{n} (x_i*, y_i*) \tag{6}$$

where n is the number of neighboring nodes of P, and it is also the number of intersecting points.

4.2 Angle-Based Smoothing for Tetrahedral Meshes

Similar to angle-based smoothing for triangular mesh, herein we compute the intersecting point of the dihedral angular bisectors for tetrahedral meshes. As shown in Fig. 4, triangle $P_i P_j P_k$ is one facet formed by adjacent nodes of P. The edge $P_i P_j$ is one edge of the facet, and its orientation vector is v_i. The orientation vectors of the facets formed by neighboring nodes of P, adjacent to edge $P_i P_j$ are v_{i-1} and v_{i+1}. The dihedral angular bisector through edge $P_i P_j$ can be written as:

$$(x - x_i, y - y_i, z - z_i)\left(\frac{v_{i-1}}{||v_{i-1}||} - \frac{v_{i+1}}{||v_{i+1}||}\right) = 0, \text{ if } \left(\frac{v_{i-1}}{||v_{i-1}||} - \frac{v_{i+1}}{||v_{i+1}||}\right) \neq 0 \tag{7}$$

$$(x - x_i, y - y_i, z - z_i)\left(\frac{v_{i-1}}{||v_{i-1}||} \times v_i\right) = 0, \text{ if } \left(\frac{v_{i-1}}{||v_{i-1}||} - \frac{v_{i+1}}{||v_{i+1}||}\right) = 0 \tag{8}$$

where (x_i, y_i, z_i) is the coordinate of node P_i. Then, the computation of the intersecting point of the dihedral angular bisectors around a facet can be seen as the solution of a system of linear equations of three unknowns. Supposing (x_i*, y_i*, z_i*) is the i-th interesting point's coordinate, the new coordinate of P is:

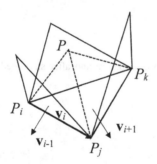

Fig. 4. Angle-based smoothing for tetrahedral mesh.

$$(x*, y*, z*) = \frac{1}{n} \sum_{i=1}^{n} (x_i*, y_i*, z_i*) \tag{9}$$

where n is the number of neighboring facets of P, and it is also the number of intersecting points.

5 Examples and Analysis

In this section, three examples including two triangular meshes and one tetrahedral mesh are selected to test the effectiveness of the proposed method. The qualities of these initial meshes are very poor. GETMe, Laplacian smoothing, smart Laplacian smoothing, angle-base smoothing, and smart angle-based smoothing are conducted to improve the quality of the mesh. The quality measures of smart Laplacian smoothing and smart angle-based smoothing in this paper are minimum angle and maximum angle. Only when the minimum angle is increased and the maximum angle is decreased in the local mesh, the methods relocate the node P to the new coordinate. In the practice of angle-based smoothing, intersecting points of angular bisectors or dihedral angular bisectors of a ring may be far away from the free node P, and even not be existing. Hence, if there are more than one angle of the adjacent edges or adjacent facets is larger than 150° in the ring, we use the Laplacian smoothing method instead.

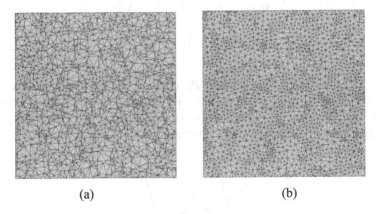

(a)   (b)

Fig. 5. Result of the first mesh case before and after smoothed by our angle-base smoothing method.

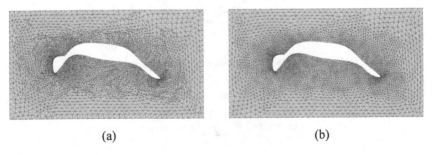

(a)   (b)

Fig. 6. Result of second mesh case before and after smoothed by our angle-base smoothing method.

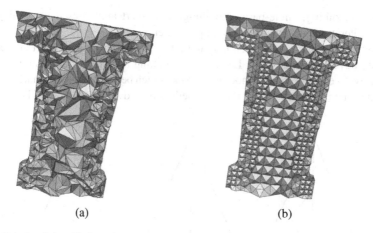

(a)   (b)

Fig. 7. Result of the third mesh case before and after smoothed by our angle-base smoothing method.

Table 1. Comparison with other smoothing methods.

Mesh	Method	Nodes	Min. Angle/(°)	Max. Angle/(°)
First	Original	2200	0.220	179.543
	GETMe	2200	8.938	153.721
	Laplacian	2200	7.870	160.976
	Smart Laplacian	2200	8.809	152.609
	Angle-based	2200	10.387	146.400
	Smart angle-based	2200	11.485	143.274
Second	Original	4244	2.571	172.860
	GETMe	4244	18.788	128.254
	Laplacian	4244	11.123	132.441
	Smart Laplacian	4244	16.170	128.110
	Angle-based	4244	15.236	127.221
	Smart angle-based	4244	18.732	123.012
Third	Original	45533	0.010	179.780
	GETMe	45533	22.839	129.334
	Laplacian	45533	20.123	135.230
	Smart Laplacian	45533	22.895	131.236
	Angle-based	45533	23.556	127.665
	Smart angle-based	45533	24.112	127.002

5.1 Examples

Figures 5, 6 and 7 show the meshes before and after smoothed by our angle-base smoothing method. Table 1 shows the results of GETMe, Laplacian smoothing, smart

Laplacian smoothing, angle-base smoothing, and smart angle-based smoothing. The number of iterations on mesh quality improvement is 10.

From the results, it can be seen that angle-based smoothing and smart angle-based smoothing have advantages over Laplacian smoothing and smart Laplacian smoothing respectively. Even angle-based smoothing gives a much better result than smart Laplacian smoothing. The results also show that our method has advantages than GETMe.

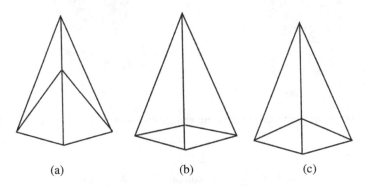

(a)   (b)   (c)

Fig. 8. The performance of different methods in an intuitive case. (a) Original mesh; (b) Laplacian smoothing; (c) Angle-based smoothing.

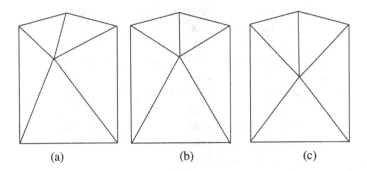

(a)   (b)   (c)

Fig. 9. The performance of different methods in another intuitive case. (a) Original mesh; (b) Laplacian smoothing; (c) Angle-based smoothing.

5.2   Analysis

The purpose of the Laplacian smoothing method is to improve the mesh element shape directly, while our angle-based smoothing method is obviously different, it can improve the mesh element angle or dihedral angle directly. Therefore, our method is better in increasing the minimum angle and decreasing the maximum angle than the Laplacian smoothing method. Its performance has been validated by the above examples. There are also some intuitive cases as shown in Fig. 8 and Fig. 9. These cases show that angle-based smoothing is more suitable for improving the special meshes than Laplacian smoothing.

6   Conclusions and Future Work

In this paper, an angle-based smoothing method has been presented. It can be used for triangular mesh smoothing and tetrahedral mesh smoothing. Some examples proved that our method generates a higher-quality mesh than Laplacian smoothing while keeping the same computational complexity.

We showed that our method can handle some special cases than Laplacian smoothing, which is beneficial to increase the minimum angle and decrease the maximum angle of the mesh.

In the future, we will extend our angle-based smoothing method to other types of meshes to show its generalizability.

References

1. Turk, G.: Re-Tiling Polygonal Surfaces. ACM SIGGRAPH Computer Graphics. **26**(2), 55–64 (2001)
2. Rivara, M.C.: New longest-edge algorithms for the refinement and/or improvement of unstructured triangulations. Int. J. Numer. Meth. Eng. **40**(18), 3313–3324 (1997)
3. Escobar, J.M., Montenegro, R., Montero, G., Rodríguez, E., González-Yuste, J.M.: Smoothing and local refinement techniques for improving tetrahedral mesh quality. Comput. Struct. **83**(28–30), 2423–2430 (2005)
4. Guo, Y., Shang, F., Liu, J.: Surface adaptive mesh generation for STL models based on ball-packing method. Journal of Computer-Aided Design & Computer Graphics. **30**(4), 549–556 (2018)
5. Chen, J., Zheng, J., Zheng, Y., Xiao, Z., Si, H., Yao, Y.: Tetrahedral mesh improvement by shell transformation. Eng. Comput. **33**(3), 393–414 (2016). https://doi.org/10.1007/s00366-016-0480-z
6. Liu, J., Chen, Y.Q., Sun, S.L.: Small polyhedron reconnection for mesh improvement and its implementation based on advancing front technique. Int. J. Numer. Meth. Eng. **79**(8), 1004–1018 (2009)
7. Peng, C.H., Zhang, E., Kobayashi, Y., Wonka, P.: Connectivity editing for quadrilateral meshes. ACM Trans. Graph. **30**(6), 141 (2011)
8. Cheng, K.S.D., Wenping, W., Hong, Q., Wong, K.Y.K., Huaiping, Y., Yang, L.: Design and analysis of optimization methods for subdivision surface fitting. IEEE Trans. Visual Comput. Graphics **13**(5), 878–890 (2007)
9. Chen, L., Zheng, Y., Chen, J.: Topological improvement for quadrilateral finite element meshes. J. Comput.-Aided Des. Comput. Graph. **19**(1), 78–83 (2007)
10. Field, D.A.: Laplacian smoothing and delaunay triangulations. Commun. Appl. Num. Meth. **4**(6), 709–712 (1988)
11. Lo, S.H.: A new mesh generation scheme for arbitrary planar domains. Int. J. Numer. Meth. Eng. **21**(8), 1403–1426 (1985)
12. Vollmer, J., Mencl, R., Muller, H.: Improved laplacian smoothing of noisy surface meshes. Comput. Graphics Forum. **18**(3), 131–138 (1999)
13. Blacker, T.D., Stephenson, M.B.: Paving: a new approach to automated quadrilateral mesh generation. Int. J. Numer. Meth. Eng. **32**(4), 811–847 (2010)

14. Vartziotis, D., Athanasiadis, T., Goudas, I., Wipper, J.: Mesh smoothing using the geometric element transformation method. Comput. Methods Appl. Mech. Eng. **197**(45–48), 3760–1767 (2008)
15. Vartziotis, D., Wipper, J.: The geometric element transformation method for mixed mesh smoothing. Eng. Comput. **25**(3), 287–301 (2009)
16. Vartziotis, D., Wipper, J.: Fast smoothing of mixed volume meshes based on the effective geometric element transformation method. Comput. Methods Appl. Mech. Eng. **201–204**, 65–81 (2012)
17. Sun, S., Zhang, M., Gou, Z.: Smoothing algorithm for planar and surface mesh based on element geometric deformation. Math. Prob. Eng. **2015**, 1–9 (2015)
18. Ahmed, A.G.M., Guo, J., Yan, D.M., Franceschia, J.Y., Zhang, X., Deussen, O.: A simple push-pull algorithm for blue-noise sampling. IEEE Trans. Vis. Comput. Graph. **23**(12), 2496–2508 (2017)
19. Zhou, T., Shimada, K.: An Angle-Based Approach to Two-Dimensional Mesh Smoothing. Louisiana, New Orleans (2000)
20. Lin, T., Guan, Z., Chang, J.: An efficient method for unstructured dynamic mesh deformation-vertex-ball spring smoothing. J. Comput.-Aided Des. Comput. Graph. **25**(11), 1651–1657 (2013)
21. Leordeanu, M., Hebert, M.: Smoothing-based optimization. In: IEEE Conference on Computer Vision & Pattern Recognition (2008)
22. Xu, H., Newman, T.S.: An angle-based optimization approach for 2D finite element mesh smoothing. Finite Elem. Anal. Des. **42**(13), 1150–1164 (2006)
23. Hongtao, X.U., NEWMAN, Timothy S.: 2D FE quad mesh smoothing via angle-based optimization (2005)
24. Xu, K., Gao, X., Chen, G.: Hexahedral mesh quality improvement via edge-angle optimization. Comput. Graph. **70**, 17–27 (2018)
25. Kodiyalam, P.S.: A constrained optimization approach to finite element mesh smoothing. J. Finite Elements Anal. Des. **9**(4), 309–320 (1991)
26. Lo, S.H.: Optimization of tetrahedral meshes based on element shape measures. Comput. Struct. **63**(5), 951–961 (1997)
27. Lo, S.H.: Generating quadrilateral elements on plane and over curved surfaces. Comput. Struct. **31**(3), 421–426 (1989)
28. Brewer ML, Diachin LF, Knupp PM, Leurent T, Melander DJ.: The mesquite mesh quality improvement toolkit. In: Proceedings International Meshing Roundtable (2003)
29. Freitag, L.A.: On combining laplacian and optimization-based mesh smoothing techniques. AMD Trends in Unstructured Mesh Generation (1997)
30. Chen, Z., Tristano, J.R., Kwok, W.: Combined Laplacian and Optimization-based Smoothing for Quadratic Mixed Surface Meshes. IMR (2008)

Virtual Reality and Human-Computer Interaction

Rendering Method for Light-Field Near-Eye Displays Based on Micro-structures with Arbitrary Distribution

Jiajing Han[1], Weitao Song[2(✉)], Yue Liu[1,3], and Yongtian Wang[1,3]

[1] Beijing Engineering Research Center of Mixed Reality and Advanced Display,
School of Optics and Photonics, Beijing Institute of Technology,
Beijing 100081, China
muc_hanjiajing@163.com, {liuyue,wyt}@bit.edu.cn
[2] School of Electrical and Electronic Engineering, Nanyang Technological
University, S2.1 B6-02 50 Nanyang Dr., Singapore, Singapore
wtsong@ntu.edu.sg
[3] AICFVE of Beijing Film Academy, 4, Xitucheng Road, Haidian,
Beijing 100088, China

Abstract. It is a hot topic to realize high-immersion three-dimensional (3D) virtual reality near-eye displays without the accommodation-convergence conflict. Light-field near-eye display method is considered as one of the most potential solutions to achieve this goal. To obtain virtual image for light-field near-eye displays, many researchers have focused on the light field rendering method. In this paper, we propose a universal rendering method to generate light-field pattern for light-field near-eye displays based on OpenGL computer graphics. Each micro-structure is treated as a virtual camera, which can easily be parallelly computed and employed in light-field near-eye displays using multiple micro-structures with arbitrary distribution including a micro-lens array, a pinhole array and random holes. To verify the proposed method, the rendering system based on micro-structures with arbitrary distribution is constructed. Experimental results have shown that the rendered images can reconstruct the virtual scene and be focused on different depth in the light-field near-eye displays system.

Keywords: Virtual reality · Light field · Near-eye displays · Computer graphics · Accommodation-convergence conflict

1 Introduction

In recent years, near-eye displays have been used in virtual reality and augmented reality applications such as entertainment [1], medical treatment [2], education [3, 4] and other fields. In most of the commercial products, binocular parallax has been used to provide the user three-dimensional sense, which may lead to accommodation-convergence conflict. Studies have proved that such conflict may be the main reason in visual discomfort during the usage of near-eye displays.

© Springer Nature Singapore Pte Ltd. 2020
Y. Wang et al. (Eds.): IGTA 2020, CCIS 1314, pp. 237–247, 2020.
https://doi.org/10.1007/978-981-33-6033-4_18

To solve this problems, researchers have proposed a variety of solutions, including multi-focal plane near-eye displays [5], holographic near-eye displays [6, 7], and light-field near-eye displays [8–10] and so on. Among these methods, light-field near-eye display method is considered as a most promising solution by reconstructing sufficient light rays into the user's pupils. Light-field near-eye displays system is constructed based on micro structures. Micro structures including micro-lens array, pinhole array, and random holes, are most commonly employed to modulate the display device to generate light field. However, the light-field near-eye displays system based on a micro-structure array may have the repeated zones and Moiré fringes problems. It has been proved that random holes can be applied to solve these problems in light field displays with periodic structures [11]. To realize the light-field near-eye display, it is necessary to study how to render light-field pattern for universal light-field near-eye displays based on micro-structures.

Wilburn et al. [12, 13] built a camera array consisting of 100 simple cameras to capture the light field, which acquired 3D scenes information from different views and generated the light-filed image. However, the light field rendering method based on camera array has many limitations, for example, the high cost and high requirement for experimental system. Chen et al. [14] developed the system to render the light field, which composed of a normal DSLR camera and a mask (a pinhole array) printed using Kodak LVT technique. With the development of computational optics, the computational light field rendering methods have become increasingly popular. Li et al. [15] proposed a computational rendering method for rendering a 3D scene with one-shot orthographic projection using combing multiple orthographic frustum. Similarly, Aksit et al. [16] realized the light field rendering using the orthographic projection algorithm and developed the rendering system based on a pinhole array. In our previous work [17], a light field rendering method was proposed based on ray-tracing algorithm, and the light-field near-eye display simulation system was developed based on a pinhole array in which the overlapping area of the elemental images is different due to different pupil sizes, so the uncertain net-shaped bright-dark pattern will occur in the refocusing image. Yao et al. [18] optimized the rendering method based on pinhole array using the gradual virtual apertures in the rendering system, but computational rendering methods have low calculation efficiency due to the large amount of calculation of light field rendering. Lanman et al. [19, 20] developed a light field rendering system using GPU-accelerated light-field renderers to realize real-time light field rendering, including a general ray tracing method and a "backward compatible" rasterization method. However, the above method of light field rendering can only be used for light-field near-eye displays system based on a matrix-structure micro-structure array. A universal rendering method is required in light-field near-eye displays.

In this paper, a light field rendering method has been proposed based on OpenGL computer graphics, which is universally applicable to the light-field near-eye displays using micro-structures. The rendering system is developed using the Unity 3D engine, and experiments have also been set up to verify the proposed rendering method. The rest of this paper is organized as follows. The proposed method is discussed in Sect. 2. Section 3 introduces the experiments based on the proposed method, and the experimental results are discussed in this section. Finally, Sect. 4 summarizes and concludes this paper.

2 Method

In order to realize the light-field near-eye displays based on OpenGL graphics, we first create 3D virtual scenes, and render the light field of the 3D virtual scenes to obtain the light-field image. The rendering method for light-field near-eye displays using micro-structures with arbitrary distribution is described in detail in this Section. To make it clear, micro-structure array method is firstly analyzed, and then the universal method has been presented in Sect. 2.2.

2.1 Light Field Rendering

Study on light field rendering technology is based on the 4D light field theory, and researchers have proposed many algorithms to render the light field, such as ray-tracing algorithm [17], multiple orthographic frustum combing algorithm [15] etc. In this paper, the proposed method of light field rendering utilizes the perspective projection algorithm of the OpenGL computer graphics, and each micro-structure is treated as a virtual camera.

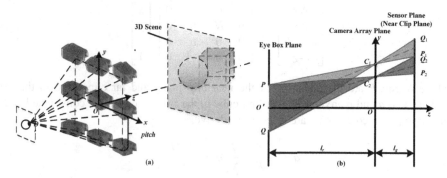

Fig. 1. The diagram of light field rendering system. (a) The diagram of light field rendering system based on camera array. (b) Light field rendering system profile diagram; the eye box plane is introduced to demonstrate the calculation process. The 3D virtual scene (z) is located behind the sensor plane (z_1), $z > z_1$.

As shown in Fig. 1(a), a 3D Cartesian coordinate system O-xyz is set with the x-y plane on the virtual camera array plane and positive z-axis points to the 3D scene. The camera array is generated with the reverse extension of the optical axis converging at O', and a 3D virtual scene is built using Unity 3D engine. Each virtual camera acquires 3D virtual scene between the near clip plane and the far clip plane. As shown in Fig. 1 (b), the projection plane of the cameras is set on the near clip plane, and in fact it can be set on any plane parallel to the near clip plane. If the camera coordinate position C_k is given, the range of elemental image can be derived according to the principle of similar triangles as shown in Eq. (1).

$$\left\{ P_1 = (x', y') \middle\| P_1 - C_k \cdot \frac{l_r + l_g}{l_g} \middle| \le \frac{l_e \cdot l_g}{l_r} \right\} \tag{1}$$

where l_e is the size of eye box, l_g is the gap between camera array plane and sensor plane, l_r is the eye relief of the system.

Every camera in the camera array is the same and the projection method used by the camera is perspective projection [21], which conforms to the characteristics of human visual system [22]. The camera array is placed on the x-y plane, and it is assumed that there is a camera at coordinate origin O. $P(x, y, z)$ is a random point between the near clip plane and the far clip plane of the camera, and $P'(x', y', z')$ is the intersection of OP and the near clip plane. According to the principle of similar triangle, the coordinate of P' can be derived as:

$$P' = \left(-N\frac{x}{z}, -N\frac{y}{z}, -\frac{az + b}{z} \right) \tag{2}$$

Therefore, the equation from P to P' is:

$$\begin{bmatrix} Nx \\ Ny \\ az + b \\ -z \end{bmatrix} = \begin{bmatrix} N & 0 & 0 & 0 \\ 0 & N & 0 & 0 \\ 0 & 0 & a & b \\ 0 & 0 & -1 & 0 \end{bmatrix} \begin{bmatrix} x \\ y \\ z \\ 1 \end{bmatrix} \tag{3}$$

where N and F are the distances of the near clip plane and far clip plane to the x-y plane, and a and b are constant. A unit cube in the z direction can be constructed and the appropriate coefficients are selected to derive:

$$a = -\frac{F + N}{F - N}, b = -\frac{2FN}{F - N} \tag{4}$$

Finally, according to the principle of linear interpolation, the projection plane is mapped into the unit cube to derive the projection matrix:

$$M = \begin{bmatrix} \frac{2N}{x_{max} - x_{min}} & 0 & \frac{x_{max} + x_{min}}{x_{max} - x_{min}} & 0 \\ 0 & \frac{2N}{y_{max} - y_{min}} & \frac{y_{max} + y_{min}}{y_{max} - y_{min}} & 0 \\ 0 & 0 & -\frac{F + N}{F - N} & -\frac{2FN}{F - N} \\ 0 & 0 & -1 & 0 \end{bmatrix} \tag{5}$$

where x_{max}, x_{min}, y_{max}, and y_{min} are the boundary of the projection plane.

The light field rendering system acquires the light field information of the 3D virtual scene, and generates a 2D light-field image with a resolution of $M \times N$ pixels. Based on the light-field image, the 3D virtual scene content is displayed on the display screen by applying the bilinear interpolation algorithm. After the parameters of the display panel including pixel size and screen size are given, we can derive the resolution of the display screen. Each sampled ray in the 3D scene corresponds to a pixel on the rendering image which is shown on the display screen. By determining the corresponding position of each ray on the 2D light-field image and the rendering image, the gray value of the pixel on the light-field image is interpolated to derive the gray of each pixel on the rendering image.

2.2 Light Field Rendering System Based on Cameras with Arbitrary Distribution

The proposed rendering method can also be applied universally in light-field near-eye displays using random structures. It can be seen from [11] that light field displays with random micro structures can solve the repeated zones and Moiré fringes problems. To construct the rendering system based on random structures, the virtual cameras can be set at the positions corresponding to micro structures in the light-field near-eye displays system, as shown in Fig. 2(a). Supposing p is the minimum distance of any two holes, l_r is the eye relief, and g is the gap between micro-structure plane and the display panel, the exit pupil size can be expressed as:

$$D = \frac{l_r^2}{l_r g + g^2} p \qquad (6)$$

For a given exit pupil size and eye relief, the minimum distance of random pinholes is determined according to Eq. (6). In light-field near-eye displays using pinhole array or a micro-lens array, the positions of micro structures are given in advance according to the fixed pitch. Before rendering a light field image for random holes, the position of random holes should be obtained firstly.

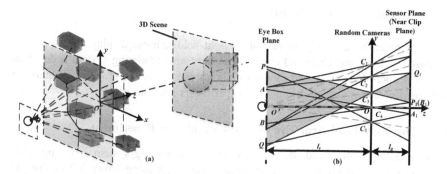

Fig. 2. The diagram of light field rendering system. (a) The diagram of light field rendering system based on random cameras. (b) Light field rendering system profile diagram; the eye box plane is introduced to demonstrate the calculation process and the random cameras are labeled C_i.

Random number generators are applied to generate a number of holes whose numbers are the number of holes should be much larger than the area can afford. For example, for a given display area, the micro structure array with a given pitch has N micro structure in total, thus, the number of generated holes can be more than 10 times of N at first. We give all the holes a serial number from one to the number of generated holes, and then traversal all the holes. For the kth hole, if all the distance between this hole and the first to the $(k-1)th$ one is less than the given value, this hole is retained; otherwise, it should be deleted. These retained holes are the required ones which will be manufactured as a random micro-structure to modulate the display.

Given the position of random holes, the camera can be located at the same position of random holes to obtain the elemental images for each hole. As mentioned in Sect. 2.1, the micro-structure is periodic, and the image size of each camera can be easily determined. In terms of random holes, things go different, and the next step will be how to determine the elemental image size. The random holes can be marked using another serial number one to the number of generated holes. For each pixel on the display panel, we first trace the lines between this pixel and all the micro-structures, and find the intersection point on the eye relief plane. We then find the closest intersection with the center of exit pupil, and this pixel will be labeled as the serial number of corresponding hole with the closest intersection. Thus, all the pixels have been labeled by different number. Such process can be performed in advance, and this label number will not be calculated again, which will not affect the rendering speed. The pixels with the same labeled number are the projected area of the responding camera, as shown in Fig. 2(a). The proposed rendering method can be employed universally for light-field near-eye display using micro-structures with arbitrary distribution.

3   Experiments and Results

We implement the rendering method of the light-field near-eye displays by adopting the proposed method into Unity 3D engine. The light field rendering system is developed based on micro-lens array and random pinholes. To verify the rendering method, the light-field near-eye displays simulation experiments are constructed.

3.1   The Light-Field Near-Eye Displays Simulation Experiment

According to the proposed method, a light field rendering system is set up with a camera array based on a 3D virtual scene constructed using Unity 3D engine. The relevant parameters of the light-field near-eye simulation experiment are shown in Table 1. The near clip plane of the camera is set at the distance 'Gap' from the camera plane. The 3D virtual scene 'Cube-3' and 'Cube-D' are set at the depth of 300 mm and 2000 mm respectively.

Table 1. Parameters of the light-field near-eye displays simulation system.

Parameters	Specifications
Eye relief	30 mm
Eye box	10 mm
The pixel size	0.04725 mm
The micro-lens pitch	1 mm
Number of micro-lens	57
Gap	3.3 mm
Resolution of elemental image	100×100 pixels

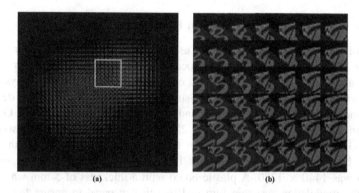

(a)   (b)

Fig. 3. (a) Rendered image on the display panel. (b) Enlarged rendered image.

The 3D virtual scene is rendered by adopting the proposed method, and generates the rendered contents on the display screen as shown in Fig. 3(a). The magnified light-field image is shown in Fig. 3(b). In order to verify the accuracy of the rendered contents, the experiment of near-eye display performance is conducted. The characteristics of human visual system are taken account into the near-eye displays simulation system. The actual eye pupil is usually 2-8 mm and the eye box in the simulation system is 10 mm, so the eye pupil of the eye model is set as 8 mm so that eye model can focus on different depth. The rendered image is shown on the display panel and can be refocused on the "Cube-3" and "Cube-D", as shown in Fig. 4(a) and (b) respectively.

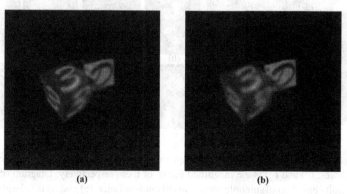

(a)   (b)

Fig. 4. Light-field near-eye displays simulation results focusing on the depth (a) 300 mm and (b) 2000 mm.

Table 2. Parameters of the light-field near-eye displays system.

Parameters	Display using a pinhole array	Display using random holes
Eye relief	50 mm	50 mm
Eye box	8 mm	8 mm
The pixel size	0.04725 mm	0.04725 mm
The micro-structure pitch	0.57 mm	No less than 0.57 mm
Number of micro-lens	57	
Focal depth	1000 mm	1000 mm
Gap	3.356 mm	3.356 mm

3.2 Near-Eye Displays Using Random Holes and a Pinhole Array

In terms of light-field near-eye displays using micro-structures with arbitrary distribution, the projected cameras should be located at the positions of micro-structures as discussed above in Sect. 2. To perform a comparison between near-displays using a micro-structure array and random micro-structures, a prototype has been developed using a random-hole film, a LCD panel and a plastic board (see Fig. 5(a)). One LCD display panel with a resolution of 1440 × 2560 and a refresh rate of 60 Hz were used, and the RGB pixel size was 0.0475 mm × 0.0475 mm with full color. As the display panel can be used for both eyes, half of LCD panel can be utilized, the rendering resolution was 1440 × 1280. A plastic board with a thickness of 5 mm was inserted between the modulation structure film and the display panel to ensure the gap. The equivalent gap in air can be treated as 3.356 mm because the refractive index of the plastic board is 1.49 for visible wavelength. A random-hole film and a pinhole array film were employed respectively for different setup. To obtain an exit pupil of 10 mm at eye-relief of 50 mm, the closest hole distance of random holes was set as 15 pixels, while the pitch of the pinhole array was also set as 15 pixels in near-eye displays using a pinhole array. The pinhole-array, random-hole film pattern and the enlarged pattern are shown in Fig. 5(b–c) and Fig. 5(f–g), respectively. The specifications of light-field near-eye displays using random holes and a pinhole array are shown in Table 2.

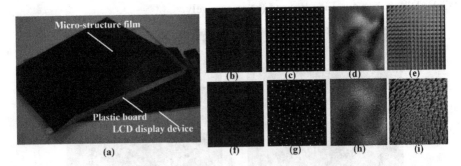

Fig. 5. (a) Diagram of light-field near-eye displays prototype. Diagram of (b) pinhole array and (f) random holes; (c) and (g) is the magnified image of them respectively. Diagram of light field rendering results based on (d) pinhole array and (h) random holes; (e) and (i) is enlarged image of them respectively. (Color figure online)

As described in Sect. 2, the elemental images for these two displays using the developed method are shown in Fig. 5(d) and (h), respectively. Enlarged images have also been presented in Fig. 5(e) and (i), which show the periodicity and the randomness of these two displayed images. Due to the modulation function of micro-structures, light field of virtual objects can be seen in the light-field near-eye displays system. Here, one camera was used around the exit pupil to imitate the process that we observe the displays. When the camera is placed inside the view zone, the required light field using the developed method can be observed as shown in Fig. 6(a) and (d). By moving the position of camera, the process that the users utilize the light-field near-eye displays can be obtained. When the camera is placed near the border of view zone, images from the repeated zone can be seen at one time in the display using a pinhole array (see Fig. 6(b)), while the signal-noise ratio of virtual image decreases in the system using random holes, which displays an unclear image (see Fig. 6(e)). Figure 6(c) and Fig. 6(f) presents the display results outside of view zone which shows that image from wrong repeating zone can be presented in the pinhole-array method, and no clear image can be observed in random-hole method. The experimental results meet the principle of light-field near-eye display, which can verify the proposed rendering method.

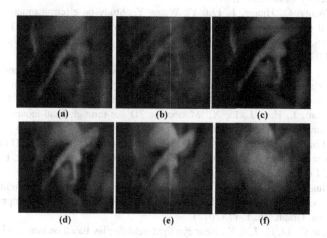

Fig. 6. Diagram of light-field near-eye displays results based on (a) pinhole array and (b) random holes when the eyes are inside the view zone. (b) and (c) are the images of light-field near-eye displays results based on pinhole array when the eyes are outside the view zone. (b) and (c) are the images of light-field near-eye displays results based on random holes when the eyes are outside the view zone.

4   Conclusion

This paper proposes a universal light field rendering method based on OpenGL computer graphics, which can be applied in light-field near-eye displays based on micro-structures with arbitrary distribution. In the proposed light field rendering system, micro-structures including a micro-lens array, a pinhole array and random holes are treated virtual cameras. To verify the rendering results, a simulation experimental system and a prototype have been developed, realizing the virtual scene reconstruction and refocusing. Future work includes optimizing and speeding up the rendering method, which make it more suitable for light-field near-eye displays.

Acknowledgments. This work was supported by This work was supported by the National Key Research and Development Program of China (No. 2018YFB1403901) and the National Natural Science Foundation of China (No. 61727808) and the 111 Project (B18005).

References

1. Martel, E., Muldner, K.: Controlling VR games: control schemes and the player experience. Entertain. Comput. **21**, 19–31 (2017)
2. Willaert, W., Aggarwal, R., Van Herzeele, I., Cheshire, N.J., Vermassen, F.: Recent advancements in medical simulation: patient-specific virtual reality simulation. World J. Surg. **36**(7), 1703–1712 (2012)
3. Stefan, P., et al.: An AR edutainment system supporting bone anatomy learning. In: IEEE Virtual Reality Conference (2014)
4. Liu, X., Wang, C., Huang, J., Liu, Y., Wang, Y.: Study on electromagnetic visualization experiment system based on augmented reality. In: Wang, Y., Huang, Q., Peng, Y. (eds.) IGTA 2019. CCIS, vol. 1043, pp. 699–712. Springer, Singapore (2019). https://doi.org/10.1007/978-981-13-9917-6_65
5. Huang, F., Chen, K., Wetzstein, G.: The light field stereoscope: immersive computer graphics via factored near-eye light field displays with focus cues. In: International Conference on Computer Graphics and Interactive Techniques, vol. 34, no. 4, p. 60 (2015)
6. Gao, Q., Liu, J., Han, J., Li, X.: Monocular 3D see-through head-mounted display via complex amplitude modulation. Opt. Express **24**(15), 17372–17383 (2016)
7. Gao, Q., Liu, J., Duan, X., Zhao, T., Li, X., Liu, P.: Compact see-through 3D head-mounted display based on wavefront modulation with holographic grating filter. Opt. Express **25**(7), 8412–8424 (2017)
8. Shi, L., Huang, F.C., Lopes, W., Matusik, W., Luebke, D.: Near-eye light field holographic rendering with spherical waves for wide field of view interactive 3D computer graphics. ACM Trans. Graph. **36**, 1–17 (2017)
9. Liu, M., Lu, C., Li, H., Liu, X.: Near eye light field display based on human visual features. Opt. Express **25**(9), 9886–9900 (2017)
10. Yao, C., Cheng, D., Yang, T., Wang, Y.: Design of an optical see-through light-field near-eye display using a discrete lenslet array. Opt. Express **26**(14), 18292–18301 (2018)
11. Song, W., et al.: Design of a light-field near-eye display using random pinholes. Opt. Express **27**(17), 23763–23774 (2019)
12. Wilburn, B., Joshi, N., Vaish, V., Levoy, M., Horowitz, M.: High speed video using a dense array of cameras. In: Proceedings of CVPR (2004)
13. Wilburn, B., et al.: High performance imaging using large camera arrays. In: International Conference on Computer Graphics and Interactive Techniques, vol. 24, no. 3, pp. 765–776 (2005)
14. Chen, C., Lu, Y., Su, M.: Light field based digital refocusing using a DSLR camera with a pinhole array mask. In: International Conference on Acoustics, Speech, and Signal Processing (2010)

15. Li, S.L., Wang, Q.H., Xiong, Z.L., Deng, H., Ji, C.C.: Multiple orthographic frustum combing for real-time computer-generated integral imaging system. J. Disp. Technol. **10**(8), 704–709 (2014)
16. Akşit, K., Kautz, J., Luebke, D.: Slim near-eye display using pinhole aperture arrays. Appl. Opt. **54**(11), 3422 (2015)
17. Han, J., Song, W., Liu, Y., Wang, Y.: Design of simulation tools for light-field near-eye displays with a pinhole array. In: SID Symposium Digest of Technical Papers, vol. 50, pp. 1370–1373 (2019)
18. Yao, C., Cheng, D., Wang, Y.: Uniform luminance light field near eye display using pinhole arrays and gradual virtual aperture. In: 2016 International Conference on Virtual Reality and Visualization. IEEE (2016)
19. Lanman, D., Luebke, D.: Near-eye light field displays. ACM Trans. Graph. **32**(6), 1 (2013)
20. Maimone, A., Lanman, D., Rathinavel, K., Keller, K., Luebke, D., Fuchs, H.: Pinlight displays: wide field of view augmented reality eyeglasses using defocused point light sources. ACM Trans. Graph. **33**(4), 1–11 (2014)
21. Brigui, F., Ginolhac, G., Thirion-Lefevre, L., Forster, P.: New SAR target imaging algorithm based on oblique projection for clutter reduction. IEEE Trans. Aerosp. Electron. Syst. **50**(2), 1118–1137 (2014)
22. Groot, S.G.D., Gebhard, J.W.: Pupil size as determined by adapting luminance. J. Opt. Soc. Am. **42**(7), 492–495 (1952)

AUIF: An Adaptive User Interface Framework for Multiple Devices

Chen Dong[1,2], Bin Li[1,2], Li Yang[1], Liang Geng[1], and Fengjun Zhang[1,3(✉)]

[1] Institute of Software, Chinese Academy of Sciences, Beijing, China
dongchenchn@126.com,
delialamb@163.com, {yangli2017,lianggeng,fengjun}@iscas.ac.cn
[2] University of Chinese Academy of Sciences, Beijing, China
[3] State Key Laboratory of Computer Science, Institute of Software,
Chinese Academy of Sciences, Beijing, China

Abstract. Nowadays, it is becoming increasingly common for users to use multiple devices in their daily lives. Smartphones, tablets, laptops and other devices constitute a multi-device environment for users. Making full use of the multi-device environments so that users can work conveniently and efficiently has been a challenging problem in the field of user interface (UI) research. Here, we present AUIF, a web-based application framework for dynamically creating distributed user interfaces (DUIs) across multiple devices. We propose multi-device collaboration patterns and UI components allocation algorithm and combine them in our implementation. A prototype system based on AUIF is designed as an illustration of concept. Finally, user experiments based on the prototype system are carried out, and the result verifies the effectiveness of AUIF in actual use.

Keywords: Cross-device interaction · Distributed user interface · Multi-device environment

1 Introduction

With the rapid development of electronic information technology, an increasing number of users use more than one electronic device in daily life and work. These devices are located in the same physical space and connected through the network to form an interactive space which is called multi-device environments [2] that provides users with information resources. Unlike common single-device usage scenarios, in a multi-device environment, application state and data are shared, and DUIs are applied instead of traditional UIs, taking advantage of multiple input and output channels.

There are some theoretical frameworks providing design guidance and principles for multi-device applications [20,25]. Some of the existing works discuss users' workflows across devices. They explore multi-device use patterns [1,9,17,24] and create UIs for multi-devices based on these patterns [1,16]. Some

© Springer Nature Singapore Pte Ltd. 2020
Y. Wang et al. (Eds.): IGTA 2020, CCIS 1314, pp. 248–263, 2020.
https://doi.org/10.1007/978-981-33-6033-4_19

works utilize the input and output characteristics of different devices to automatically [6,19,27] or manually [7,10,18] distribute UI components. Among these works, pattern-based approaches seem to fit the users' habits, but they are not flexible enough to deal with the changing interaction scenarios. The methods of automatic distribution cleverly utilize the characteristics of different devices, but without considering the usage habits and personalized needs of different users. The manual distribution methods can be applied in all scenarios, but it's cumbersome for users and developers to operate the interface tools, and there is some learning cost for users. As far as we know, no existing method can both meet the requirements of the user's cross-device interactive workflow and make full use of the input and output characteristics of different devices.

Our goal is to provide a UI allocation solution that can automatically assign UI components to the most suitable devices while adapting to the users' cross-device interaction habits in a multi-device environment. Thus, we propose the framework AUIF, which includes *i)* data synchronization mechanisms for different cross-device collaboration patterns, *ii)* a novel UI allocation algorithm for adaptive devices based on integer linear programming, *iii)* a DUI generation framework capable of dynamically adjusting DUIs in changing multi-device environments, so that users can seamlessly switch between different workflows. Similar to Panelrama [27], we calculate the optimal solution of an integer programming formula as the output of the UI allocation algorithm. We not only consider the compatibility of UI components and devices, but also the ease of use. And different from AdaM [19], factors such as user access, privacy, and roles are beyond our scope.

In the remainder of this paper, firstly we introduce the existing research. Then we present the details of AUIF, including cross-device patterns and corresponding data synchronization mechanisms, the UI allocation algorithm, and the DUI generation framework. Next, we implement a prototype system with AUIF to validate the feasibility. After that, we present a user study to evaluate AUIF and collect user feedback. We end by discussing the pros and cons of AUIF and ideas of future work.

2   Related Work

To investigate user's usage habits in multi-device environments, an in-depth user survey must be performed first. Santosa and Wigdor [24] research on the user's workflow and observe single-user use patterns of multiple devices. Besides, they find devices play different roles in workflows. Jokela *et al.* [9] divide patterns of multi-device use into parallel use and serial use. It can be found after comparison that Santosa and Wigdor's research focuses on the function of devices in workflows, while Jokela *et al.*'s research focuses on the way of multi-devices performing tasks. Nebeling and Dey use the end-user customization tool, XDBrowser [17], to explore the user's usage habits and summarize seven multi-device design patterns. The disadvantages in the above works include that the proposed patterns are not highly differentiated, the scenario combination is not comprehensive enough, and it does not support combinations of more than three devices.

Inspired by these works, we propose four cross-device patterns according to whether the data is synchronized and whether the interface is the same. There is no overlap between these patterns, and these patterns cover all situations from the perspective of data and interface.

Model-based, manual, and automatic methods are proposed for generating DUIs. The model-based methods [12,15,21] describe the UI and its distribution in a modeling language, and then systems generate executable code based on the authored model. The manual methods [7,8,18] migrate UI components by operating manually. They not only allow users to share entire or partial web pages (the push operation) and select web pages from other devices (the pull operation) by sending URLs [7], but also allow peripherals (such as Bluetooth keyboards) to be shared between devices [8]. The automatic methods use automatic mechanisms [16] and algorithms [6,19,27] to generate UIs for different devices. They solve the interfaces generation problem as an optimization problem, consider user traces [6], user roles [19], and maximize the adaptability of the panel and the device [27]. In AUIF, we apply the integer linear optimization algorithm similar to Panelrama. In the definition of the objective function. We take the compatibility of UI components and devices into consideration as well as the difficulty of user interaction with UI components on the devices.

An important issue that cross-device tools need to address is synchronizing data between different platforms. Gaia [23] encapsulates the underlying information of different physical devices and provides a unified programming interface for applications. Gummy [14] abstracts the UI into a platform-independent representation so that when an update occurs, the UI remains consistent across all platforms. D-Macs [13] uses the method of demonstration programming. It records macros of operations on one device and plays them back on other devices. At present, most cross-device toolkits [1,10,22,27] are built based on web applications. Because they have almost no migration costs across platforms. It's easy and free for developers to manipulate the HTML DOM. PolyChrome [1] uses point-to-point connections between clients and adds a server connected to each device to store interaction logs, so that newly added devices can quickly synchronize with the old devices in the session. Webstrates [10] implements web pages as nested *webstrates*, users can directly change the HTML and JavaScript code, and then the server synchronizes the changes on DOM to other devices. In view of mature web technologies and the good compatibility of web applications, we build AUIF using JavaScript and implement our prototype as a web application.

3   Design of AUIF

We conduct systematic research on distributed UIs in a multi-device environment to build an adaptive interface tool. Our design mainly includes the following three parts:

1) Establish multi-device collaboration patterns that can meet different user interaction behaviors.

2) Propose an integer linear optimization algorithm adapted to the characteristics of different devices to guide the distribution of UI components across multiple devices.

3) Construct a DUI generation framework that makes users to work smoothly in a multi-device environment.

We will present these three parts in detail in this section.

3.1 Multi-device Collaboration Patterns

Through the introduction of cross-device workflows in Sect. 2, we know that users have different interaction needs in multi-device use scenarios. Multiple devices can either access data synchronously to complete tasks collaboratively, or access data asynchronously to complete each part of the task separately. Multiple devices can either have the same UI components, or have different UI components with different functions depending on the characteristics of each device. We divide the multi-device collaboration patterns into the following four types:

Duplicate: Devices access data synchronously and display the same UI components.

Extend: Devices access data asynchronously and display the same UI components.

Joint: Devices access data synchronously and display distributed UIs.

Split: Devices access data asynchronously and display distributed UIs (Fig. 1).

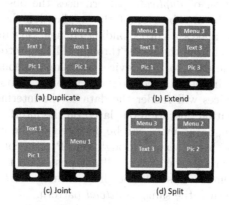

Fig. 1. Four multi-device collaboration patterns

Our UI components allocation algorithm is responsible for calculating the best allocation of UI components, which will be introduced later. We design and implement the data interaction process in each pattern under the B/S architecture, as shown in Fig. 2. The server is connected to each device through

WebSocket[1], and the devices that access data synchronously are in the same group.

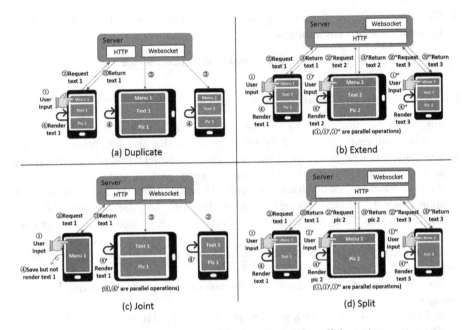

Fig. 2. Data interaction process of four multi-device collaboration patterns

The implementation of *duplicate* pattern uses the observer pattern. Each device is equivalent to an observer of the application data. When the data changes, each device makes a corresponding change. Specifically, when a user requests a piece of text from the server through HTTP protocol with his device, the server returns the data to this device, and simultaneously sends the same data to other devices in the same device group through WebSocket. After receiving the data, all devices will render the data to the interface to complete the real-time synchronization operation displayed on the interface.

In *extend* pattern, each device can be operated by the user in parallel, and data is exchanged with the server through HTTP protocol. And because each device is in its own group, the server will not push data to other devices, and operations on different devices will not affect each other. The data interaction process in *split* pattern is the same as *extend* pattern.

In *joint* pattern, the data interaction process is similar to *duplicate* pattern, the difference is that when the server returns the same data to each device, because different UI components are distributed on different devices, if the device has no UI components that can display data, the data will be saved on the browser but not rendered on the interface. Other devices with corresponding UI components still render the data.

---

[1] https://www.w3.org/TR/2012/CR-websockets-20120920/.

3.2   UI Components Allocation Algorithm

In order to distribute UI components with different functions to the appropriate devices, we design an automatic allocation algorithm. Two optimization targets are considered: maximizing the compatibility of UI components and devices, and minimizing the difficulty of user interaction with UI components on the devices.

Compatibility. Let UI components and devices be in the same attribute space, c_e is the attribute vector of UI component e, and c_d is the attribute vector of device d. For example, in Table 1, the attribute space can be summarized from function, input channels, and visual size into four attributes: controllability (also the portability of devices), keyboard function, touch screen function, and area. The value of the attribute is an integer with a maximum value of 5 and a minimum value of 0, used to indicate the degree. The keyboard function values of laptops, smartphones, and tablets are 5, 1, and 2, respectively, indicating that laptops are very suitable for keyboard input, smartphones are not suitable for keyboard input, and keyboard input functions of tablets are better than smartphones but worse than laptops. The text editor in UI components requires keyboard input, so the attribute value of its keyboard function is 5, and the sketch canvas does not require keyboard input, and its corresponding attribute value is 0.

Table 1. Example of attribute values of UI components and devices

		Controllability (Portability)	Keyboard function	Touch screen function	Area
UI components	Online devices bar	5	0	0	2
	Sketch toolbar	5	0	0	2
	Sketch canvas	0	0	3	5
	Text editor	0	5	0	5
Devices	Laptop	2	5	0	5
	Tablet	4	2	5	4
	Smartphone	5	1	5	2

We use the euclidean distance between the attribute vector c_e of the UI component e and the attribute vector c_d of the device d to measure whether these two match. That is, larger the value of $\|c_e - c_d\|$, less the UI component e is suitable for allocation to the device d. Then on all device sets D and all UI component sets E, the overall mismatch value M is calculated by the following formula, where x_{de} indicates whether UI component e is distributed to device d.

The value of x_{de} is 0 or 1.

$$M = \sum_{d=1}^{D} \sum_{e=1}^{E} \|c_e - c_d\| x_{de} \tag{1}$$

Difficulty. Inspired by Sketchplore [26], we use Fitts' law [11] to measure the difficulty of user interaction with UI components on the devices. Fitts' law quantitatively examines factors that affect the time MT it takes for a pointing device to reach a target. Given the distance D between device and target, and the target width W, MT is:

$$MT = a + b \log_2(\frac{D}{W} + 1) \tag{2}$$

We optimize the total time T for the user to operate pointing interaction with UI components on each device. Assuming that UI components are located at the corner of the device screen, D in formula 2 is the screen size s_d of device d and W is the width w_e of the UI component e. According to the concept of responsive web design[2], in the actual generated UI, the size of the UI components should be changed within a certain range according to the screen size. To simplify the problem, we use the minimum size of UI components. In this way, we consider the worst case when optimizing T (in the actual generated interface, UI components are generally not placed in the corner of the screen, and their sizes are not necessarily the lower bound of the range). T is calculated as:

$$T = \sum_{d=1}^{D} \sum_{e=1}^{E} [a + b \log_2(\frac{s_d}{w_e} + 1)] x_{de} \tag{3}$$

Parameters a and b in formula 3 depends on the input channels of devices. For mouse-input devices, the values of a and b are available from the original study [11] of Fitts' law; for devices with touch screen, we use values from the study [5].

We aim to optimize both the compatibility of UI components and devices, and the difficulty of user interaction with UI components on the devices, means minimizing the sum of M in formula 1 and T in formula 3. In addition, there are two constraints: $i)$ each UI components must be distributed to only one device, $ii)$ each device must get at least one UI component. In summary, the automatic allocation algorithm of UI components can be described as:

$$\min \sum_{d=1}^{D} \sum_{e=1}^{E} [a + b \log_2(\frac{s_d}{w_e} + 1) + \|c_e - c_d\|] x_{de} \tag{4}$$

$$s.t. \quad \begin{matrix} \sum_{d=1}^{D} x_{de} = 1, & e = 1, 2, 3 \dots, E \\ \sum_{e=1}^{E} x_{de} \geq 1, & d = 1, 2, 3 \dots, D \end{matrix}. \tag{5}$$

---

[2] http://alistapart.com/article/responsive-web-design/.

In the above formula, parameter s_d can be obtained from the *window* object in JavaScript, w_e, c_e, c_d are determined by developers, and x_{de} is the variable to be solved. When implementing the algorithm, we use the GNU Linear Programming Kit[3].

3.3 DUI Generation Framework

In AUIF, there are four phases to build DUIs: preparation, device connection, UI components allocation, and user interaction. The four phases are shown in Fig. 3.

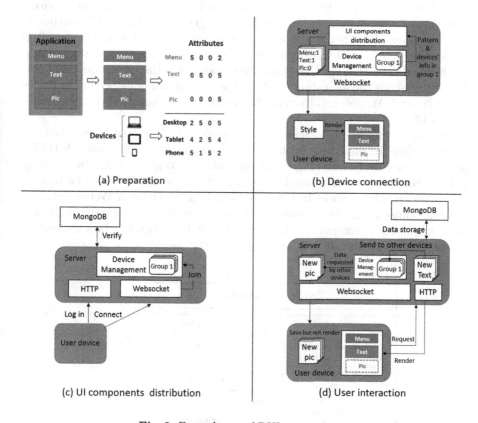

Fig. 3. Four phases of DUI generation

In the preparation phase, developers must first determine the functions and business logic of the application and separate the relatively independent functions as UI components. Then the types of devices that might interact with the application are determined. Finally, the developer gives the attribute space of

---

[3] http://hgourvest.github.com/glpk.js/.

UI components and devices, and sets the attribute value for each UI component and device.

In the device-connection phase, the user device first logs in to the application and then connects to the server through WebSocket protocol. Users can choose to join in existing device groups or create their groups. Device groups can be nested, but they cannot be intersected referencing. Each device group needs to apply a collaboration pattern, which defaults to be *duplicate* pattern. The device manager on the server side maintains information about all connected devices in device groups.

In the UI-components allocation phase, the server runs the UI components allocation algorithm based on collaboration patterns and device type information of each device group. There is no need to assign UI components in *duplicate* pattern and *extend* pattern, and each device gets all of them. The output of the allocation algorithm is the allocation result of each device. The result is dictionary-type data, of which the key is UI component id, and the value is 1 or 0, which indicates that the UI component is assigned to the current device or not. The server sends the allocation result to the corresponding device in the device group through WebSocket. Devices then render the assigned UI components to form a UI.

In the user-interaction phase, users request data from the server then the server fetches the data from MongoDB and returns it to users through HTTP protocol. User devices re-render the page after getting the data. In *duplicate* pattern and *joint* pattern, the data requested by one user will also be sent by the server to other devices in the same device group through WebSocket, and the user himself will also receive the data requested by other devices. If user devices have no UI components that can display data, then the data is saved on the browser, otherwise, it will be rendered on the page.

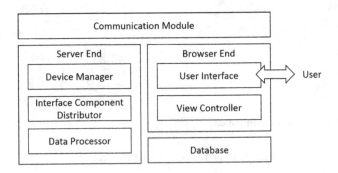

Fig. 4. The architecture of AUIF

AUIF is built with B/S architecture, as shown in Fig. 4. The User interface is the interface displayed on the user device. The view controller assembles the UI components, UI styles, and data into the user interface. The device manager manages a multi-device environment and handles device connection and disconnection. The interface component distributor runs the UI components allocation algorithm to distribute UI components for different devices. The data processor handles the data operations and makes changes to the database. The database stores the data. And the communication module is the communication pipe between server and browser.

4   Prototype System

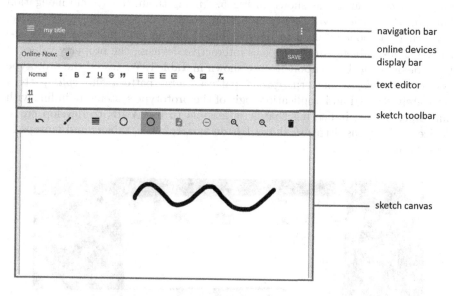

Fig. 5. UI components of the prototype system

In order to verify the feasibility of AUIF in real applications, we design a multi-person collaborative meeting scenario. The main task for the users in this scenario is to participate in the conference discussions together. During the discussion, people record the conference information and inspire through sketches. Finally, people save the records and sketches of each meeting. In this scenario, users viewing the same record are in the same device group. If the content of the viewed document is updated, it needs to be synchronized on the interface in real-time. Multiple devices of a single user can also be regarded as the same device group, and users can view or edit different notes at the same time. There are four main types of interactions:

1) Multiple users view or edit the same note at the same time, which corresponds to *duplicate* pattern in collaboration patterns.

2) Single user uses multiple devices for different functions when viewing or editing the same note, which corresponds to *joint* pattern.
3) Multiple users view or edit different notes, which corresponds to *extend* pattern.
4) Single user uses multiple devices to view or edit different notes, which corresponds to *split* pattern.

We use Node.js runtime environment to build a JavaScript-based server-side framework, and the front-end libraries React[4], HTML, and CSS to build a browser-side framework. UI components of the prototype system include the navigation bar, the online devices display bar, the text editor, the sketch toolbar, and the sketch canvas, as shown in Fig. 5. Among them, the system navigation bar provides devices with a logout option, so it is displayed on every device, and other components need to be distributed among multiple devices. We use smartphones, tablets, and laptops as interactive devices. The property values of these devices and UI components are shown in Table 1. We tested the prototype system in *duplicate* pattern, *extend* pattern, *joint* pattern and *split* pattern separately. The UI and application logic of the prototype system are in line with our expectations, indicating that it is feasible to use AUIF to develop cross-device applications. Figure 6 shows UIs on smartphones, tablets and laptops in *joint* pattern.

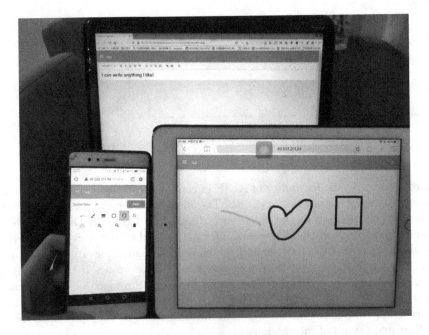

Fig. 6. UIs on smartphones, tablets and laptops in *joint* pattern

---

[4] https://reactjs.org/.

Table 2. Statements in the 7-point Likert scale

No	Statements
Q1	I often use at least two of my phone, tablet, and laptop at the same time
Q2	I think the user interface of the prototype system on different devices is pretty
Q3	I think the interface components assigned on different devices are reasonable
Q4	I think the multi-device task is difficult to complete
Q5	I can adapt to the multi-device usage of the prototype system
Q6	Compared to cross-device usage, I prefer single-device usage

5   User Study

To confirm the effectiveness of AUIF, we conducted user experiments based on the prototype system and analyzed the results.

5.1   Design of the Experiment

We set up a single-device task and a multi-device task in user experiments. In the single-device task, we asked the participants to create a document, edit texts and sketches and save them by mobile smartphone, tablet and laptop separately on the same account. In the multi-device task, we asked the participants to use the combination of smartphone + laptop, smartphone + tablet, laptop + tablet, smartphone + laptop + tablet, editing texts and sketches at the same time. In the multi-device task, participants will experience that the UI components are distributed on different devices, so that different devices will have different functions when editing documents together.

We had a total of 14 participants (5 female), aged from 24 to 30 years old. Nine of them did not own a tablet, and they were unable to perform tablet-related tasks. To facilitate the conducting of the experiment, we deployed the prototype system on a cloud server. The participants only need to log in to the deployed address using their own devices.

Before the experiment, each participant needs to fill out a questionnaire on basic information, including the name, age, gender, and experience of using the device. Each participant then had 10 min to become familiar with the prototype system. During the experiment, we asked the participants to use their smartphones, tablets and laptops to complete the single-device task and the multi-device task. After the experiment, each participant must fill in the System Usability Scale (SUS) [4], the seven-point Likert scale based on the statement in Table 2, and questions to collect user suggestions.

5.2   Result

In terms of system availability, The average SUS score of the prototype system is 79.82, which is above the SUS score threshold(70) [3], indicating the good usability of the prototype system.

Table 3 summarizes the result of the seven-point Likert scale, presenting the relative frequency of the level of agreement (1 for Strongly Disagree to 7 for Strongly Agree), the mean of the answers ($\bar{x}$), and the standard deviation (s). From the closer relative frequency and the low standard deviation, we can see that most users are close to the statement, so the conclusion of our experiment is valid.

Table 3. Summary of the 7-point Likert scale

	1	2	3	4	5	6	7	$\bar{x}$	s
Q1	0.0%	0.0%	7.1%	7.1%	7.1%	14.3%	64.3%	6.21	1.31
Q2	0.0%	7.1%	7.1%	0.0%	21.4%	42.9%	21.4%	5.50	1.45
Q3	0.0%	0.0%	21.4%	7.1%	7.1%	28.6%	35.7%	5.50	1.61
Q4	14.3%	35.7%	35.7%	7.1%	0.0%	7.1%	0.0%	2.64	1.28
Q5	0.0%	0.0%	0.0%	0.0	21.4%	50%	28.6%	6.07	0.73
Q6	7.1%	14.3%	28.6%	21.4%	14.3%	14.3%	0.0%	3.64	1.50

From the results of Q1, we know that most people have multiple device usage behaviors in daily life. The results of Q2 and Q3 tell us that the UI components allocation algorithm proposed in AUIF can effectively obtain a DUI with a reasonable distribution of UI components. The result of Q4 shows that the tasks are easy to complete, and the prototype system is interaction-friendly. The results of Q5 and Q6 show that the participants were able to adapt to the re-assigned UI, and tended to use the DUI.

6   Conclusions

We focus on solving the problem of DUI construction in a multi-device environment, presenting an adaptive UI framework that enables traditional single-person single-device applications to be used in multi-person multi-device scenarios. The multi-device collaboration patterns we proposed meet the requirements of the user's cross-device interactive workflow and the UI allocation algorithm we designed makes full use of the input and output characteristics of different devices. The user experiments based on the prototype system verify the effectiveness of AUIF in actual use.

However, there are still improvements that we wish to make. In the future, we will take user personalization into consideration in the process of UI component allocation. And we will apply the key algorithms and techniques in AUIF to traditional desktop applications.

Acknowledgments. This work was supported by the National Key Research and Development Program of China (No. 2018YFB1005002), the National Natural Science Foundation of China (No. 61572479), the Strategy Priority Research Program of Chinese Academy of Sciences (No. XDA20080200).

References

1. Badam, S.K., Elmqvist, N.: PolyChrome: a cross-device framework for collaborative web visualization. In: Proceedings of the Ninth ACM International Conference on Interactive Tabletops and Surfaces, ITS 2014, pp. 109–118. Association for Computing Machinery, New York (2014)
2. Balakrishnan, R., Baudisch, P.: Introduction to this special issue on ubiquitous multi-display environments. Hum.-Comput. Interact. **24**(1–2), 1–8 (2009)
3. Bangor, A., Kortum, P., Miller, J.: Determining what individual SUS scores mean: adding an adjective rating scale. J. Usability Stud. **4**(3), 114–123 (2009)
4. Brooke, J.: SUS - a quick and dirty usability scale, pp. 189–194, January 1996
5. El Batran, K., Dunlop, M.D.: Enhancing KLM (keystroke-level model) to fit touch screen mobile devices. In: Proceedings of the 16th International Conference on Human-Computer Interaction with Mobile Devices & Services, MobileHCI 2014, pp. 283–286. Association for Computing Machinery, New York (2014)
6. Gajos, K., Weld, D.S.: SUPPLE: automatically generating user interfaces. In: Proceedings of the 9th International Conference on Intelligent User Interfaces, IUI 2004, pp. 93–100. Association for Computing Machinery, New York (2004)
7. Ghiani, G., Paternò, F., Santoro, C.: Push and pull of web user interfaces in multi-device environments. Association for Computing Machinery, New York (2012)
8. Hamilton, P., Wigdor, D.J.: Conductor: enabling and understanding cross-device interaction. In: Proceedings of the SIGCHI Conference on Human Factors in Computing Systems, CHI 2014, pp. 2773–2782. Association for Computing Machinery, New York (2014)
9. Jokela, T., Ojala, J., Olsson, T.: A diary study on combining multiple information devices in everyday activities and tasks. In: Proceedings of the 33rd Annual ACM Conference on Human Factors in Computing Systems, CHI 2015, pp. 3903–3912. Association for Computing Machinery, New York (2015)
10. Klokmose, C.N., Eagan, J.R., Baader, S., Mackay, W., Beaudouin-Lafon, M.: Webstrates: shareable dynamic media. In: Proceedings of the 28th Annual ACM Symposium on User Interface Software & Technology, UIST 2015, pp. 280–290. Association for Computing Machinery, New York (2015)
11. MacKenzie, I.S.: Fitts' law as a research and design tool in human-computer interaction. Hum.-Comput. Interact. **7**(1), 91–139 (1992)
12. Melchior, J., Vanderdonckt, J., Van Roy, P.: A model-based approach for distributed user interfaces. In: Proceedings of the 3rd ACM SIGCHI Symposium on Engineering Interactive Computing Systems, EICS 2011, pp. 11–20. Association for Computing Machinery, New York (2011)
13. Meskens, J., Luyten, K., Coninx, K.: D-Macs: building multi-device user interfaces by demonstrating, sharing and replaying design actions. In: Proceedings of the 23rd Annual ACM Symposium on User Interface Software and Technology, UIST 2010, pp. 129–138. Association for Computing Machinery, New York (2010)

14. Meskens, J., Vermeulen, J., Luyten, K., Coninx, K.: Gummy for multi-platform user interface designs: shape me, multiply me, fix me, use me. In: Proceedings of the Working Conference on Advanced Visual Interfaces, AVI 2008, pp. 233–240. Association for Computing Machinery, New York (2008)

15. Miao, G., Hongxing, L., Songyu, X., Juncai, L.: Research on user interface transformation method based on MDA. In: 2017 16th International Symposium on Distributed Computing and Applications to Business, Engineering and Science (DCABES), pp. 150–153 (2017)

16. Nebeling, M.: XDBrowser 2.0: semi-automatic generation of cross-device interfaces. In: Proceedings of the 2017 CHI Conference on Human Factors in Computing Systems, CHI 2017, pp. 4574–4584. Association for Computing Machinery, New York (2017)

17. Nebeling, M., Dey, A.K.: XDBrowser: user-defined cross-device web page designs. In: Proceedings of the 2016 CHI Conference on Human Factors in Computing Systems, CHI 2016, pp. 5494–5505. Association for Computing Machinery, New York (2016)

18. Nebeling, M., Mintsi, T., Husmann, M., Norrie, M.: Interactive development of cross-device user interfaces. In: Proceedings of the SIGCHI Conference on Human Factors in Computing Systems, CHI 2014, pp. 2793–2802. Association for Computing Machinery, New York (2014)

19. Park, S., et al.: Adam: adapting multi-user interfaces for collaborative environments in real-time. In: Proceedings of the 2018 CHI Conference on Human Factors in Computing Systems, CHI 2018, Association for Computing Machinery, New York (2018)

20. Paternò, F., Santoro, C.: A logical framework for multi-device user interfaces. In: Proceedings of the 4th ACM SIGCHI Symposium on Engineering Interactive Computing Systems, EICS 2012, pp. 45–50. Association for Computing Machinery, New York (2012)

21. Paterno', F., Santoro, C., Spano, L.D.: MARIA: a universal, declarative, multiple abstraction-level language for service-oriented applications in ubiquitous environments. ACM Trans. Comput.-Hum. Interact. 16(4), 1–30 (2009)

22. Roels, R., De Witte, A., Signer, B.: INFEX: a unifying framework for cross-device information exploration and exchange. Proc. ACM Hum.-Comput. Interact. 2(EICS), 1–26 (2018)

23. Roman, M., Campbell, R.H.: Gaia: enabling active spaces. In: Proceedings of the 9th Workshop on ACM SIGOPS European Workshop: Beyond the PC: New Challenges for the Operating System, EW 9, pp. 229–234. Association for Computing Machinery, New York (2000)

24. Santosa, S., Wigdor, D.: A field study of multi-device workflows in distributed workspaces. In: Proceedings of the 2013 ACM International Joint Conference on Pervasive and Ubiquitous Computing, UbiComp 2013, pp. 63–72. Association for Computing Machinery, New York (2013)

25. Sørensen, H., Raptis, D., Kjeldskov, J., Skov, M.B.: The 4C framework: principles of interaction in digital ecosystems. In: Proceedings of the 2014 ACM International Joint Conference on Pervasive and Ubiquitous Computing, UbiComp 2014, pp. 87–97. Association for Computing Machinery, New York (2014)

26. Todi, K., Weir, D., Oulasvirta, A.: Sketchplore: sketch and explore layout designs with an optimiser. In: Proceedings of the 2016 CHI Conference Extended Abstracts on Human Factors in Computing Systems, CHI EA 2016, pp. 3780–3783. Association for Computing Machinery, New York (2016)
27. Yang, J., Wigdor, D.: Panelrama: enabling easy specification of cross-device web applications. In: Proceedings of the SIGCHI Conference on Human Factors in Computing Systems, CHI 2014, pp. 2783–2792. Association for Computing Machinery, New York (2014)

Applications of Image and Graphics

Control and on-Board Calibration Method for in-Situ Detection Using the Visible and Near-Infrared Imaging Spectrometer on the Yutu-2 Rover

Jia Wang[1], Tianyi Yu[1], Kaichang Di[2], Sheng Gou[2], Man Peng[2], Wenhui Wan[2(⊠)], Zhaoqin Liu[2], Lichun Li[1], Yexin Wang[2], Zhifei Rong[1], Ximing He[1], Yi You[1], Fan Wu[1], Qiaofang Zou[1], and Xiaohui Liu[1]

[1] Beijing Aerospace Control Center, Beijing 100094, China
15210106156@139.com, yuty.bacc@foxmail.com,
lichunmail@163.com, 15011442144@139.com,
heximing15@nudt.edu.cn, ywyykathy@yeah.net,
wufanlqaz@163.com, zouqiaofang@163.com,
LXH910803@126.com

[2] State Key Laboratory of Remote Sensing Science, Aerospace Information Research Institute, Chinese Academy of Sciences, Beijing 100101, China
{dikc,gousheng,pengman,wanwh,liuzq,wangyx716}@radi.ac.cn

Abstract. The visible and near-infrared imaging spectrometer (VNIS) carried by the Yutu-2 rover is mainly used for mineral composition studies of the lunar surface. It is installed in front of the rover with a fixed pitch angle and small field of view (FOV). Therefore, in the process of target detection, it is necessary to precisely control the rover to reach the designated position and point to the target. Hence, this report proposes a vision-guided control method for in-situ detection using the VNIS. During the first 17 lunar days after landing, Yutu-2 conducted five in-situ scientific explorations, demonstrating the effectiveness and feasibility of this method. In addition, a theoretical error analysis was performed on the prediction method of the FOV. On-board calibration was completed based on the analysis results of the multi-waypoint VNIS images, further refining the relevant parameters of the VNIS and effectively improving implementation efficiency for in-situ detection.

Keywords: Yutu-2 rover · Visible and near-infrared imaging spectrometer · In-situ detection · Control method · On-board calibration

1 Introduction

The Chang'e-4 (CE-4) probe successfully landed in the Von Kármán crater of the South Pole–Aitken basin on the far side of the Moon at 10:26 (UTC + 8) on January 3, 2019 [1]. Its landing site, which is named Station Tianhe, was centered at 177.5991°E, 45.4446°S with an elevation of −5935 m with respect to the radius of the Moon [2–4].

© Springer Nature Singapore Pte Ltd. 2020
Y. Wang et al. (Eds.): IGTA 2020, CCIS 1314, pp. 267–281, 2020.
https://doi.org/10.1007/978-981-33-6033-4_20

Three scientific payloads were installed on the Yutu-2 rover, including a panoramic camera, lunar penetrating radar, and visible and near-infrared imaging spectrometer (VNIS), which were mainly used to study the morphologies, shallow structures, and mineralogical compositions of the traversed areas, respectively [5]. The VNIS instrument consisted of a visible/near-infrared (VIS/NIR) imager with an effective pixel size of 256 × 256 pixels (∼ 1 mm/pixel) and a shortwave infrared (SWIR) single-pixel detector [6]. The VIS/NIR imager operated in the range 450 − 945 nm with a spectral resolution of 2 to 7 nm, and the SWIR detector worked in the range 900 − 2395 nm, with a spectral resolution of 3 to 12 nm [7–9].

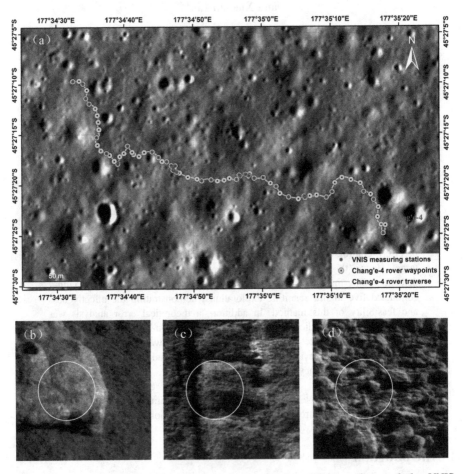

Fig. 1. Traverse map of the Yutu-2 rover during the first 17 lunar days and the VNIS measurements. (*a*) Base map is a high-resolution (0.9 m/pixel) digital orthophoto map (DOM) [12] generated from Lunar Reconnaissance Orbiter Camera Narrow Angle Camera imagery [13]; green lines are Yutu-2 center lines. (*b*), (*c*), and (*d*) 750-nm-band images of waypoints LE00303, LE00702 and LE00910, obtained by the VNIS, respectively showing stone, ruts, and the impact crater. The white circles show the field of view (FOV) of the SWIR detector.

During the first 17 lunar days after landing, the Yutu-2 rover performed a series of scientific exploration experiments, such as "stone detection," "rutting detection," and "impact pit detection," using the VNIS, and obtained spectral data for the lunar soil on the far side of the Moon, providing the fundamental data for the study of the mineral composition and space weathering of the lunar soil. The results based on the exploration data further revealed the material composition of the far side of the Moon, facilitating the improvement of formation and evolution models for the Moon [10, 11] (Fig. 1).

The VNIS was mounted at the front of the Yutu-2 rover at a 45° angle at a height of 0.69 m above the lunar surface [14]. The FOVs of the VIS/NIR and SWIR are narrow (8.5° × 8.5° and Φ 3.58°, respectively) [15]. As a result, approaching and pointing at a specific science target depends completely on the precise control of the moving rover [16]. In addition, because of the complex terrain and many impact craters in the CE-4 landing area, it is necessary not only to ensure the safety and reliability of the movement of the Yutu-2 rover but also to implement in-situ detection with the appropriate solar altitude and azimuth angles. These factors render detection experiments challenging, particularly when approaching a small rock or an impact crater and pointing at the desired location.

The stereo cameras of the Yutu-2 rover used for environment perceptions, including the navigation camera (Navcam), panoramic camera (Pancam) and hazard avoidance camera (Hazcam), acquired 3-D spatial information for the lunar surface [16], which could be employed in target approaching operations. However, the precise installation calibration of VINS was not conducted before launching [17]. The position and altitude of the VINS mounted on the rover were needed for accurate approaching and target pointing operations.

In the CE-4 mission, we developed a control method for VNIS in-suit detection based on visual guidance to approach detection targets safely and accurately, and in the process of implementation, we predicted the FOV of the VNIS and assisted the ground remote operators in evaluating and adjusting the detection azimuth. Furthermore, we performed error analysis using the VNIS spectral images from multiple waypoints, completed the on-board calibration based on the analysis results, and further refined the installation parameters of the VNIS. Now, the developed control method and the calibration result have been routine using in the mission operations of the Yutu-2 rover.

2 Control Method and Engineering Applications

Conducting in-situ detection using the VNIS in the complex terrain environment on the far side of the Moon presents various challenges. To overcome these challenges, we formulated an implementation scheme of "aiming from a distance, pausing, and fine-tuning when approaching." Based on the vehicle-mounted multiple imaging systems, we used visual guidance to realize the precise control of in-situ detection using the VNIS. Section 2.1 introduces the proposed control method in detail, which includes moving and turning, FOV prediction, and adjustment. Section 2.2 describes the implementation of in-situ detection of the VNIS in the CE-4 mission in detail.

2.1 Control Method

The proposed method of controlling in-situ detection using VNIS based on visual guidance consists of two main steps. The first step involves controlling the rover movement near the detection area and controlling the rover to turn toward the detection target according to the location of the detection target and the lunar rover so that it aims at the detection target. The second step involves controlling the rover to approach the target using the "short distance straight line" approach, during which the course angle is precisely adjusted using the predicted FOV of the VNIS. Figure 2 presents the overall control flowchart for in-situ detection using VNIS during the CE-4 mission, with the work introduced in this study highlighted in blue.

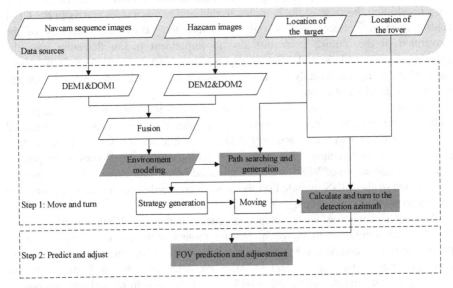

Fig. 2. Overall flowchart of the vision-based control method for the VNIS in-situ detection. (Color figure online)

Movement and Turning. As mentioned above, owing to the limitations of the installation angle, location, and FOV, we adopt a "segment" control method when implementing in-situ detection using the VNIS. Before moving, stereo images are acquired at every waypoint using the Navcam, and sometimes using the Hazcam. Local digital elevation models (DEMs) and digital orthophoto maps (DOMs) of 0.02 m grid spacing are routinely generated at each waypoint after the images are downlinked [18]. It is considered that only using the image data of the Navcam produces a blind area close to the rover body, which results in security risks when the rover is moving. In the CE-4 mission, we utilized a multi-source image processing method, which used the terrain data derived from the Hazcam to fill in the blind spots in the Navcam DEM. For the detailed process, see reference [19]. Based on the integrated DEM, we extracted several statistical indices to quantify the terrain features, including slope, roughness, and height gradient. The slope cost, roughness cost, height gradient cost, illumination, and communication shadow are weighted and superposed to generate a comprehensive cost map for path searching [20, 21]. Based on the established comprehensive cost map, the A* algorithm [22] is used to obtain a sequence of path points. The corresponding movement control command is then generated by combining the geometric parameters of the lunar rover, the motion parameters of the wheel system and mechanism, and the operating capability of the lunar rover [23].

When we control the rover such that it moves to a position close to the detection target, the "detection start point" is set for target detection. The following factors should be considered when selecting the "detection start point," which mainly considers distance, slope, and safety. Additionally, because the VNIS is a passive detection device, it is necessary to adjust the azimuth of the rover appropriately according to the change rule of the solar altitude and azimuth angles to avoid excessive shadow in the image [24].

FOV Prediction and Adjustment. Figure 3 shows the flowchart of the method of FOV prediction of the VNIS using Hazcam images. In this method, we chose Hazcam images for the following reasons. First, the installation mode of the Hazcam is similar to that of the VNIS. It is fixed in front of the lunar rover. The changes in the external orientation elements of the Hazcam and VNIS are only caused by changes in the altitude of the rover and are not affected by the control accuracy of other moving mechanisms, such as the Navcam, which is affected by the control accuracy of the mast yaw and pitch, among others. Second, the Hazcam uses a fish-eye lens with a large FOV, enabling observation of the terrain environment of the rover at a short distance. At such a short distance (e.g., within 2 m), the terrain is out of the FOV of the Navcam [18].

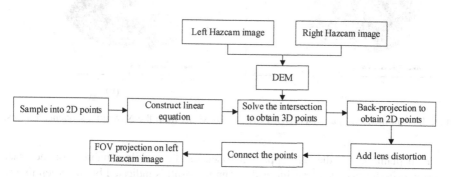

Fig. 3. Flowchart for prediction of the FOV of the VNIS on Hazcam image.

The FOV prediction can reflect whether the current azimuth satisfies the detection implementation requirements. If not, it can be realized by turning slightly to the left or right.

2.2 Engineering Applications

This section introduces the control process for VNIS in-situ detection, using stone detection on the 12th lunar day as an example. At the dormant waypoint, we found that approximately 7 m southwest of the rover, there was an unusually shaped stone near several successive impact craters. According to the analysis, these impact craters are relatively new and of significant scientific value.

According to the preliminary analysis of the terrain, the terrain near the stone was complex, and there were several small pits on both sides, making it difficult for the rover to approach. Combined with the change of the solar azimuth, to meet the requirements of the VNIS regarding the angle of the incident light, detection could only be conducted toward the southwest. Additionally, during the implementation process, the stone was likely to become shaded with changes in the rays from the Sun. In such a case, time is of urgent importance. The exposed volume of the stone was quite small, with a diameter of approximately 0.25 m and a height of only 0.06 m. Detection control in this case can be very challenging, as it is necessary to control the movement of the lunar rover precisely so that the VNIS FOV points to the stone as intended, under several strict constraints.

In this operation, the DEM and DOM with 0.02 m grid spacing were automatically generated as the basis for obstacle analysis and path searching using the sequence images from the Navcam. The targets to be detected were then selected in DEM&DOM data, as shown in Fig. 4.

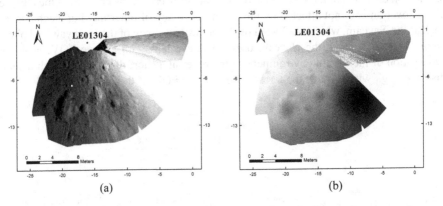

Fig. 4. Automatically generated DOM (*a*) and DEM (*b*) from the Navcam's sequence images.

Path planning for the safe approach based on the mobile capability of the lunar rover was performed. The specific path-planning result is indicated by the green line in Fig. 5(a), while the yellow sector represents the original turning. Due to the wheel design of the Yutu-2 lunar rover, every effort should be made to prevent the sieve-like wheel from being stuck by rock [18]. Therefore, the searched path should be checked by back-projection on the image mosaic stitched from Navcam images, which has a finer texture, as shown in Fig. 5(b). The planned path first avoids a conspicuous stone to the right front of the rover, as shown by arrow 7. Secondly, it bypasses the shallow impact craters, as shown by numbers 1–6 and 8 in Fig. 5(b). Towards the end, the left wheel rolls onto the edge of an impact crater, as shown by arrow 8, at a depth of approximately 5 cm.

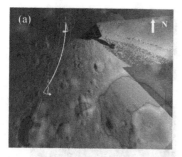
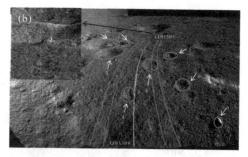

Fig. 5. Searched path projected on DOM (*a*) and Navcam image mosaic (*b*). Inset: the detection stone found in the Pancam image.

After the rover arrived at the end waypoint of the planned path, we predicted the FOV of the VNIS at this waypoint, as shown in Fig. 6. It was evident that the FOV of the VNIS is not pointing toward the target, and it was necessary to continue controlling the movement of the rover in the direction of the stone. According to terrain measurement, the distance between the mass center of the rover and the stone was approximately 1.45 m. Additionally, compared with the position of the center of the stone, the predicted FOV center deviated slightly to the left, accordingly, the rover had to turn to the right at a certain angle.

Fig. 6. Predicted FOV of the VNIS at the LE01305 waypoint (the red oval represents the FOV of the SWIR detector).

We subsequently controlled the movements of the rover several times and finally ensured that the FOV of the VNIS pointed toward the center of the stone at the LE01310 waypoint, as shown in Fig. 7.

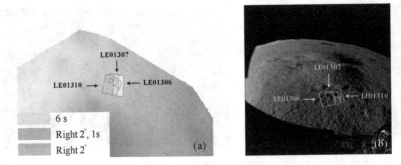

Fig. 7. Predicted FOVs of the VNIS on the DEM (*a*) and Hazcam image (*b*) corresponding to the three movements.

3   Error Analysis and on-Board Calibration

By comparing the actual VNIS images obtained using multiple waypoints with the predicted FOVs, we found that there was always a certain deviation between the two. This section presents an analysis of the theoretical error of the back-projection method based on the method of predicting the FOV of the VNIS, combined with the principle of photogrammetry and the law of error propagation. We then describe the completion of the on-board calibration of the VNIS placement matrix by using the deviation between the measured results and predicted data. We demonstrate that the deviation between the results and measured values is significantly reduced when the corrected parameters are used to predict the FOV.

3.1   Error Analysis

According to the method of predicting the FOV of the VNIS described in Sect. 2.1, the entire back-projection procedure involves three error propagation processes. These involve (1) the error of the DEM generated by the stereo Hazcam images, (2) the error of projecting the 3-D points onto the 2-D Hazcam image, (3) the insufficient accuracy of the placement matrix of the VNIS.

DEM Generation. The DEM of the Hazcam image can be obtained through bilinear interpolation of matched 3-D point clouds, which produce errors, mainly in the forms of (1) matching errors and (2) interpolation errors.

According to the principle of photogrammetry, the measurement error in the object space of the matching point can be obtained as follows [25]:

$$\sigma_Y = \frac{Y_p^2}{Bf}\sigma_p,$$

$$\sigma_X = \sqrt{\left(\frac{Y_p^2}{Bf}\right)^2 \left(\frac{x}{f}\right)^2 \sigma_p^2 + \left(\frac{Y_p}{f}\right)^2 \sigma_x^2} \tag{1}$$

$$\sigma_Z = \sqrt{\left(\frac{Y_p^2}{Bf}\right)^2 \left(\frac{y}{f}\right)^2 \sigma_p^2 + \left(\frac{Y_p}{f}\right)^2 \sigma_y^2}$$

where (X_p, Y_p, Z_p) are the object coordinates, (x, y) represent the image coordinates, B denotes the baseline of the stereo camera, f is the focal length, σ_p is the matching precision, and σ_x and σ_y represent the measurement errors of horizontal and vertical image plane coordinates, respectively. According to the law of error propagation, $\sigma_p = \frac{1}{3}\sigma_x$. Using this law in formula (2), we can obtain formula (2) as follows:

$$\sigma_Y = \frac{Y_p^2}{Bf} \cdot \frac{1}{3}\sigma_x = \frac{Y_p^2}{Bf} \cdot \frac{1}{3}\sigma_y$$

$$\sigma_X = \frac{Y}{f}\sigma_x \tag{2}$$

$$\sigma_Z = \frac{Y}{f}\sigma_y$$

The detection range of the VNIS is approximately 0.7 m [6] in front of the lunar rover. Thus, a matching error within 1 m can be obtained according to the relevant parameters of the Hazcam [26] and formula (2). Please refer to Table 1 for details, where $\sigma_p = 0.3$ pixel. It can be seen from Table 1 that the error in this distance range is less than 2 mm, and that the error in the y range direction is significantly higher than those in the other two directions.

$$\sigma_{match} = \sqrt{\sigma_X^2 + \sigma_Y^2 + \sigma_Z^2}$$

Table 1. Matching error of the Hazcam image. (unit: mm).

Distance	600	700	800	900	1000
σ_X	0.37	0.43	0.49	0.55	0.62
σ_Z	0.37	0.43	0.49	0.55	0.62
σ_Y	0.74	1.01	1.32	1.66	2.05
σ_{xyz}	0.91	1.18	1.49	1.84	2.23

In the CE-4 mission, the bilinear model is used to build DEMs. According to the accuracy formula and considering the maximum and minimum elevation errors, we can obtain the accuracy of the DEM grid, as shown in formula (3) [27]:

$$\sigma_{surf}^2 = \frac{2}{9}\sigma_{zmin}^2 + \frac{2}{9}\sigma_{zmax}^2 + \frac{3}{5}\sigma_T^2, \tag{3}$$

where σ_{surf} is the average accuracy of each grid point on the DEM; σ_{zmin} and σ_{zmax} represent the minimum and maximum elevation errors, respectively, in a certain distance range; and σ_T denotes the loss of interpolation accuracy caused by the bilinear surface. Considering that the terrain of the detection area was flat, and the overall slope was small, $\sigma_T = 0.3$ m was selected in this study. Further, it is known that the interpolation accuracy is approximately 0.5 mm within a range of approximately 0.7 m.

Projecting the 3-D Points onto the 2-D Hazcam Image. We denote the 3-D coordinates in the work coordinate system as (X_c, Y_c, Z_c), and the obtained image coordinates through the imaging model of the Hazcam are (x_c, y_c), where

$$x_c = \frac{X_c}{Z_c}f, \quad y_c = \frac{Y_c}{Z_c}f. \tag{4}$$

According to the law of error propagation [28], the errors of the corresponding image coordinates can be calculated using the following formulas:

$$\sigma_{x_c} = \sqrt{\left(\frac{f}{Z_c}\right)^2 \sigma_{X_c}^2 + \left(\frac{x}{Z_c}\right)^2 \sigma_{Z_c}^2}, \quad \sigma_{y_c} = \sqrt{\left(\frac{f}{Z_c}\right)^2 \sigma_{Y_c}^2 + \left(\frac{y}{Z_c}\right)^2 \sigma_{Z_c}^2} \tag{5}$$

where $(\sigma_{x_c}, \sigma_{y_c})$ are the projection errors after the propagation effect; $\sigma_{X_c}, \sigma_{Y_c}$ and σ_{Z_c} are measurement errors calculated in the first error propagation process; and, (x, y) are the image coordinates that come from the image in the first error propagation process. It can be seen that the DEM construction error will lead to the 2-D image coordinate error of back-projection onto the Hazcam image.

Inaccurate Placement Parameters. Because of inevitable collision or vibration in the launching and landing processes, instrument installation parameters are likely to change, which will lead to a deviation of the alignment with the detection target of the VNIS. Therefore, we chose the spectral data of multiple waypoints and compared the predicted FOV with the measured results to complete the on-board calibration of the installation parameters of the VNIS.

3.2   On-Board Calibration

Based on the above analysis, we obtained the in-situ detection data at four waypoints, as shown in Figs. 8, 9, 10 and 11. In these figures, subfigures (a) represent the left Hazcam images and subfigures (b) are enlarged views of local regions of the images in (a), where the red and green dots represent the measured and predicted centers of the SWIR, respectively. Subfigures (c) show the VIS/NIR images ($\sim$ 1 mm/pixel) @ 750 nm, where the red dots represent the center of the SWIR. Subfigures (d) depict the predicted FOVs of the VNIS. The center position of the SWIR in a VIS/NIR image is marked in reference [6]. Tables 2 and 3 list specific 2-D coordinates of the image points in the left Hazcam images and 3-D coordinates of the vehicle system, respectively.

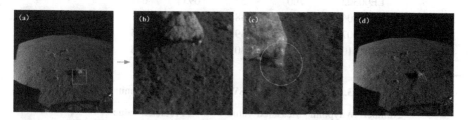

Fig. 8. First comparison between the predicted and measured results of the FOV of the VNIS, at waypoint LE01302.

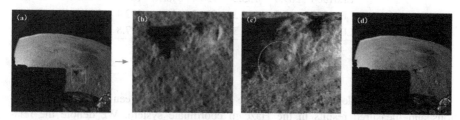

Fig. 9. Second comparison between the predicted and measured results of the FOV of the VNIS, at waypoint LE01306.

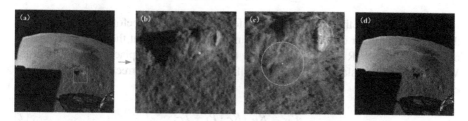

Fig. 10. Third comparison between the predicted and measured results of the FOV of the VNIS, at waypoint LE01307.

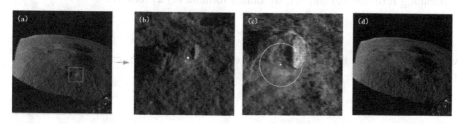

Fig. 11. Fourth comparison between the predicted and measured results of the FOV of the VNIS, at waypoint LE01310.

Table 2. Comparison of the SWIR center position (unit: pixel).

Number	Waypoint	Measured value		Predicted value		Deviation	
		x	y	x'	y'	Δx	Δy
1	LE00302	700	637	693	641	7	−4
2	LE01306	691	638	678	641	13	−3
3	LE01307	694	639	681	645	13	−6
4	LE01310	659	630	645	635	14	−5

Table 3. Comparison of the SWIR center position (unit: mm).

Number	Waypoint	Measured value			Predicted value		
		x	y	z	x'	y'	z'
1	LE01302	293.4	−190.8	−761	280.2	−194.7	−764.4
2	LE01306	269.0	−185.2	−750.4	248.7	−189.6	−748.8
3	LE01307	269.0	−184.3	−729.7	245.6	−192.4	−724.5
4	LE01310	215.4	−169.3	−745.7	189.6	−177.8	−746.0

$$\Delta x = x - x', \quad \Delta y = y - y'$$

Tables 2 and 3 demonstrate that there are deviations between the predicted positions and measured results in the Hazcam coordinate system. We denote the relationship between the VNIS and Hazcam as follows:

$$P_H = R_C \cdot P_I + T_C, \tag{6}$$

where R_C represents the positioning attitude matrix calibrated before launch, and T_C is the translation vector calibrated before launch. Considering that the position deviation of the center of SWIR is mainly caused by the deviation of the placement matrix of the VNIS and the translation effect is small, the following correction model can be established:

$$P'_H = R_M \cdot R_C \cdot P_I + T_C, \tag{7}$$

where P'_H is the coordinate after correction, R_M is the rotation matrix for correction. Combining formulas (6) and (7), we obtain formula (8) as follows:

$$P'_H - P_H = (R_M - E) \cdot R_C \cdot P_I, \tag{8}$$

where R_M represents the orthogonal matrix to be solved. $R_C \cdot P_I$ can be obtained from formula (6):

$$R_C \cdot P_I = P_H - T_C. \tag{9}$$

Finally, the following formula can be obtained:

$$P' = (R_M - E) \cdot C. \tag{10}$$

Because redundant observation results were obtained, the least squares method can be used to optimally estimate the calibration parameters. Simultaneously, to maintain the orthogonality of the matrix, we introduced the Euler angle parameter and established the following error formula:

$$v = P' - (R_M(o, p, k) - E) \cdot C. \tag{11}$$

where $R_M(o, p, k)$ represents the rotation matrix obtained using the Euler angle o, p, k. Based on the least squares principle, the results of attitude correction and the Euler angle parameters were obtained to be $(1.5604°, -2.2279°, -3.0516°)$. The root-mean-square deviation (REMS) of the Euclidean distance residuals of the corrected 3-D coordinates is 8.257 mm, and Table 4 lists the residual values in each direction. The above comparison demonstrates that the accuracy of the prediction results is signifi-

Table 4. Comparison of the calibrated results of the SWIR center position (unit: mm).

Number	Waypoint	Residual before optimization			Residual after optimization		
		Δx	Δy	Δz	$\Delta x'$	$\Delta y'$	$\Delta z'$
1	LE01302	13.2	3.9	3.4	−5.551	−2.051	−2.231
2	LE01306	20.3	4.4	−1.6	1.816	−2.827	−6.186
3	LE01307	23.4	8.1	−5.2	6.003	1.361	−9.561
4	LE01310	25.8	8.5	0.3	6.669	−1.836	−2.421

cantly improved after calibration, providing reliable measurement and positioning results for efficient detection of the VNIS.

4 Conclusion

This paper presented a successful method of the VNIS detection for the Yutu-2 rover based on visual guidance. By using vision-based measurement technologies, the driving path of the Yutu-2 rover was accurately controlled to ensure that the FOV of the VNIS was focused on the detected target. The method was verified in the CE-4 mission. The detection results successfully revealed the material composition on the far side of the Moon and deepened human understanding of its formation and evolution [29]. Furthermore, we conducted error analysis on the prediction method, deduced the accuracy model, and analyzed the sources of the prediction errors using photogrammetry and error propagation theories. We also established a prediction optimization model and solved for the parameters using the measured results of the multiple waypoints of the VNIS images. Additionally, we further refined the prediction results and on-board calibration of the VNIS installation parameters based on the analysis results. After verification, the optimized parameters were used for prediction, and the results were found to be much closer to the measured values. These results demonstrate the accuracy and efficiency of the VNIS in-situ detection and the usability of this approach in the ongoing CE-4 mission.

In the existing methods, the lunar rover is regarded as a rigid structure, and the impacts of the vehicle dynamics model (including the suspension structure) and the attitude on the planning are not considered. In the future, we will continue to study the lunar rover path-planning algorithm to further enhance the control precision of the lunar rover.

Acknowledgments. The Chang'e-4 mission was conducted by the Chinese Lunar Exploration Program. This study was supported by the National Natural Science Foundation of China (grant Nos. 41771488, 11773002, and 61573049).

References

1. Li, C.L., Liu, D., Liu, B., et al.: Chang'E-4 initial spectroscopic identification of lunar far-side mantle-derived materials. Nature **569**, 378–382 (2019)
2. Di, K., Liu, Z., Liu, B., et al.: Chang'e-4 lander localization based on multi-source data. J. Remote Sens. **23**(1), 177–180 (2019)
3. Liu, J.J., Ren, X., Yan, W., et al.: Descent trajectory reconstruction and landing site positioning of Chang'E-4 on the lunar farside. Nat. Commun. **10**, 4229 (2019)
4. Wang, J., Wu, W.R., Li, J., et al.: Vision Based Chang'e-4 Landing Point Localization (in Chinese). Sci Sin Tech **2020**(50), 41–53 (2020)
5. Jia, Y., Zou, Y., Ping, J., et al.: Scientific objectives and payloads of Chang'E-4 mission. Planet. Space Sci. **162**, 207–215 (2018)
6. Li, C.L., Wang, Z., Xu, R., et al.: The Scientific Information Model of Chang'e-4 Visible and Near-IR Imaging Spectrometer (VNIS) and In-Flight Verification. Sensors **19**(12), 2806 (2019)
7. He, Z., Li, C., Xu, R., et al.: Spectrometers based on acoustooptic tunable filters for in-situ lunar surface measurement. Appl. Remote Sens. **13**(2), 027502 (2019)
8. Gou, S., Di, K., Yue, Z.Y., et al.: Lunar deep materials observed by Chang'e-4 rover. Earth Planet. Sc. Lett. **528**, 115829 (2019)
9. Gou, S., Yue, Z.Y., Di, K., et al.: In situ spectral measurements of space weathering by Chang'e-4 rover. Earth Planet. Sc. Lett. **535**, 116117 (2020)
10. Gou, S., Di, K., Yue, Z.Y., et al.: Forsteritic olivine and magnesium-rich orthopyroxene materials measured by Chang'e-4 rover. Icarus **345**, 113776 (2020)
11. Huang, J., Xiao, Z., Xiao, L., et al.: Diverse rock types detected in the lunar south pole-Aitken Basin by the Chang'E-4 lunar mission. Geology **48**(7), 723–727 (2020)
12. Liu, B., Niu, S., Xin, X., et al.: High precision DTM and DOM generating using multi-source orbital data on Chang'e-4 landing site. In: International Archives Photogrammetry Remote Sensinh Spatial Information Sciences Copernicus Publications, Enschede, The Netherlands, pp. 1413–1417 (2019)

13. Robinson, M.S., Brylow, S.M., Tschimmel, M., et al.: Lunar Reconnaissance Orbiter Camera (LROC) instrument overview. Space Sci. Rev. **150**, 81–124 (2010)
14. Hu, X., Ma, P., Yang, Y., et al.: Mineral abundances inferred from in-situ reflectance measurements of Chang'E-4 landing site in South Pole-Aitken basin. Geophys. Res. Lett. **46**, 943–9447 (2019)
15. Li, C.L., Xu, R., Lv, G., et al.: Detection and calibration characteristics of the visible and near-infrared imaging spectrometer in the Chang'e-4. Rev. Sci. Instrum. **90**, 103106 (2019)
16. Liu, Z., Di, K., Li J., et al.: Landing Site topographic mapping and rover localization for Chang'e-4 mission. J. Sci. China Inform. Sci., **63**, 140901:1–140901:12 (2020)
17. Xu,R., Lv, G., Ma, Y., et al.: Calibration of Visible and Near-infrared Imaging Spectrometer (VNIS) on lunar surface. In: Proceedings of SPIE 9263, Multispectral, Hyperspectral, and Ultraspectral Remote Sensing Technology, Techniques and Applications V. 926315. SPIE Asia-Pacific Remote Sensing, Beijing (2014)
18. Wang, Y., Wan, W., Gou, S., et al.: Vision-based decision support for rover path planning in the Chang'e-4 Mission. Remote Sens. **12**, 624 (2020)
19. Wang, J., Li, J., Wang, S., et al.: Computer vision in the teleoperation of the Yutu-2 rover. In: Remote Sensing and Spatial Information Sciences ISPRS Geospatial Week, pp. 595–602. ISPRS, Gottingen (2020)
20. Chen, X., Yu, T. C.: Design and implementation of path planning algorithm based on the characteristics of complex lunar surface. In: 6th National Symposium on Flight Dynamics Technology, pp. 248–53. Kunming (2018)
21. Yu, T., Fei, J., Li, L., et al.: Study on path planning method of lunar rover. Deep Space Explore. **6**, 384–390 (2019)
22. Wu, W., Zhou, J., Wang, B., et al.: Key technologies in the teleoperation of Chang'E-3 Jade Rabbit rover. Sci. China Inf. Sci. **44**, 425–440 (2014)
23. Shen, Z., Zhang, W., Jia, Y., et al.: System design and technical characteristics analysis of Chang'e-3 lunar rover. Spacecr Eng. **24**, 8–13 (2015)
24. Yu, T., Liu, Z., Rong, Z., et al.: Implementation strategy of visible and near-infrared imaging spectrometer on Yutu-2 rover based on vision measurement technology. In: Remote Sensing and Spatial Information Sciences ISPRS Geospatial Week (2020)
25. Peng, M., Wan, W., Wu, K., et al.: Topographic mapping capability analysis of Chang'e-3 Navcam stereo images and three-dimensional terrain reconstruction for mission operations. Remote Sens. **18**, 995–1002 (2014)
26. Chen, J., Xing, Y., Teng, B., et al.: Guidance, navigation and control technologies of chang'E-3 lunar rover. Sci. China Technol. **44**, 461–469 (2014)
27. Li, Z., Zhu, Q., Gold, C.: Digital elevation model. Wuhan University Press, Wuhan (2001)
28. Wang, Z.: Principles of Photogrammetry (with Remote Sensing), Publishing House of Surveying and Mapping: Beijing pp. 312–313. (1990)
29. Gou, S., Yue, Z., Di, K., et al.: Impact melt breccia and surrounding regolith measured by Chang'e-4 rover. Earth Planet. Sci. Lett. **544**, 116378 (2020)

Deep Attention Network for Remote Sensing Scene Classification

Yuchao Wang[1(✉)], Jun Shi[2], Jun Li[3], and Jun Li[2]

[1] Space Star Technology Co., Ltd, Beijing 100086, China
wangyuchao503@sina.com
[2] School of Software, Hefei University of Technology, Hefei 230601, China
juns@hfut.edu.cn, junl@mail.hfut.edu.cn
[3] Jiangxi Aerospace Pohuyun Technology Co., Ltd., Nanchang 330096, China
lijun@htphy.com

Abstract. Scene classification plays a crucial role in understanding high-resolution remote sensing images. In this paper, we introduce a novel scene classification method based on deep attention network. It aims to integrate the attention mechanism and Convolutional Neural Network (CNN) to learn the more discriminative features from the scene content. Concretely we use three branches (main, spatial attention and channel attention branches) to explore the scene features. The main branch is the feature maps of the backbone CNN which are used to describe the high-level semantic features of scene. The spatial attention branch aims to investigate the long-range contextual dependencies through non-local operation. The channel attention branch intends to exploit the key semantic response. Finally, the three branches are fused to extract the more representative features and suppress irrelevant features. Experimental results on AID and NWPU-RESISC45 remote sensing scene datasets demonstrate the effectiveness of our method.

Keywords: Remote sensing · Scene classification · Feature learning · Attention

1 Introduction

Scene classification is a significant task for remote sensing image understanding, which aims to assign a semantic label to a remote sensing image according to its scene content. However, there exist the high intraclass diversity and interclass similarity. Therefore, it is necessary to extract the discriminative feature representation from the remote sensing images which contributes to scene classification.

Traditional scene classification methods use the low-level features (e.g. color, texture, shape) to characterize the scene content. To bridge the gap between the low-level features and high-level semantics, Bag-of-Words (BoW) based methods have been applied in remote sensing images and effectively improve the performance of scene classification. However, these methods suffer from feature selection. More importantly, they cannot characterize the image content accurately due to the intraclass diversity and interclass similarity of scenes and thus generate weak feature

© Springer Nature Singapore Pte Ltd. 2020
Y. Wang et al. (Eds.): IGTA 2020, CCIS 1314, pp. 282–289, 2020.
https://doi.org/10.1007/978-981-33-6033-4_21

representation. With the advent of Convolutional Neural Network (CNN) which aims to automatically learn high-level features via hierarchical deep structure, the CNN based methods achieve the impressive performance on remote sensing scene classification [1, 2]. However, most of these methods generally extract the features from the whole image through a pre-trained CNN model. Consequently, they fail to distinguish the crucial features and irrelevant features within the image in the process of global feature learning. In the past few years, the attention mechanism [3–6] has been proposed with the aim of obtaining the crucial information and suppressing redundant information. It mainly falls into two categories: the spatial attention which aims to discover the spatial contextual information, and the channel attention which intends to explore the interdependencies between feature maps and thus boost the sematic-specific response. In the field of remote sensing scene classification, APDC-Net [7] use a spatial attention pooling layer to concentrate the local features of scene. RADC-Net [8] introduce a residual attention block to capture the local semantics in the spatial dimension. To further exploit the advantage of the spatial and channel attention modules, Attention GANSs [9] integrate both of them into the discriminator of Generative Adversarial Networks (GANs) and thus capture the contextual information. SAFF [10] use spatial-wise and channel-wise weightings to enhance the responses of the representative regions within the scene. These attention-based methods obtain the better performance of scene classification compared to the traditional CNN-based methods. However, most existing spatial attention-based methods fail to consider the long-range dependency and thus cannot describe the global context effectively.

In this paper, a novel remote sensing scene classification method based on deep attention network is proposed. We design a feature aggregation block consisting of three branches (main, spatial attention and channel attention branches) to generate more discriminative feature representation of scene, as shown in Fig. 1. The main branch preserves the feature maps of the CNN which can characterize the high-level features of scene. Motivated by the SENet block [4], the channel attention branch uses the Global Average Pooling (GAP) and Global Max Pooling (GMP) to aggregate the spatial information of feature maps and then fuse the aggregated information to explore the channel-wise relationships. Consequently, it can effectively learn the importance of each feature map. The spatial branch applies the simplified the non-local (NL) block [5, 6] to capture the long-range dependencies. Therefore, it can obtain the importance of each spatial position. Finally, the outputs of three branches are fused to produce the more representative attention-based feature maps for improving the classification performance. Experimental results on two public datasets (i.e. AID and NWPU-RESISC45) validate the effectiveness of our method for remote sensing scene classification.

The rest of the paper is arranged as follows: Sect. 2 present the proposed method. Section 3 shows the experimental results and Sect. 4 draws the conclusion.

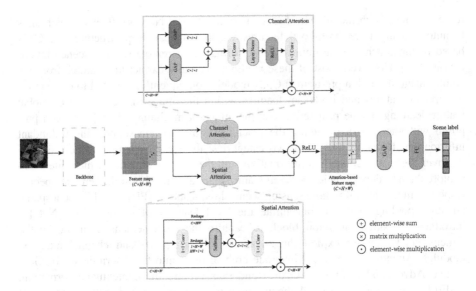

Fig. 1. Overview of the proposed method for remote sensing scene classification.

2   The Proposed Method

2.1   Overview

As shown in Fig. 1, the pre-trained DenseNet-121 [11] is taken as the backbone CNN and the feature maps of the last conv-layer with the size $C \times H \times W$ which denote the number of channels, height and width can be obtained. Then the subsequent part after the backbone contains three branches: main branch, channel attention branch and spatial attention branch. The main branch retains the original feature maps. Furthermore, the feature maps are fed into the channel and spatial attention branches in parallel. For the channel attention branch, the spatial information of feature maps is aggregated through GAP and GMP and then the aggregated features with the size $C \times 1 \times 1$ are inputted into the sigmoid function. After 1×1 convolution, layer normalization and activation function, the importance weights of each feature map can be learned. The final channel-wise attention features can be obtained by multiplying the learned weights and the original feature maps. On the other hand, the spatial attention branch groups the features of all positions together and then uses 1×1 convolution and softmax to obtain the global contextual features in terms of the simplified the non-local block [5, 6]. Consequently, the importance weights of each spatial position can be gained. Similar to the final channel-wise attention features, the final spatial attention features are produced by multiplying the learned weights and the original feature maps. The features from the main branch are fused with the features from the other two branches thus the more discriminative attention-based feature maps are generated. Cross entropy loss is used to train the whole network after GAP and fully connected (FC) layer.

2.2 Channel Attention

The feature map from each channel can represent the scene semantic-specific response. Therefore, it can capture the meaningful scene semantics through mining the interdependencies between feature maps. Inspired by the SENet [4] and GCNet [6], we use the channel attention branch to explore the inter-channel dependencies and thus enhance the sematic-specific representational ability.

Given the feature maps $U = [u^1, \cdots, u^C]$ of the backbone CNN with the size $C \times H \times W$, we aggregate the spatial features of feature map through two pooling operations (GAP and GMP). The interdependencies of channels $W_C = [w_C^1, \cdots, w_C^C] \in \mathbb{R}^C$ can be obtained through 1×1 convolution, layer normalization and activation function:

$$W_C = W_v^2\big(\mathrm{ReLU}\big(\mathrm{LN}\big(W_v^1(U_{GAP} \oplus U_{GMP})\big)\big)\big) \tag{1}$$

where $U_{GAP}, U_{GMP} \in \mathbb{R}^C$ denote the output generated by GAP and GMP, W_v^1 and W_v^2 are the weight matrices of 1×1 convolution. $\mathrm{LN}(\cdot)$ and $\mathrm{ReLU}(\cdot)$ denote the layer normalization and ReLU activation function respectively. Note that layer normalization is used as a regularizer in favor of generalization [6]. The final channel-wise attention features $U_C = [u_C^1, \cdots, u_C^C]$ can be gained:

$$u_C^c = w_C^c \cdot u^c, \ 1 \le c \le C \tag{2}$$

2.3 Spatial Attention

Considering there exist the high intraclass variation, the representative parts within the scene is significant to classifying different scene semantics. Therefore, we use the spatial attention branch to exploit the importance of each spatial position and thus discover the spatial contextual features. Particularly we apply the simplified the non-local block derived from GCNet [6] to further capture the long-range dependencies and boost the perception capability of global context.

Given the c-th feature map $u^c = \{u_i^c\}_{i=1}^N$ where $N = H \times W$, $1 \le c \le C$, according to the GCNet [6], a 1×1 convolution W_t^1 and softmax are used to generate the attention weights and then use the attention pooling to gain the global contextual features. Then a 1×1 convolution W_t^2 is applied for feature transformation. Finally, we use the element-wise multiplication to achieve the feature aggregation from the global contextual features to the features of each spatial position. The simplified NL block can be expressed as follows:

$$\tilde{u}_i^c = u_i^c \cdot \left(W_t^2 \sum\nolimits_{j=1}^N \frac{\exp\big(W_t^1 u_j^c\big)}{\sum_{k=1}^N \exp\big(W_t^1 u_k^c\big)} u_j^c \right) \tag{3}$$

where $\tilde{u}^c$ denotes the output of the simplified NL block and thus the final spatial attention features can be described as $U_S = [\tilde{u}^1, \cdots, \tilde{u}^C]$.

2.4   Branch Fusion

The features of the main branch characterize the high-level features of the original feature maps and they are enriched by fusing the final channel-wise attention features and spatial attention features. The final attention features of remote sensing scene image can be described:

$$\tilde{U} = ReLU(U \oplus U_C \oplus U_S) \tag{4}$$

where $\oplus$ indicates element-wise sum. Afterwards, cross entropy loss function is applied to train the proposed deep attention network after the operations of GAP and FC.

3   Experimental Results

To evaluate the scene classification performance of our method, two public remote sensing scene datasets (AID [1] and NWPU-RESISC45 [2]) are used in the experiments. The AID dataset contains 30 scene classes and 10000 images with the size of 600×600 pixels. The images are captured under different countries and imaging conditions, and thus there exist the high intraclass variation. The NWPU-RESISC45 dataset consists of 700 scene images from 45 scene classes with the size of 256×256 pixels. Compared with AID, NWPU-RESISC45 dataset has more complex scene variation, higher intraclass diversity and higher interclass similarity.

According to the experimental protocol [1, 2], the training ratio of the AID dataset is 50% and the ratio of the NWPU-RESISC45 dataset is 20%. The overall accuracy and standard deviations on ten rounds are taken as the final classification results. Furthermore, the confusion matrix is also to assess the scene classification performance. We compare the proposed method with the classic CNN-based scene classification methods (e.g. CaffeNet [1, 2], VGG-VD-16 [1, 2] and GoogLeNet [1, 2]) and the attention-based methods (e.g. APDC-Net [7], RADC-Net [8], Attention GANs [9] and SAFF [10]).

Table 1 shows the classification results of these methods. Compared with CaffeNet [1, 2], VGG-VD-16 [1, 2] and GoogLeNet [1, 2] which only use the pre-trained CNN model for scene classification, the attention-based methods have better classification performance. It indicates that the attention mechanism is conductive to obtaining the more representative features. Particularly the methods based on spatial and channel fusing attention (e.g. Attention GANs [9] and SAFF [10]) outperforms the methods based on a single attention mechanism (e.g. APDC-Net [7] and RADC-Net [8]). It verifies that the spatial-wise and channel-wise attention can better enhance the representational power of features. In contrast with other methods, our method is more effective. It may be explained that the spatial attention branch guided by the simplified NL block can better capture the long-range dependencies within feature map, and the improved channel attention branch boost the scene sematic-specific representational capability. More importantly, the fused three branches make full use of the high-level, channel-wise and spatial information.

Table 1. Overall accuracies (%) and standard deviations on AID and NWPU datasets

Methods	AID (50% train)	NWPU-RESISC45 (20% train)
CaffeNet [1, 2]	89.53 ± 0.31	85.16 ± 0.18
VGG-VD-16 [1, 2]	89.64 ± 0.36	90.36 ± 0.18
GoogLeNet [1, 2]	86.39 ± 0.55	86.02 ± 0.18
APDC-Net [7]	92.15 ± 0.29	87.84 ± 0.26
RADC-Net [8]	92.35 ± 0.19	87.63 ± 0.28
Attention GANs [9]	94.93 ± 0.21	90.58 ± 0.23
SAFF [10]	93.83 ± 0.28	87.86 ± 0.14
Our method	**95.22 ± 0.34**	**91.78 ± 0.19**

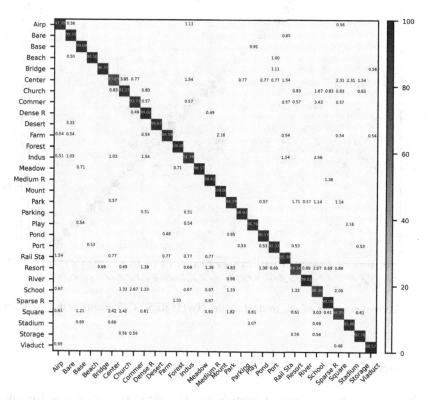

Fig. 2. Confusion matrix of our method on the AID dataset with 50% training ratio.

Figures 2 and 3 illustrate the confusion matrices of our method on these two datasets. For the results of AID dataset with 30 classes shown in Fig. 2, our method achieves the superior classification results for most of scene classes. Particularly there are 21 classes with the classification accuracy over 95% and the accuracies of challenging scene classes (e.g. dense residential, medium residential, sparse residential) reach 99.02%, 98.62% and 98%. As for the results of NWPU-RESISC45 dataset, our method basically has the better classification performance. The accuracies of 33 scene classes exceed 90%. It indicates that our method achieves the separability of different scene classes through fusing the high-level, channel-wise and spatial attention features.

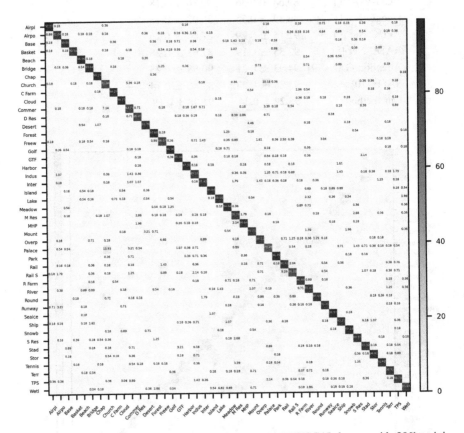

Fig. 3. Confusion matrix of our method on the NWPU-RESISC45 dataset with 20% training ratio.

4　Conclusion

In this paper, a novel remote sensing scene classification method based on deep attention network is introduced. It uses the spatial attention branch guided by the simplified NL block to explore the long-range contextual features and applies the channel attention branch to capture the crucial semantic response. Finally, the high-level features from main branch are fused with the other two attention branches and further obtain the more discriminative features and suppress irrelevant features. Experimental results on remote sensing scene classification verify our method has better classification results and the attention mechanism contributes to exploring more representational features.

References

1. Xia, G., Hu, J., Hu, F., Shi, B., Bai, X., Zhong, Y., Zhang, L., Lu, X.: AID: a benchmark data set for performance evaluation of aerial scene classification. IEEE Trans. Geosci. Remote Sens. **55**(7), 3965–3981 (2017)
2. Cheng, G., Han, J., Lu, X.: Remote sensing image scene classification: benchmark and state of the art. Proc. IEEE **105**(10), 1865–1883 (2017)
3. Vaswani, A.: Attention is all you need. In: Advances in Neural Information Processing Systems (NIPS), pp. 6000–6010 (2017)
4. Hu, J., Shen, L., Sun, G.: Squeeze-and-excitation networks. In: IEEE Computer Vision and Pattern Recognition (CVPR), pp. 7132–7141 (2018)
5. Wang, X., Girshick, R., Gupta, A., He, K.: Non-local neural networks. In: Proceedings of the IEEE Conference on Computer Vision and Pattern Recognition, pp. 7794–7803 (2018)
6. Cao, Y., Xu, J., Lin, S., Wei, F., Hu, H.: Gcnet: Non-local networks meet squeeze-excitation networks and beyond. In: Proceedings of the IEEE International Conference on Computer Vision Workshops (2019)
7. Bi, Q., Qin, K., Zhang, H., Xie, J., Li, Z., Xu, K.: APDC-Net: Attention pooling-based convolutional network for aerial scene classification. In: IEEE Geoscience and Remote Sensing Letters (2019)
8. Bi, Q., Qin, K., Zhang, H., Li, Z., Xu, K.: RADC-Net: A residual attention based convolution network for aerial scene classification. Neurocomputing **377**, 345–359 (2020)
9. Yu, Y., Li, X., Liu, F.: Attention GANs: unsupervised deep feature learning for aerial scene classification. IEEE Trans. Geosci. Remote Sens. **58**(1), 519–531 (2019)
10. Cao, R., Fang, L., Lu, T., He, N.: Self-attention-based deep feature fusion for remote sensing scene classification. IEEE Geoscience and Remote Sensing Letters (2020)
11. Huang, G., Liu, Z., Van Der Maaten, L., Weinberger, K.: Densely connected convolutional Networks. In: IEEE Conference on Computer Vision and Pattern Recognition (CVPR), Honolulu, HI, USA, pp. 2261–2269, July 2017

Thin Cloud Removal Using Cirrus Spectral Property for Remote Sensing Images

Yu Wang[1,2(✉)], Xiaoyong Wang[1,2], Hongyan He[1,2], and Chenghua Cao[1,2]

[1] Beijing Institute of Space Mechanics and Electricity, Beijing 100094, China
93031@163.com
[2] Key Laboratory for Advanced Optical Remote Sensing Technology of Beijing, Beijing 100094, China

Abstract. The atmosphere and cloud affects the quality of multi-spectral remote sensing images. At present, algorithms and software for atmospheric correction are widely used, but removing the effects of clouds in multi-spectral remote sensing images is still a major challenge, this paper presents a study of utilizing the cirrus band for this purpose. A cloud removal algorithm, using the spectral property of cirrus clouds, is developed for thin cloud contamination correction. Taking a Landsat-8 image affected by cloud as an example, after the cloud removal algorithm is executed, the thin cloud is removed. Use a reference image without cloud influence to further verify the cloud removal results. In a homogenous surface area, there is a certain linear relationship between the DN value of the visible band and the cirrus band. After cloud removal, the spatial correlation coefficients are all above 0.87, which is significantly higher than that before. The change of reflectance in each band and the improvement of spatial correlation coefficient fully verify the effectiveness and reliability of the algorithm.

Keywords: Thin cloud removal · Cirrus spectral property · Remote sensing images

1 Introduction

The scattering and absorption of solar energy by atmosphere have effects on acquisition of optical satellite images. The presence of cloud reduces the quality and limits the use of the remote sensing image. Due to the spatial uncertainty and temporal variability of clouds, coupled with the limited cloud bands in remote sensing, the removal of clouds for multi-band remote sensing images still faces great challenges [1]. Therefore, it is necessary to study thin cloud removal algorithms to improve data usage.

According to the spatial property of clouds, cloud removal algorithms can be divided into thin cloud removal algorithms and thick cloud removal algorithms [2]. Thick clouds cannot penetrate through the solar radiation, and the remote sensing images covered by thick clouds hardly contain any information on the ground. There is a blind zone, and the ground features can hardly be restored. Cirrus clouds are generally located at the top of the troposphere and the lower part of the stratosphere. The scattered radiation of cirrus clouds is an important interference source for space target

© Springer Nature Singapore Pte Ltd. 2020
Y. Wang et al. (Eds.): IGTA 2020, CCIS 1314, pp. 290–297, 2020.
https://doi.org/10.1007/978-981-33-6033-4_22

identification. Cirrus clouds are composed of ice crystal particles of various shapes. The scattering properties of cirrus particles will directly affect the climate model, the radiation transmission of light in the atmosphere, remote sensing detection, and atmospheric detection. Therefore, it is great significance to study the removal of cirrus clouds for remote sensing images.

The existence of thin cloud prevents the remote sensing ground observation system from correctly acquiring surface information, which affects the quality of remote sensing image [3, 4]. The cloud removal algorithm based on spectral property considers the influence of thin clouds during the radiation transmission process, and uses the difference in the spectral responses of thin clouds and ground features in different spectral ranges to establish corresponding experience or physical relationships to eliminate the influence of thin clouds. The key is to find out the spectral difference between the thin cloud and the ground features in different spectral bands.

Most remote sensing earth observation systems set the band which is a spectral range where absorption in the atmosphere is weak and energy is easier to pass. When there is a thin cloud, these bands will also capture the atmospheric information while acquiring the information of the ground [5, 6]. On the other hand, in the spectral ranges of 1.3–1.5 μm and 1.8–2.0 μm, the absorption of water vapor exceeds 80%, and in the bottom atmosphere, there is a lot of water vapor, and the electromagnetic energy in these bands is absorbed by the water vapor [7]. It cannot reach the ground, so remote sensing devices equipped with these band sensors only capture the radiant energy reflected by ice crystals and other substances in thin clouds at high altitudes.

Therefore, only the thin cloud information is recorded in these bands without being affected by the ground features, which is also the wavelength range of the band used by the remote sensing system to detect clouds. The current methods for removing clouds from remote sensing images can be divided into two categories [8–10]: The first category is based on DOS. This method may cause the image loss parts of the information. The other category is based on physical models. This method establishes a cloud radiation scattering model, understands the physical mechanism of image degradation, and inverts and restores the results. The restored image has a real effect and is close to the original scene of the degraded foreground.

2 Methodology

2.1 Cirrus Spectral Property

Cirrus clouds usually appear at altitudes above 6 km. These clouds are mainly composed of ice crystals. In the vicinity of the 1.38 μm band, it has strong absorption of water vapor. Due to the large amount of water vapor in the lower atmosphere, the main radiant energy captured in the 1.38 μm band (cirrus band) comes from the reflection of ice crystals. Therefore, the 1.38 μm band can detect cirrus spectral property. It can be used to correct cirrus cloud pollution in other bands (visible and near infrared). The results show that the clouds appear bright white in the cirrus band, while other surface information appears dark gray due to the absorption of water vapor. This paper is based on the cirrus spectral property to study the cloud removal processing.

The Landsat-8 satellite was successfully launched on February 11, 2013. The satellite is equipped with OLI and TIRS sensors. Added the coastal zone bands that are mainly used for coastal zone observation and the cirrus cloud bands that can be used for cloud detection. As an example, the below image in Fig. 1 is the true color composite image of an area near Shanghai, China recorded on May 17, 2016. The central position of the image is the result of multiple surface types and cloud coverage. The right image is a cirrus band. Compared with the picture on the left, the distribution intensity information of cirrus clouds can be obtained intuitively.

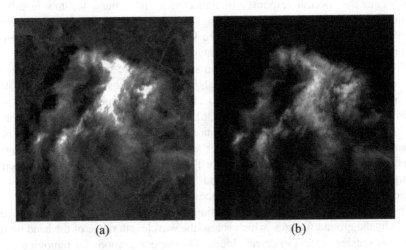

(a)   (b)

Fig. 1. (a) True color image (b) Corresponding image in the cirrus band

2.2 Removal of Thin Cloud Effect

Theoretical Derivation
The general radiative transfer equation affected by thin clouds is

$$r_{\lambda_0} = T_{w,\lambda_0} r_{c,\lambda_0} + T_{w,c,\lambda_0} \frac{r_{s,\lambda_0}}{1 - r_{s,\lambda_0} \bullet r_{c,\lambda_0}} \tag{1}$$

For the band with a center wavelength of 1.38 μm, there is:

$$r_{1.38} = T_{w,1.38} r_{c,1.38} \tag{2}$$

In the visible band, the absorption of water vapor can be ignored. Therefore, the transmittance of water vapor is a fixed value. If you ignore the geometric relationship between the sun and the observation, it will have $T_{\lambda_0,1.38} = 1$.

The apparent reflectance of a visible band can be expressed as:

$$r_{\lambda_0} = r_{c,\lambda_0} + \overline{r_{\lambda_0}}$$ (3)

The study found that the reflectivity of the thin cloud in the 1.38 μm band is linearly related to the reflectivity of the thin cloud in the 0.4–1.0 μm band:

$$r_{c,1.38} = \Gamma r_{c,\lambda_0} + \delta$$ (4)

To find the conversion coefficient of thin cloud Γ becomes the key to remove the influence of thin cloud. First, consider a simple case when the area acquired by a remote sensing image is a homogenous area, such as a calm ocean surface. Then, the variable δ will be constant, and the $r_{1.38}$ and r_{c,λ_0} scatterplot will be shown in Fig. 2(a). The slope value of the straight line is the thin cloud conversion coefficient Γ. The coefficient δ is composed of two parts, one part is the scattering effect from the atmosphere under the cloud layer, and the other part is the energy of the uniform surface. The x-axis intercept is the mixture of the minimum reflectivity of the ground surface under the clouds and the atmospheric scattering under the clouds.

For more complicated land surfaces, the surface reflectance will change with the ground features. The reflected energy of the complex surface affect the measured value of the sensor, but has no effect on the 1.38 μm band. Therefore, the data points are distributed on the right side of the straight line. In this case, the scatter plot of 1.38 μm is shown in Fig. 2(b). Therefore, the straight line slope value is the thin cloud conversion coefficient Γ.

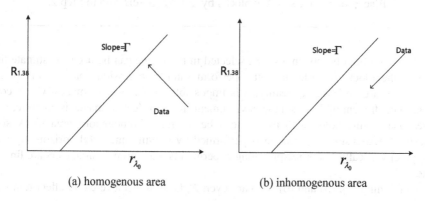

(a) homogenous area   (b) inhomogenous area

Fig. 2. Schematic diagram of calculating the thin cloud conversion factor Γ

The Effect of Thin Cloud Model

The Landsat-8 OLI cirrus band image provides a new opportunity to remove the thin cloud effect. Let $L_i(x, y)$ be the digital number (DN) value recorded by the OLI at band i, $i = 1, 2 \ldots 7$, for the pixel at location (x, y). It can be written as the following formula:

$$L_i(x, y) = N_i^o(x, y) + N_i^c(x, y) \tag{5}$$

Where $N_i^o(x, y)$ is the true DN associated with the land cover, and $N_i^c(x, y)$ is the contribution from cirrus cloud that is a function of cloud density. The purpose of cirrus cloud removal is to recover $N_i^o(x, y)$.

According to the previous description, $N_i^c(x, y)$ is linearly related to the image recorded in the cirrus band $c(x, y)$,

$$N_i^c(x, y) = \Gamma[c(x, y) - minc(x, y)] \tag{6}$$

Firstly, we select an area with a single surface type and no cloud coverage in this area, then it can be considered that the DN value of the pixels in this area is not affected by cirrus clouds. The transmittance value Γ can be obtained by the scatter plot of the DN value of the cloud-containing pixel and the cirrus band image, and then performing linear regression calculation.

In order to implement this method for a given band, the following procedure was developed to search for homogeneous blocks and use them to estimate Γ.

Step 1: $\tau = 0$; Divide the imaged scene into N by N blocks.

Step 2: Run regression for each block between the data to correct and the data in cirrus band.

Step 3: Find the highest R^2, R_{max}^2.

   If $R_{max}^2 > f$,

   Output the corresponding correlation coefficient Γ.

   Else $\tau = \tau + 1$; shift the blocks by $\tau * M$ pixels, and to Step 2.

End.

The maximum homogenous area selected in the image can be used to estimate the size of the block. In order to better perform sample regression analysis, the block should be large enough. In reality, it is impossible to obtain two images with a correlation coefficient of 1. When the environment in the collected images is complicated, a reasonable threshold (such as 0.9) can be set for a homogenous area block size threshold. Exhaustive contraction is performed by moving one pixel horizontally and one pixel vertically. After reciprocating experiments, the optimal parameters are finally determined.

Combining Eq. (5) and (6), we can revert $N_i^o(x, y)$ to achieve cloud effect removal using

$$N_i^o(x, y) = L_i(x, y) - \Gamma_i[c(x, y) - minc(x, y)] \tag{7}$$

There is no need to set the threshold for generating a cloud coverage, because the DN value of the no cloud image pixels will be close to the minimum value in cirrus band and $c(x, y) - minc(x, y)$ will be about zero, so these pixels will not change.

3 Experimental Results and Analysis

In this paper, a scene of Landsat-8 satellite remote sensing image is selected as the experimental data. The remote sensing data acquisition time is May 17, 2016, and the row number is P118/R040. A sub-image is cropped from the image as the research image, and part of it is covered by clouds. The area where the image is located is near Shanghai. It can be seen from the image that different kinds of thin clouds are distributed in the upper left side of the image and the l upper middle part of the image. In the study area, there are different types of features, such as water bodies and buildings, but the forest covers the most area. At the same time, another remote sensing image data of the same area at different times is used as a reference image for evaluation. There is no cloud in this image, and it was acquired on July 20, 2016. The changes in the features are also the smallest.

In order to obtain more accurate cloud removal correction parameters, this paper selects four cloud-covered water blocks, and selects the most homogenous blocks from them. Table 1 shows the slope Γ and R^2 results obtained by using the blue spectrum and cirrus spectrum scatter plots. The slope value 1.962 with the highest R^2 of 0.956 is selected as the final regression model. Table 2 shows the R^2 values before and after cloud removal in different surface areas, which can illustrate the effect of cloud removal.

As shown in Fig. 3, by visual evaluation, it can be seen that the correction effect of this method is better. Comparing the corrected image with the reference image, the correlation coefficient R^2 is used as the evaluation index. The results obtained are listed in Table 2. The increased value indicates that the correction effect is obvious.

Table 1. Linear regression results between Band 2 and Cirrus Band

Water blocks	Slope	R^2
1	1.542	0.655
2	1.962	0.956
3	1.864	0.886
4	1.782	0.943

Table 2. Averaged R^2 of the visible bands to the reference image

Aera	Landscape	Before removal	After removal
1	Soil and farmland	0.72	0.91
2	Urban and lake	0.78	0.87
3	Mountain and grassland	0.69	0.89

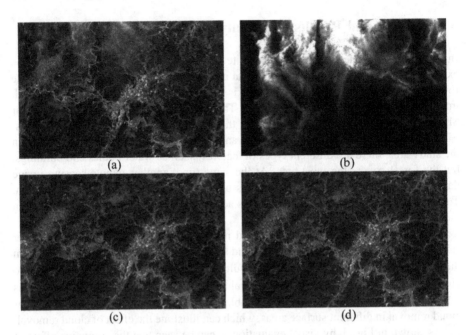

Fig. 3. (a) True colour image, (b) the cirrus band, (c) after cloud removal, (d) the reference.

In order to compare with traditional image correction methods, DOS and HOT methods are tested. When the image is corrected by DOS, the original lower limit relative to the reference image remains unchanged after correction. For the HOT method, the R^2 value of area 3 is increased from 0.69 to 0.80, which is lower compared to the proposed method, for areas 1 and 2, the R^2 value is increased from 0.72 and 0.78 to 0.03 and 0.04, respectively, which indicates that the correction has no significant effect.

4   Conclusion

In this paper, the 1.38 μm band is used to accurately detect cirrus cloud. The visible light wave infrared bands have different penetrating capabilities for thin clouds. A cloud removal algorithm using spectral property is proposed. Cloud removal based on the cirrus band is less affected by the bright surface (snow or desert), thereby reducing the requirements for the ground type. Experimental results show the effectiveness of the algorithm quantitatively.

Using this algorithm has a good cloud removal effect, but it ignores the effect on the short wave infrared band when the thin cloud has a certain thickness. It should be noted here that this algorithm will no longer valid when the cloud cover is thicker and obscure the ground information. In the following research, we should focus on the problems in the algorithm and provide a more stable and widely applicable cloud removal algorithm.

References

1. Vermote, E.F., Kotchenova, S.: Atmospheric correction for the monitoring of land surfaces. J. Geophys. Res.: Atmos. 113 (2008)
2. Shen, H., Li, H., et al.: An effective thin cloud removal procedure for visible remote sensing images. ISPRS J. Photogramm. Remote Sens. **96**, 224–235 (2014)
3. Richter, R., Wang, X., et al.: Correction of cirrus effects in Sentinel-2 type of imagery. Int. J. Remote Sens. **32**(10), 2931–2941 (2011)
4. van der Meer, F.J., et al.: Potential of ESA's Sentinel-2 for geological applications. Remote Sens. Environ. **148**, 124–133 (2014)
5. Parmes, E., et al.: Automatic cloud and shadow detection in optical satellite imagery without using thermal bands - application to Suomi NPP VIIRS images over Fennoscandia. Remote. Sens. **9**, 806 (2017)
6. Ji, C.: Haze reduction from the visible bands of Landsat TM and ETM + images over a shallow water reef environment. Remote Sens. Environ. **112**, 1773–1783 (2008)
7. Gao, B.-C., Li, R.-R.: Removal of thin cirrus scattering effects in Landsat 8 OLI images using the cirrus detecting channel. Remote. Sens. **9**, 834 (2017)
8. Mustak, Sk.: Correction of atmospheric haze in Resourcesat-1 Liss-4 MX data for urban analysis: an improved dark object subtraction approach. Remote Sens. Spat. Inf. Sci. **XL**, 283–287 (2013)
9. Li, J.: Haze and thin cloud removal via sphere model improved dark channel prior. IEEE Geosci. Remote Sens. Lett. **16**, 472–476 (2019)
10. Makarau, A., Richter, R., Schläpfer, D.: Combined haze and cirrus removal for multispectral imagery. IEEE Geosci. Remote Sens. Lett. **13**(5), 1–8 (2016)

A Multi-line Image Difference Technique to Background Suppression Based on Geometric Jitter Correction

Guoliang Tian[1,2(✉)], Hongchen Tang[1,2], Minpu Qiu[1,2],
and Jinping He[1,2]

[1] Beijing Institute of Space Mechanics and Electricity, Beijing 100094, China
tglmingbai@163.com
[2] Beijing Key Laboratory of Advanced Optical Remote Sensing Technology,
Beijing 100094, China

Abstract. With the development of space optical remote sensing technology, in order to meet the needs of large-scale moving target detection and recognition, the research of multi-line detector remote sensing system has become a key development direction. However, after using multi-line push-broom imaging, the background suppression is performed by image difference, and its accuracy depends largely on the suppression effect of geometric jitter in the imaging process. Therefore, this paper proposes a multi-line image difference technique to background suppression based on geometric jitter correction. First, an adaptive algorithm is used to select a template with a suitable size for image registration. Secondly, the gray correlation method is used to detect the geometric displacement error between the images, the geometric jitter correction is performed on the image, and finally the multi-line image is processed differentially to obtain the background. Suppress the image. Simulation and experiments show that this method can quickly and accurately perform geometric jitter correction on the remote sensing image sequence of multi-line push-broom imaging. The correction accuracy is 0.3 pixels. After the difference, the moving target in the image is obvious, and the effectiveness can reach 98%.

Keywords: Geometric registration · Image difference · Gray correlation · Background suppression

1 Introduction

With the development of space optical remote sensing technology, space-based optical remote sensing technology is widely used in the search and recognition of moving targets due to its wide detection range and flexible detection time [1]. However, due to factors such as small detection target, complicated background, and large amount of image information data, the traditional image recognition algorithm based on target features does not meet the needs of space-based remote sensing detection. Using the physical characteristics that the image background does not change and the moving target is displaced within a short time interval, by performing differential background suppression processing on two images with a short time interval, the result can be a

© Springer Nature Singapore Pte Ltd. 2020
Y. Wang et al. (Eds.): IGTA 2020, CCIS 1314, pp. 298–309, 2020.
https://doi.org/10.1007/978-981-33-6033-4_23

moving target image shift image, greatly reducing The noise of target recognition is reduced, and the target recognition characteristics are increased [2]. However, during system imaging, platform jitter will cause geometric image shift in the image, so the effectiveness of the image difference algorithm depends on the geometric registration accuracy between the images before the difference. How to achieve high-precision and rapid image geometric registration, and then effectively suppress the image background noise through the image difference algorithm, and improve the accuracy of target detection is the main research content of this article [3].

In recent years, domestic and foreign research institutions have carried out a lot of research on image geometric flutter correction technology. Among them, Wang Mi et al. of Wuhan University analyzed the impact of the tremor geometric accuracy of high-resolution optical satellite image platforms in 2018 [4]. The platform tremor in seconds will cause the imaging error of the sub-satellite point to reach 2.4 m (@500 km), and the maximum image distortion will reach 4.8 pixels (@0.5mGSD); in 2015, Zhang Xiao et al., Shanghai Institute of Technology, Chinese Academy of Sciences The dual-line detector push-sweep image registration technology was studied, and the geometric registration algorithm based on target features was used to achieve offline sub-pixel registration accuracy [5]; In 2014, Chen Jinwei of Zhejiang University matched the agile satellite remote sensing image Preparing for research, using a variety of geometric registration algorithms based on target features to achieve geometric image restoration [6]; in 2017, Sutton et al. of the Lunar and Planetary Laboratory at the University of Arizona in the United States conducted spatial jitter in high-resolution scientific experimental images For the correction, the algorithm of Fourier analysis is used to realize the sub-pixel registration and correction of the image [7]. At present, the research on geometric flutter correction at home and abroad has basically achieved sub-pixel level image registration, but the timeliness is difficult to meet the requirements of online correction.

Aiming at the application requirements of large-scale moving target detection and recognition in space-based space optical remote sensing system, this paper proposes a differential background suppression method for multi-line images based on geometric jitter correction. The core algorithm is based on adaptive correlation geometric jitter correction method. First, in order to improve the timeliness of the algorithm, an adaptive algorithm is used to select a template of appropriate size for image registration, and secondly, a bilinear interpolation algorithm is used to preprocess the registered image, and then the geometric displacement between the images is detected by the gray correlation method Error, geometric jitter correction of the image, and finally differential processing of the multi-line image to obtain a background suppression image.

2   A Multi-line Image Difference Technique to Background Suppression Based on Geometric Jitter Correction

Satellite platform tremor refers to the disturbance caused by the satellite platform's attitude adjustment, pointing control, solar windsurfing adjustment, periodic motion of the moving parts of the satellite during orbits of the satellite, etc. Vibration response. The transfer of the tremor of the satellite platform to the image surface causes image image shift, which mainly includes image blur and geometric distortion [8]. At present, most of the satellite platforms at home and abroad have a vibration amplitude of 1–5 Hz, and the maximum vibration amplitude is between 50–100 μrad; when 5–10 Hz, the vibration amplitude is between 10–50 μrad; within the frequency of 10–100 Hz, The amplitude is within 1–10 μrad; within the frequency of 100–200 Hz, the amplitude is within 1 μrad. High-frequency vibrations cause image-level or even sub-pixel-level image shifts, and low-frequency vibrations cause image shifts of dozens or even tens of pixels. Low-frequency vibration can be combined with satellite pointing information to correct it based on image features, and this article focuses on how to correct image movement caused by high-frequency vibration [9].

2.1   Geometric Jitter Correction Based on Adaptive Correlation

Geometric jitter correction is the core and premise of image differential background suppression. The core of geometric jitter correction based on adaptive correlation is cross-correlation registration based on line pixel templates [10]. However, before performing geometric registration, it is necessary to pre-process the registration image to give play to the timeliness and accuracy of the core algorithm.

2.1.1   Geometric Correction Preprocessing

Geometric registration is to locate the template in the image to be registered, so the preprocessing of geometric correction includes the resampling of the image to be registered and the selection of the template. Resampling is to improve the spatial resolution of the image. During the imaging process, the restoration target is located between adjacent pixels or the target image with low resolution, so that the matching accuracy enters the sub-pixel level. Adaptive template selection can not only improve the registration efficiency and avoid the "black hole effect"(The black hole effect means that when the template is smaller than the connected area of the same pixel value in the image, the algorithm fails and the registration fails), but also improve the registration accuracy and reduce the processing time.

Compared with other resampling algorithms, bilinear interpolation algorithm can not only avoid the discontinuity of local pixel grayscale, but also reduce the amount of calculation and increase the processing speed. Therefore, on the premise of ensuring registration accuracy, bilinear interpolation is used as the resampling algorithm in the geometric correction method (Fig. 1).

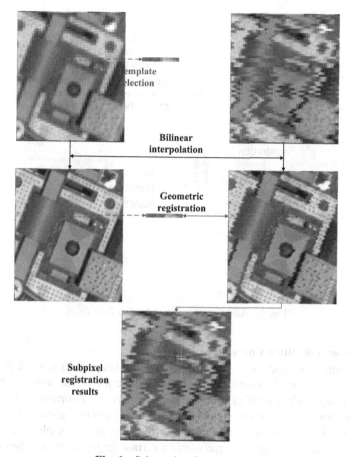

Fig. 1. Schematic of resampling

Adaptive template selection is a method to solve the balance between template size and registration accuracy. The system performs image registration given the initial size of a line template under initial conditions. After registration, the two images correspond to the line image for jitter correction, and the corresponding difference processing is performed to calculate the gray average value of the line image after the difference And the standard deviation, which is used as the threshold for discrimination. Before the next line enters the registration, the template size is changed, the registration, correction, and difference of the image in the next line are calculated. The mean and standard deviation of the image in the next line are calculated, compared, and the template size is adjusted. Increase and decrease, complete adaptive feedback closed loop, improve registration accuracy and processing efficiency (Fig. 2).

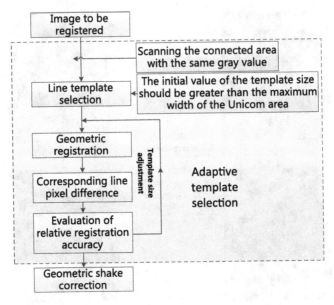

Fig. 2. Adaptive template selection flow chart

2.1.2 Geometric Jitter Correction

Geometric jitter correction is to detect the geometric image shift amount through the geometric registration algorithm, and then compensate the image accordingly according to the image shift amount to complete the geometric jitter correction. In geometric registration, considering the requirements for the timeliness of registration, at the same time, the local information of the remote sensing image is complex, and the gray information of the template is unique within a certain range. Therefore, the core-based correlation algorithm based on gray information is used. The similarity registration method can simplify the calculation amount of the registration process and ensure its registration accuracy.

The cross-correlation similarity metric function is shown below:

$$CC(u, v) = \frac{\sum_x \sum_y T(x, y) I(x - u, y - v)}{\sqrt{\sum_x \sum_y I^2(x - u, y - v)}} \tag{1}$$

I(x, y) is the reference image, and T(x, y) is a module area on the image to be registered (template, which is set to a certain size according to the actual situation). In the actual calculation process, T(x, y) performs a traversal search on I(x, y). When CC (u, v) gets the maximum value, (u, v) is the translation between the two images the amount. When searching for large-size images, the cross-correlation algorithm can perform FFT calculation through convolution theory to improve the efficiency of the algorithm (Fig. 3).

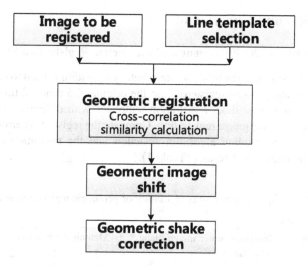

Fig. 3. Geometric Jitter correction flow chart

2.2 A Multi-line Image Difference Technique to Background Suppression Based on Geometric Jitter Correction

For the multi-line array imaging system, differential processing is performed on the image corrected by geometric flutter to complete the suppression of complex background clutter and achieve the purpose of improving the signal-to-noise ratio. In the push-broom imaging process, according to the arrangement relationship of the two linear array detectors, each push-broom process can form a multi-line array detection image with a certain time interval. For the images collected by two linear arrays, including background, target and noise, they can be expressed by the following formulas:

$$I_1(t_1) = B(t_1) + T(t_1) + N_1(t_1)$$
$$I_2(t_2) = B(t_2) + T(t_2) + N_2(t_2)$$

$$(2)$$

B(t) is the background image at time t, T(t) is the target at time t, and N(t) is the image noise at time t. Since the target is moving at a high speed and the background changes slowly, the distance traveled by the target in Δt time is $L = v \cdot \Delta t$. Two images with a stationary background are formed, and only the target moves by a few pixels. Taking one of the linear array detector images as a reference, after the difference is completed, a positive and negative target point pair is formed on the image, which is an important feature information for moving target extraction.

3 Simulation and Experiment

3.1 Experiment on the Effectiveness of Geometric Registration Strategy

In the grassland test scene, the bilinear interpolation algorithm is used to reconstruct the resolution of the image to be registered into the original 1.5 times, 2 times, 2.5 times, and 3 times. At the same time, registration templates with different sizes at random positions are selected for image registration. Compare the registration error to verify the template size, the resampling algorithm selection and the resolution reconstruction multiples on the registration results (Table 1).

Table 1. Experimental results of validity of geometric registration strategy

Bilinear interpolation					
Template width (pixel)	Template center image plane coordinates	Matching error(pixel)			
		1.5 times	2 times	2.5 times	3 times
70	156,187	0.3	0.5	0. 4	0.3
140	164,59	0.3	0.5	0.2	0.3
210	72,296	0	0	0. 2	0.3
280	170,166	0.2	0.5	0.4	0.6
280	170,166	0.3	0.5	0.2	0. 3

It can be known from experiments that the larger the template is within a certain range, the higher the matching accuracy. However, when the template size has been increasing, the amount of calculation increases, the registration speed becomes lower, and more errors will be introduced to reduce the registration accuracy; in the selection of the upsampling image multiples, it is known through experiments that bilinear The interpolation method reconstructs the original image resolution to twice the original, the geometric registration accuracy is the highest.

3.2 Image Jitter Correction Effect Verification Experiment

The images of different resolutions are superimposed line by line with different random displacements (random displacement std: 5 pixels, mean: 0 pixels) to generate two simulated jitter images, and the geometric jitter correction method based on adaptive correlation is used to simulate each group of two jitters. The image is geometrically corrected, and the correction time is recorded. The difference between the flutter displacement of each line detected in the geometric registration and the superimposed random displacement is taken as the registration error, and the mean and standard deviation are calculated to judge the correction effect of the correction algorithm (Figs. 4 and 5).

Fig. 4. Jitter simulation diagram

Fig. 5. Corrected image

Through experimental verification, it can be known that the correction method based on the adaptive correlation-based geometric jitter correction effect is good, and the correction time becomes larger as the corrected image becomes larger (Table 2).

Table 2. Image jitter correction effect

Image size/pixel	Registration error /pixel		Time/s
	Mean	STD	
200 * 390	0	0	3.3
500 * 975	0	0	4.5
1000 * 1950	0	0	7.5

3.3 Resampling Algorithm Simulation

To grassland test scenarios, using nearest neighbor interpolation, the bilinear interpolation and double three different resampling algorithm will stay registration, such as image resolution to 1.5 times, 2 times, 2.5 times, 3 times, at the same time different sizes of random position registration template image registration, registration error, to verify that the template size, resampling algorithm is chosen and the resolution of the reconstructed multiples on the result of registration (Table 3).

Table 3. Resampling algorithm simulation results

Bilinear interpolation

Template width (pixels)	Template image surface coordinates	Matching error (pixel)			
		1.5 times	2 times	2.5 times	3 times
70	156,187	0.3	0.5	0. 4	0.3
140	164,59	0.3	0.5	0.2	0.3
210	72,296	0	0	0. 2	0.3
280	170,166	0.2	0.5	0.4	0.6

Bicubic interpolation

Template width (pixels)	Template image surface coordinates	Matching error (pixel)			
		1.5 倍	2 倍	2.5 倍	3 倍
70	156,187	0.3	0.5	0.4	0. 5
140	164,59	0.2	0.5	0.2	0.3
210	72,296	0	0	0.1	0.3
280	170,166	0.3	0.5	0.2	0. 3

According to the test, the larger the template is within a certain range, the higher the matching accuracy will be. However, when the template size keeps increasing, the calculation amount will increase, and the registration speed will become lower. Meanwhile, more errors will be introduced to make the registration accuracy become lower. Compared with the resampling algorithm, the registration accuracy of the bicubic interpolation is higher than that of the bilinear interpolation, but with the increase of the template, the calculation speed is much slower than that of the bilinear interpolation. As for the selection of multiple of the upsampled image, the experiment shows that no matter what resampling method is adopted, the geometric registration accuracy is the highest when the resolution of the original image is changed to 2 times of the original (Table 4).

3.4 Differential Detection Effect Verification Experiment

Select several groups of images with different resolutions (superimposed jitter error), the two images in each group have obvious time shift, most of the background is stable, and slight changes occur locally. The moving target in the figure has obvious time shift. The method in one experiment superimposes the same degree of random jitter, performs differential processing on the two images corrected using the adaptive correlation-based geometric flutter correction method, analyzes the resulting differential image, and verifies the differential detection algorithm (Figs. 6 and 7).

Fig. 6. Images with obvious time-shift features

Fig. 7. Overlay different dithered images

Through experiments, it can be known that under certain time shift conditions, the geometric registration accuracy in differential detection remains unchanged, regardless of the resolution. After analyzing the effect map after the difference, the target image has high contrast and the differential detection is effective (Fig. 8).

Fig. 8. Difference result image

Table 4. Geometric registration analysis in difference processing

Num	Corrected lines	Lines of registration errors	Geometric correction accuracy	Time/s
1	216	4	98.15%	4
2	540	7	98.70%	5
3	1080	21	98.06%	8

4   Conclusion

This paper introduces the application requirements of large-scale moving target detection and recognition for space-based space optical remote sensing system, and proposes a differential background suppression method for multi-line array images based on geometric flutter correction. This method is based on the imaging principle of multi-line space optical remote sensing imaging system, image flutter analysis and other factors. Through experimental verification, the multi-line images difference technique to background suppression based on geometric jitter correction can achieve multi-line array image registration accuracy of 0.3 pixels, and effectively achieve image geometric jitter correction. Through the image differential background suppression method, a large range of image targets can be quickly searched, and the detection target image pair has high contrast and is easy to detect. Further, the geometric correction processing is carried out simultaneously with the linear array push-broom imaging, so as to improve the timeliness of the algorithm and finally achieve the purpose of on-board processing.

References

1. Wei, H.Q.J.: A study of the new direction of space-borne hi-resolution optical remote sensor. J. Spacecraft Recovery Remote Sens. **28**(4), 48–50 (2007)
2. Zhang, X.: The effect analysis and simulation of platform motion on image quality of spaceborn TDI CCD Camera. Harbin Inst. Technol. Harnbin (2011)
3. Sun, D., Yang, J., et al.:Design and random vibration analysis of calibration mechanism of space infrared camera. J. Infrared Laser Eng. 42, 323–328 (2013)
4. Wang M., Zhu Y., et al.: Review of research on analysis and processing of the impact of geometry precision of high resolution optical satellite image platform. J. Geomatics Inf. Sci. Wuhan Univ. **43**(12), 1899–1906 (2018)
5. Xiao, Z., Hongbo, W., et al.: Research on push-broom image registration technology of long wave infrared dual line detector. J. Infrared Technol. **37**(2), 139–146 (2015)
6. Chen, J.: Research on Registration and Stitching Technology of Agile Satellite Remote Sensing Images. D. Zhejiang University (2014)
7. Sutton, S.S., Boyd, A.K., et al.: Correcting spacecraft jitter in hirise images. In: International Symposium on Planetary Remote Sensing and Mapping. Hong Kong (2017)

8. Janesick, J., Elliott, T., Andrews, J., Tower, J., Bell, P., Teruya, A., Kimbrough, J. and Bishop.: Mk x Nk gated CMOS imager. In: Proceedings of SPIE, vol. 9211, pp. 921106–921113 (2014)
9. Acton, C.: Ancillary data services of NASA's navigation and ancillary information facility. Planetary and Space Sci. **44**(1), 65–70 (1996)
10. Tong, X., Ye, Z., Xu, Y., et al.: Framework of jitter detection and compensation for high resolution satellites. J. Remote Sens. **6**(5), 3944–3964 (2014)

Other Research Works and Surveys Related to the Applications of Image and Graphics Technology

Image Recognition Method of Defective Button Battery Base on Improved MobileNetV1

Tao Yao[⊠], Qi Zhang, Xingyu Wu, and Xiuyue Lin

School of Automation, Guangdong University of Technology,
Guangzhou 510006, Guangdong, China
921507176@qq.com

Abstract. For the problem that the traditional image algorithm is challenging to design parameters in the application of defective button battery image recognition, and the complex convolutional neural network cannot meet its high real-time demand. This paper proposed an image recognition method of defective button battery based on improved MobileNetV1. MobileNetV1's deep separable convolutional layer is an improvement of the standard convolutional layer structure. MobileNetV1 has fewer parameters, which is very suitable for mobile and embedded vision applications. To further improve the classification accuracy of MobileNetV1, reduce the amount of calculation, and meet the industrial scene's real-time nature. The deep convolution layer of MobileNetV1 was improved, and the Relu activation function was replaced by the more stable activation function named TanhExp. The experimental results show that the proposed method has an excellent performance in recognizing defective button battery images. The accuracy of the test set can reach 99.05%, which can be applied to the identification of imperfect button battery images.

Keywords: MobileNetV1 · Asymmetric convolution · Activation function

1 Introduction

In the electronics industry, button batteries have been widely used in daily life because of their small size, long life, and stable discharge current. However, in the production process, due to some unforeseen factors, it will cause defects in the battery, and it is not easy to find it with the naked eye. Therefore, the need to automatically identify defective button batteries is increasingly pressing, and machine vision technology is the key technology to meet this demand. There are two main image processing algorithms: one is a traditional image processing algorithm, and the other is a deep learning image processing algorithm based on convolutional neural networks.

Traditional image processing algorithms need to be designed for specific applications. Still, in complex industrial environments, the interference of external factors will lead to the adjustment of algorithm parameters, the later maintenance costs will be high. For example, literature [1] first masked the fonts on the button battery through template matching and then used the gradient feature method to locate the scratch location. Although there is an excellent recognition effect on a specific sample set, the change in light of the light source will seriously influence the visual system's classification effect. It is necessary to adjust the parameters repeatedly, which is not conducive to later maintenance.

The deep learning algorithm based on a convolutional neural network can learn the mapping relationship between the input layer and the output layer through a large amount of image data to obtain the ability to extract image feature information. It has achieved remarkable results in image recognition and has been widely used. For example, literature [2] proposed detecting solar cells using deep neural networks. Literature [3] proposed the detection of the steel surface defect method based on a deep learning network. It uses BP neural network and image processing technology to achieve the rapid identification of surface defects of the cold-rolled strip. However, when performing defect detection in the production process of button batteries in the actual industry, there is a high real-time requirement. Industrial PC has low hardware computing power and storage. It is not suitable for complex convolutional neural network structures, as proposed in the literature [4]. The finger vein recognition of the improved residual network has an excellent performance in recognition. Still, its network structure is about 650,000 neurons and 60 million parameters, and the time complexity and space complexity are enormous, which is not suitable for industrial applications.

In response to the above problems, this paper uses MobileNetV1 proposed in the literature [5] as the underlying network. While maintaining high recognition accuracy, it dramatically reduces the parameter amount of the network module. The author of the literature [5] uses MobileNetV1 and GoolgeNet [6], VGG16 [7] to compare the classification performance on the dataset ImageNet, and their accuracy is almost the same. But the parameter amount of GoogleNet and VGG16 is 1.6 times and 32.8 times that of MobileNetV1.

Because MobileNetV1 can achieve excellent results in image recognition, this paper proposes a recognition method of the defect button battery image basing on improved MobileNetV1. To further reduce the amount of calculation and parameters and enhance generalization, MobileNetV1's deep separable convolutional layer is improved. At the same time, the activation function of MobileNetV1 is replaced by the TanExp activation function proposed in the literature [8], which avoids the "inactivation" of neurons in the training process of MobileNetV1 and accelerates the training convergence speed.

2   Improved MobileNetV1

2.1   Basic Network

MobileNetV1 is a new type of convolutional neural network. Because of the characteristics of high efficiency and low power consumption, MobileNetV1 maintains a high accuracy rate in image classification and recognition. MobileNetV1 optimizes the standard convolution layer and decomposes the standard convolution operation into two part: depthwise convolution, and pointwise convolution collectively called depthwise separable convolution. Depthwise Convolution is to apply a single convolution kernel to the single input channel of each feature map for convolution calculation. In contrast, pointwise convolution is a standard convolution with a convolution kernel of 1×1. DepthWise Separable Convolution reduces the redundancy of standard convolution and computation. If the size of input feature map and the size of the output feature map is S × S, the input channel is N, and the output channel is M, the calculation amount of the standard convolution is:

$$C1 = S \times S \times W \times H \times N \times M \tag{1}$$

The calculation of the depth separable convolution is:

$$C2 = S \times S \times W \times H \times N + W \times H \times N \times M \tag{2}$$

The comparison of the calculated amount is:

$$\frac{S \times S \times W \times H \times M}{S \times S \times W \times H \times N + W \times H \times N \times M} = \frac{S \times S \times M}{S \times S + M}$$

It can be seen that the depth separable convolutional layer of MobileNetV1 is less computational than the standard convolutional layer. Its network structure is shown in Table 1, and the input dimension is 400 × 128, Where Conv represents standard convolution, Conv dw represents depthwise convolution, AvgPool is average pooling [9], FC is fully connected layer [10], and Soft is Softmax classifier [11].

Table 1. The network structure of MobileNetV1

Type/stride	Filter shape	Input size
Conv1/S2	3 × 3 × 1 × 32	400 × 128 × 1
Conv2 dw/S1	3 × 3 × 32	200 × 64 × 32
Conv/S1	1 × 1 × 32 × 64	200 × 64 × 32
Conv3 dw/S2	3 × 3 × 64	200 × 64 × 64
Conv/S1	1 × 1 × 64 × 128	100 × 32 × 64
Conv4 dw/S1	3 × 3 × 128	100 × 32 × 128
Conv/S1	1 × 1 × 128 × 128	100 × 32 × 128
Conv5 dw/S2	3 × 3 × 128	100 × 32 × 128
Conv/S1	1 × 1 × 128 × 256	50 × 16 × 128
Conv6 dw/S1	3 × 3 × 256	50 × 16 × 256
Conv/S1	1 × 1 × 256 × 256	50 × 16 × 256
Conv7 dw/S1	3 × 3 × 256	50 × 16 × 256
Conv/S1	1 × 1 × 256 × 512	50 × 16 × 256
Conv(8–12)dw/S1(×5)	3 × 3 × 512	50 × 16 × 512
Con/S1	1 × 1 × 512 × 512	50 × 16 × 512
Conv13 dw/S2	3 × 3 × 512	50 × 16 × 512
Conv/S1	1 × 1 × 512 × 1024	25 × 8 × 512
Conv14 dw/S2	3 × 3 × 1024	25 × 8 × 1024
Conv/S1	1 × 1 × 1024 × 1024	23 × 4 × 1024
Avg pool	Pool 23 × 4	23 × 4 × 1024
FC	1024 × 2	1 × 1 × 1024
Soft	Classifier	1 × 1 × 2

2.2   Double-Layer Asymmetric Convolution

If the given input image size is 400×128, MobileNetV1 has a total of 14 layers of separable convolutional layers, the feature image size of the input image after the 13 layers of convolutional layer calculation is 50×16, the number of input channels is 256, and the number of output channels is 512, and the size of the convolution kernel is 3×3. From Eq. (2), the calculation amount of the convolutional layer with separable depth is:

$$3 \times 3 \times 50 \times 16 \times 256 + 50 \times 16 \times 256 \times 512 = 106700800$$

Although the depth separated convolutional layer has a significant reduction in the number of parameters and calculations compared to the standard convolutional layer, a single layer has 106700800 calculations. The count of the entire network is still huge for industrial computers with poor industrial site performance. So the deep separable convolutional layer of MobileNetV1 is optimized to meet the real-time requirements.

Because double-layer asymmetric convolution has fewer network parameters and calculations than single-layer convolution, replacing single-layer convolution with double-layer asymmetric convolution can make the network deeper and have one more layer of nonlinear changes. There will be more robust feature extraction capabilities. Therefore, this paper decomposes the deep convolution 3×3 convolution kernel into two one-dimensional convolutions 1×3 and 3×1. The convolution process is shown in Fig. 1.

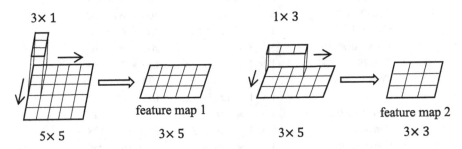

Fig. 1. Convolution process of double asymmetric convolution, the dimension of the input feature map is 5×5, the dimension of the output feature map is 3×3

The dimension of the input feature map in Fig. 1 is 5×5. First, the convolution kernel of 3×1 is used to perform the convolution operation on the input feature map, and the dimension of the output feature map 1 is 3×5. After the feature map 1 is convolved by the convolution kernel 1×3, the feature map 2 can be obtained, and its dimension is 3×3.

The two-layer asymmetric convolution is used to convolve the feature map with a dimension of 5×5. A total of 72 operations are required to obtain the feature map 2, while if the traditional convolution is used, 81 operations are required to obtain the feature map 2. In general, the input feature map has a larger dimension, more convolution times, and a double-layer asymmetric convolution reduces the more calculations.

Through experiments, it is found that when the resolution of the feature map is too high, it is not suitable to use a continuous asymmetric convolution structure, which will cause information loss. When the feature map's decision is too low, the acceleration effect of the constant asymmetric convolution structure is no longer evident. However, it has a good impact on the medium-sized feature maps. In the process of designing the network structure, the application range of the continuous asymmetric convolution structure is obtained: suitable for feature maps with a resolution of 10 × 10–50 × 50. Therefore, this paper optimizes the 6th to 12th layers of MobileNetV1 in Table 1.

2.3 Activation Function Replacement

The activation function of MobileNetV1 is Relu, and the graph is shown in Fig. 2, the expression is:

$$f(x) = \max(0, x) \tag{3}$$

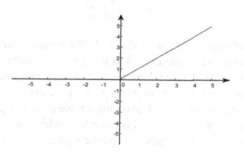

Fig. 2. Graph of the Relu activation function.

In the convolutional neural network's training process, if x_i is the input of a neuron, z is the output of the neuron, w_i is the weight, b_i is the bias, and f is the activation function. The calculation form of the forward propagation of the neuron is:

$$k = \sum_{i=1}^{n} x_i w_i + b_i \tag{4}$$

$$z = f(k) \tag{5}$$

Suppose the loss function is E, and the learning rate is η, the optimization process of weights is:

$$w_i = w_i - \eta \frac{\partial E}{w_i} \tag{6}$$

From Eqs. (4), (5), and (6), it can be seen that in the optimization process of the weights of convolutional neural networks, when the gradient is large, the weights suddenly decrease after the update, which will cause the neuron has a negative k value after the next round of forwarding propagation calculations, z is 0 after the Relu activation function, the neuron is "inactivated," and the weight cannot be updated. Therefore, the Relu activation function is sensitive to the setting of the learning rate. A high learning rate means that weights vary greatly when the weights are updated, which will cause most neurons of the network model to "inactivate," affecting the accuracy of the final recognition, a low learning rate is not conducive to the convergence of the network. It can be seen that the Relu activation function is not stable during the MobileNetV1 training process. Therefore, this text replaces the Relu activation function with the TanExp activation function in the separable convolutional layer, and its expression is:

$$f(x) = x\tanh(e^x) \tag{7}$$

Where tanh represents the double tangent function:

$$\tanh(x) = \frac{e^x - e^x}{e^x + e^x} \tag{8}$$

The curve of TanhExp is shown in Fig. 3. In the positive part, when the input is greater than one, TanhExp is almost equal to the linear transformation. The change of the output value and the input value does not exceed 0.01, consistent with the positive part of the Relu activation function; in the negative part, the minimum amount of the TanhExp activation function is −0.353, allowing a more extensive gradient flow. At the same time, it has a steep gradient when it is close to 1, which can speed up the training process's convergence rate. The Derivative gradient diagram is shown in Fig. 4.

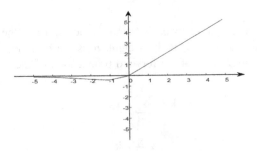

Fig. 3. Graph of the TanExp activation function.

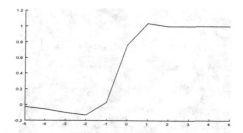

Fig. 4. Derivative gradient diagram of the TanExp activation function.

3 Experiment Procedure

3.1 Button Battery Data Set

3.1.1 Data Collection and Data Procession

In this paper, according to the button battery factory's positive and negative actual product, image data were taken at the experimental site by Daheng Industrial (model MER-310-12UC-L) camera. When acquiring images, because the defects of the button battery need to be displayed at the tilt angle, a mechanical bracket is used to fix the industrial camera at a certain tilt angle, and the lamp board is used as an auxiliary light source. The model diagram of the experimental site is shown in Fig. 5. The image samples collected by the camera are shown in Fig. 6, converted into a single channel, and preprocessed by affine changes, as shown in Fig. 7. The size of the image is 400×128, the format is still the original BMP format, and samples of good and defective products are shown in Fig. 8 and Fig. 9.

A total of 5,000 samples of good products and bad products were collected in the experiment. The experiment adopted the method of data amplification to increase the diversity of data samples; that is, the original data samples were expanded to twice the original by rotating 180°. The experiment finally selected 10000 positive and negative samples, 80% of which were used as the training set, and 20% as the test set.

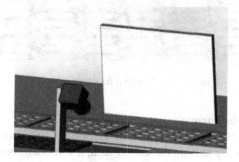

Fig. 5. The model diagram of the experimental site.

Fig. 6. The image samples collected by Daheng Industrial camera.

Fig. 7. The result after processing.

Fig. 8. One hundred positive samples.

Fig. 9. One hundred negative samples.

3.1.2 Tfrecord Data Format

Tfrecord is a unified data format provided by Tensorflow (the training environment for this experiment). Directly reading the sample data, it will often consume much memory and slow down the program. Tfrecord is a binary data encoding scheme. It occupies only one memory block and only needs to load one binary file at a time. Therefore, this experiment converts the data set to Tfrecord's data format.

3.2   Experimental Comparison and Analysis

The experiment uses the accuracy rate to evaluate the performance of the model; its formula is as follows:

$$Accuracy = \frac{TP + TN}{TP + TN + FP + FN} \tag{9}$$

TP is the correct judgment of positive samples. FP is a wrong judgment of positive samples; TN is the correct judgment of negative samples; FN is a wrong judgment of negative samples.

In the following experiment, the same training method is adopted, that is, the recognition model weights are initialized with a normal distribution, using L2 regular terms, the learning rate is 0.01, the optimizer is Adam [12], the Epoch is 10000 steps, and the batch size is 100.

3.2.1   Experimental Comparison of Different Activation Functions

In this experiment, Relu, Swish, Mish, TanExp are used as the activation function of the improved Mobilenetv1, respectively, as the experimental comparison. Swish [13] and Mish [14] are useful activation functions proposed in recent years, which joined the experiment to enhance the contrast of the experiment; their graph is shown at 15. The experimental results are shown in Table 2 and Fig. 10.

Table 2. The performance of classification results with different activation functions.

Activation function	Accuracy/%
Relu	97.89
Swish	98.75
Mish	98.64
TanExp	99.05

It can be seen from Table 2 that the accuracy of the Swish, Mish, and TanExp activation functions on the improved MobileNetV1 is close, which is 1% away from the Relu activation function, which is mainly due to the continuous smoothness of the three activation functions. These activation functions allow better information to penetrate deep into the neural network, resulting in better accuracy and generalization.

As can be seen from Fig. 11, as Epoch increases, the loss values of all activation functions gradually decrease. However, TanhExp outperforms Mish, Swish, and Relu in convergence speed. At the 1000th Epoch, the loss value of TanExp is 1.83, while Mish, Swish, and Relu are 2.41, 3.24, and 3.52, respectively, which indicates that TanhExp can quickly update the parameters and fit in the most effective direction.

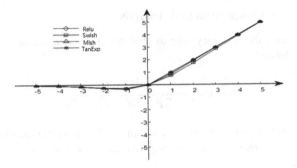

Fig. 10. Graph of four activation function.

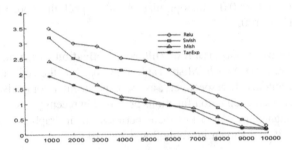

Fig. 11. Loss graph of four activation function during training. The abscissa is Epoch, and the ordinate is Loss.

3.2.2   Improve MobileNetV1 Performance

In order to verify the improved MobileNetV1 performance, this article compares it with the original MobileNetV1, and the traditional convolutional neural networks Vgg16 and ResNet18, which are close to the number of network layers. Vgg16 has 13 standard convolutional layers and three full connection layers. The convolution kernel size of each layer is 3×3, and the activation function is Relu. Resnet18 has an 18-layer structure, including a standard convolutional layer and four residual blocks. Each residual block has two standard convolutional layers. The convolution kernel size of each layer is 3×3, and the activation function is Relu.

Figure 12 shows the accuracy of the four models on the test set at different training stages. Table 3 shows the final accuracy, running time, model size of the model in the test set, and the experimental platform is GeForce GTX 1060.

Table 3. The performance of overall results with different models.

Model type	Accuracy/%	Running time/ms	Model size/M
Improve MobileNetV1	99.05	2.43	8
MobileNetV1	99.14	5.35	20
VGG16	95.51	9.08	503
ResNet18	97.38	7.34	60

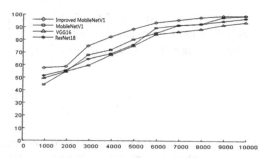

Fig. 12. Accuracy graph of four models during training. The abscissa is Epoch, and the ordinate is accuracy.

It can be seen from Fig. 12 and Table 3 that the improved MobileNetV1 has the fastest convergence speed during the training phase, the running time is 2.43 ms, the model size is 8M, and the accuracy is 99.05%, which is 0.09% lower than the original MobileNetV1, but 3.57% higher than Vgg16 and 1.67% higher than Resnet18. The reasons why the improved MobileNetV1 proposed in this article can have better performance in the image recognition of button battery defects are as follows:

1) In terms of time complexity and space complexity, as the improved MobileNetV1 improves the deep convolution layer, the single-layer convolution of the deep convolution layer is replaced by a double-layer asymmetric convolution, so that the amount of calculation and parameters decrease significantly.

2) In terms of accuracy, the improved MobileNetV1 convolutional layer structure continues the original model's deep separable convolutional layer. At the same time, the replacement of the TanExp activation function improves the robustness of the recognition model. Although the accuracy of the improved model has been reduced, the reduction accuracy is within the acceptable range, and the running time and model size is reduced by multiples, which can better meet the needs of the industry. The VGG16 and ResNet18 network structures are complicated, and the parameters are enormous. When these structures are applied to small and medium-sized data, overfitting is prone to occur.

3) In terms of convergence speed, since the TanExp activation function has a steep gradient when it is close to 1, which accelerates the training convergence speed.

In summary, the improved MobileNetV1 can be applied to image recognition of defective button batteries.

4   Conclusion

This paper proposes a method for image recognition of a defective button battery based on improved MobileNetV1. Due to the need to meet the requirements of high real-time performance in industrial production, the original MobileNetV1 deep separable convolution layer was improved. The double-layer asymmetric convolution replaced the single-layer convolution, which can significantly reduce the amount of calculation and parameters. The original Activation function of MobileNetV1 was replaced by TanExp, which improved the generalization. Experimental results show that the accuracy of this method can be up to 99.05% when it is used to identify images of defective button batteries.

References

1. Yang, Y., Sun, Z.Q., Chen, Z., Zhao, B.H.: Research on scratch detection method of button battery surface based on machine vision. Autom. Instrum. **34**(04), 50–64 (2019). (in Chinese)
2. Wang, X.B., Li, J., Yao, M.H.: Method for surface defect detection of solar cells based on deep learning. Pattern Recogn. Artif. Intell. **27**(6), 517–523 (2014). (in Chinese)
3. Xu, F.Y.: Neural network based rapid detection of steel surface defects. Nanjing University of Science and Technology (2019). (in Chinese)
4. Xiao, A.B., Yi, R., Li, Y.D., Wu, B., Zhong, L.H.: Improve the residual network refers to the vein recognition. J. Xi'an Eng. Univ. (2020). (in Chinese)
5. Howard, A.G., Zhu, M., Chen, B., et al.: MobileNets: efficient convolutional neural networks for mobile vision applications. arXiv E-print arXiv:1704.04861 (2017)
6. Simonyan, K., Zisserman, A.: Very deep convolutional networks for large-scale image recognition. In: ICLR (2015)
7. Szegedy, C., Liu, W., Jia, Y., et al.: Going deeper with convolutions. In: IEEE Conference on Computer Vision and Pattern Recognition, pp. 1–9. IEEE Computer Society (2015)
8. Liu, X.Y., Di, X.G.: TanhExp: a smooth activation function with high convergence speed for lightweight neural networks. arXiv E-prints arXiv:2003.09855 (2020)
9. Zhou, F.Y., Jin, L.P., Dong, J.: Review of research on convolutional neural networks. Acta Computerica Sinica. **6**, 1229–1251 (2017). (in Chinese)
10. Chang, L., Deng, X.M., Zhou, M.Q., et al.: Convolution neural network in image comprehension. Acta Automatica Sinica **9**, 1300–1312 (2016). (in Chinese)
11. Wan, L., Tong, X., Sheng, M.W., et al.: Overview of application of Softmax classifier in deep learning image classification. Navig. Control **6**, 1–9 (2019). (in Chinese)
12. Kingma, D.P., Ba, J.: Adam: a method for stochastic optimization. arXiv E-prints arXiv: 1412.6980 (2014)
13. Prajit, R., Barret, Z., Quoc, V.L.: Searching for activation functions. arXiv.org arXiv:1710. 05941 (2017)
14. Diganta, M.: Mish: a self regularized non-monotonic neural activation function. arXiv.org arXiv:1908.08681 (2019)

Author Index

Printed in the United States
By Bookmasters